How Chinese Acquire and Improve Mathematics Knowledge for Teaching

MATHEMATICS TEACHING AND LEARNING

Volume 4

Series Editor:
Yeping Li, Texas A&M University, College Station, USA

International Advisory Board:
Marja Van den Heuvel-Panhuizen, Utrecht University, The Netherlands
Rongjin Huang, Middle TN State University, USA
Eric J. Knuth, University of Texas-Austin, USA
Carolyn A. Maher, Rutgers – The State University of New Jersey, USA
JeongSuk Pang, Korea National University of Education, South Korea
Nathalie Sinclair, Simon Fraser University, Canada
Kaye Stacey, University of Melbourne, Australia
Günter Törner, Universität Duisburg-Essen, Germany
Catherine P. Vistro-Yu, Ateneo de Manila University, The Philippines
Anne Watson, University of Oxford, UK

Mathematics Teaching and Learning is an international book series that aims to provide an important outlet for sharing the research, policy, and practice of mathematics education to promote the teaching and learning of mathematics at all school levels as well as teacher education around the world. The book series strives to address different aspects, forms, and stages in mathematics teaching and learning both in and out of classrooms, their interactions throughout the process of mathematics instruction and teacher education from various theoretical, historical, policy, psychological, socio-cultural, or cross-cultural perspectives. The series features books that are contributed by researchers, curriculum developers, teacher educators, and practitioners from different education systems.

For further information:
http://www.brill.com/products/series/mathematics-teaching-and-learning

How Chinese Acquire and Improve Mathematics Knowledge for Teaching

Edited by

Yeping Li
Texas A&M University, USA and Shanghai Normal University, China

and

Rongjin Huang
Middle Tennessee State University, USA

BRILL
SENSE
LEIDEN | BOSTON

The Library of Congress Cataloging-in-Publication Data is available online at http://catalog.loc.gov

Names: Li, Yeping, editor. | Huang, Rongjin, editor.
Title: How Chinese acquire and improve mathematics knowledge for teaching / edited by Yeping Li (Texas A&M University, USA and Shanghai Normal University, China) and Rongjin Huang (Middle Tennessee State University, USA).
Description: Leiden : Brill Sense, c2018. | Series: Mathematics teaching and learning ; volume 4 | Includes bibliographical references and index.
Identifiers: LCCN 2018002413 (print) | LCCN 2018005275 (ebook) | ISBN 9789463512367 (E-book) | ISBN 9789463512343 (pbk. : alk. paper) | ISBN 9789463512350 (hardback : alk. paper)
Subjects: LCSH: Mathematics teachers--Training of--China. | Mathematics--Vocational guidance--China.
Classification: LCC QA10.5 (ebook) | LCC QA10.5 .H69 2018 (print) | DDC 510.71--dc23
LC record available at https://lccn.loc.gov/2018002413

ISBN: 978-94-6351-234-3 (paperback)
ISBN: 978-94-6351-235-0 (hardback)
ISBN: 978-94-6351-236-7 (e-book)

All chapters in this book have undergone peer review.

Copyright 2018 by Koninklijke Brill NV, Leiden, The Netherlands.

Koninklijke Brill NV incorporates the imprints Brill, Brill Hes & De Graaf, Brill Nijhoff, Brill Rodopi, Brill Sense and Hotei Publishing.

All rights reserved. No part of this publication may be reproduced, translated, stored in a retrieval system, or transmitted in any form or by any means, electronic, mechanical, photocopying, recording or otherwise, without prior written permission from the publisher.

Authorization to photocopy items for internal or personal use is granted by Koninklijke Brill NV provided that the appropriate fees are paid directly to The Copyright Clearance Center, 222 Rosewood Drive, Suite 910, Danvers, MA 01923, USA. Fees are subject to change.

This book is printed on acid-free paper and produced in a sustainable manner.

TABLE OF CONTENTS

Acknowledgements   vii

Part I: Introduction and Perspectives

1. Teachers' Acquisition and Improvement of Mathematical Knowledge for Teaching in China   3
   Yeping Li and Rongjin Huang

2. Educational Systems and Mathematics Teacher Education Systems in Mainland China, Hong Kong and Taiwan   9
   Huk-Yuen Law, Hak Ping Tam, Xianhan Huang and Yu-Jen Lu

3. The Conception of Mathematics Teachers' Literacy for Teaching from a Historical Perspective   37
   Feng-Jui Hsieh, Sin-Sheng Lu, Chia-Jui Hsieh, Shu-Zhi Tang and Ting-Ying Wang

4. The Conception of Mathematics Knowledge for Teaching from an International Perspective: The Case of the TEDS-M Study   57
   Martina Döhrmann, Gabriele Kaiser and Sigrid Blömeke

Part II: Acquiring and Improving Mathematics Knowledge for Teaching through Teacher Preparation

5. Elementary Mathematics Teacher Preparation in China   85
   Shu Xie, Yun-peng Ma and Wei Chen

6. Secondary Mathematics Teacher Preparation in China   109
   Yingkang Wu and Rongjin Huang

7. Pedagogical Training for Prospective Mathematics Teachers in China   137
   Zhiqiang Yuan and Rongjin Huang

8. Mathematics Teacher Preparation in China: What Do We Learn?   153
   Despina Potari

Part III: Acquiring and Improving Mathematics Knowledge for Teaching through Teaching and Professional Development

9. How Do Chinese Teachers Acquire and Improve Their Knowledge through Intensive Textbook Studies?   165
   Shuping Pu, Xuhua Sun and Yeping Li

TABLE OF CONTENTS

10. Experienced Teacher's Learning through a Master Teacher Workstation Program: A Case Study　　185
   Xingfeng Huang and Rongjin Huang

11. In-Service Mathematics Teachers' Professional Learning in Teaching Research Group: A Case Study from China　　209
   Yudong Yang and Bo Zhang

12. Learning from Developing and Observing Public Lessons　　225
   Su Liang

13. Improving Teachers' Expertise and Teaching through Apprenticeship Practice in Mainland China: Case Studies　　241
   Yeping Li and Rongjin Huang

14. Learning and Improving MKT through Teaching and Professional Development Mechanisms　　263
   Gloria Ann Stillman

Part IV: Reflection and Conclusion

15. Mathematics Knowledge for Teaching: What Have We Learned?　　277
   Tim Rowland

16. Some Remarks on the Preparation of Mathematics Teachers in China　　289
   Hung-Hsi Wu

About the Contributors　　305

Index　　313

ACKNOWLEDGEMENTS

We would like to take this opportunity to thank all 29 contributors from 9 countries or regions for their commitment and contribution. The majority of chapters are contributed by those insiders who have pertinent knowledge and experience in mathematics teaching and teacher education in China. Their contributions help provide what this book is about. With contributions from scholars in many other countries or regions, the book is also able to bring in different perspectives on this topic for the international community. This book would not have been possible without all of these important contributions. In addition, many of these contributors also volunteered their time to review chapters. Their collective efforts help ensure this book's quality.

Thanks also go to a dedicated group of external reviewers, including Chunlian Jiang (Macau), Ji-Eun Lee (USA), Aihui Peng (China), Andrzej Sokolowski (USA), Ji-Won Son (USA) and Xinrong Yang (China/Germany), for taking the time to help review chapters of the book. Their reviews and comments helped improve the quality of many chapters, and on behalf of all the contributors, we extend our appreciation to them.

Finally, we want to thank Michel Lokhorst (Brill | Sense) for his patience and support. Working with Michel has been a productive and pleasant experience, with this book's publication as the fourth volume of the book series on *Mathematics Teaching and Learning*.

PART I
INTRODUCTION AND PERSPECTIVES

YEPING LI AND RONGJIN HUANG

1. TEACHERS' ACQUISITION AND IMPROVEMENT OF MATHEMATICAL KNOWLEDGE FOR TEACHING IN CHINA

INTRODUCTION

Efforts to improve the quality of teaching have led to the ever-increasing studies on teachers' knowledge and skills that are important to have for effective teaching and how such knowledge and skills can be acquired and improved over years. Existing research has advanced our understanding about the structure and components of mathematics teachers' knowledge and skills for teaching from different perspectives and approaches (e.g., Ball, Thames, & Phelps, 2008; Blömeke, Hsieh, Kaiser, & Schmidt, 2014; Depaepe, Verschaffel, & Kelchtermans, 2013; Hill, Ball, & Schilling, 2008; Schmidt, Blömeke, & Tatto, 2011). In particular, Ball and her colleagues have built upon their work on teaching to identify and formulate specific constructs about teachers' mathematics knowledge for teaching (MKT, e.g., Ball, Thames, & Phelps, 2008). Their work has expanded our understanding of the nature and features of teachers' MKT. At the same time, another line of studies has taken an international approach to examine cross-system similarities and differences in teacher preparation programs and pre-service teachers' knowledge acquisition at the time when finishing up their program studies (e.g., Li, Ma, & Pang, 2008; Schmidt, Blömeke, & Tatto, 2011). Cross-system comparisons revealed not only differences in what mathematics teachers may know (e.g., Li, Ma, & Pang, 2008; Schmidt, Blömeke, & Tatto, 2011), but also questioned about whether mathematics teachers are actually expected to learn different knowledge and skills that are valued in different systems and cultural contexts (Depaepe et al., 2013; Döhrmann, Kaiser, & Blömeke, 2012; Hsieh, Lin, & Wang, 2012). Such questions suggest the importance of examining and understanding mathematics teachers' knowledge and skills and their acquisition and development as situated in specific systems and cultural contexts.

This book is designed to focus on Chinese teachers' acquisition and improvement of mathematics knowledge for teaching. It is a book that extends our recent work on how Chinese teach mathematics and improve teaching (Li & Huang, 2013) and the nature and forms of teaching and learning mathematics with variation in China (Huang & Li, 2017). These previous books focused on what Chinese teachers do in preparing, conducting, and improving teaching, but not on Chinese teachers' mathematics knowledge and skills for teaching

Y. Li & R. Huang (Eds.), How Chinese Acquire and Improve Mathematics Knowledge for Teaching, 3–8.
© 2018 Koninklijke Brill NV. All rights reserved.

and their acquisition and improvement. Although existing studies revealed interesting features of Chinese mathematics teachers' knowledge and skills (e.g., An, Kulm, & Wu, 2004; Li & Huang, 2008; Ma, 1999), much remains to be understood about specific knowledge and skills that Chinese mathematics teachers may need, and how they acquire and improve such knowledge and skills over years. A systematic examination of Chinese teachers' mathematics knowledge for teaching and their acquisition and improvement becomes important, especially in the current context of learning more from the high-achieving Chinese education system in their efforts for improving the quality of mathematics instruction and students' mathematics learning.

This book intends to reflect upon and extend current research efforts on this topic. In particular, this book aims to accomplish three main aspects:

1. This book builds upon existing studies to present an extended effort to examine mathematics knowledge for teaching in China. In particular, this book contains a collection of chapters that provide either comprehensive reviews of mathematics knowledge for teaching in China as well as from an international perspective, or in-depth studies of some important and distinctive features of mathematics knowledge for teaching.
2. This book includes chapters that provide system and contextual information about school education and mathematics teacher education in different regions related to China, i.e., Hong Kong, mainland China, and Taiwan. It aims to help readers to learn about some substantial differences across these regions that otherwise are often perceived as no difference by outsiders. In this way, readers can hopefully appreciate more and better about the situation in mainland China that is being focused on in this book. The book is also designed with the inclusion of such chapters that can help readers to make possible connections of teachers' mathematical knowledge for teaching in mainland China with educational policies and program structures for mathematics teacher education in that system context (e.g., Li, 2014).
3. This book views knowledge acquisition and improvement as part of teachers' life-long professional learning process. Specifically, this book not only examines teacher preparation programs to learn how mathematics teachers are prepared for their professional teaching career, but also includes such a set of chapters that examine how mathematics teachers improve their mathematics knowledge for teaching through regular school-based activities and special programs.

Taken together, these intended aspects allow the book to make important contributions to help better understand teachers' mathematics knowledge for teaching and its acquisition and improvement in mainland China. As an important case in the international context, it provides valuable literature on advancing the understanding and development of teachers' mathematics knowledge for and in teaching and promotes further research and scholarly exchange across the globe.

TEACHERS' ACQUISITION AND IMPROVEMENT OF MATHEMATICAL KNOWLEDGE

STRUCTURE OF THE BOOK

With a focus on teachers' acquisition and improvement of mathematical knowledge for teaching, this book is designed with the inclusion of such chapters that can highlight specific conceptions about teachers' mathematical knowledge for teaching, system policies and context, different stages of teachers' knowledge development along their life-long professional learning process. These chapters are thus structured in four parts, with two of these parts to parallel teachers' knowledge development from pre-service preparation to in-service teaching practice and professional development: (I) Introduction and perspectives, (II) Acquiring and improving mathematics knowledge for teaching through teacher preparation, (III) Acquiring and improving mathematics knowledge for teaching through teaching and professional development, and (IV) Reflection and conclusion.

Part I contains four chapters. The present introductory chapter provides readers with an overview of the book. Chapter 2 describes the educational systems and teacher education systems in Hong Kong, mainland China and Taiwan. In contrast to a common perception about their similarities, Law, Tam, Huang and Lu highlight the differences in social-historical and socio-political developments across these three regions and emphasize their contributions to the differences in school education, curriculum structures and mathematics teacher education. Chapter 3 (by Hsieh, Lu, Hsieh, Tang, & Wang) presents a historical account of the conception and acquisition of mathematics teachers' literacy and its evolvement in China. In Chapter 4, Döhrmann, Kaiser, and Blömeke provide an overview of different descriptions and conceptualizations about mathematical knowledge for teaching, and further situate the discussion in a large-scale international study, Teacher Education and Development Study in Mathematics (TEDS-M, 2008).

Part II focuses on the stage of pre-service teacher preparation when teachers receive mathematical and pedagogical training through university-based program studies. Scholarly inquires and related research about teachers' knowledge acquisition at this stage have not received much attention until recently. Empirical studies revealed important differences across educational systems in terms of not only teacher preparation program differences but also the status of pre-service teachers' knowledge acquisition through their program studies (e.g., Li, Ma, & Pang, 2008; Schmidt, Blömeke, & Tatto, 2011). Three chapters are included in this part to present specific accounts about mathematical and pedagogical training provided in elementary and secondary teacher preparation programs in mainland China. In Chapter 5, Xie, Ma, and Chen discuss the current state of pre-service elementary teachers' mathematical knowledge for teaching and possible contributing sources to their knowledge acquisition and development. They collected data from pre-service teachers, teacher educators, teacher mentors, and administrators through surveys and interviews, in addition to the analyses of teacher preparation program curriculum documents. Based on the results, they emphasize the importance of further improving mathematical and pedagogical training to pre-service elementary teachers. In the next Chapter, Wu and Huang focus on secondary mathematics teacher preparation programs in China. They present and analyze such program documents to examine what are expected for pre-service secondary teachers, and also report what were found from a large survey study about pre-service secondary teachers' knowledge of algebra for teaching. Yuan and Huang (Chapter 7) highlight the importance of pedagogical training provided through pre-service teacher preparation programs. In particular, three types of practice-based pedagogical training (field observations, microteaching, and student teaching) are described, and a new type of Teaching Research Group activity called Same Content Different Designs (SCDD) activity is discussed. In her commentary Chapter (Chapter 8) for this part, Potari provides important account of all three chapters included in this part, discusses them with identified issues and challenges, and further offers insight that can be gained from reading these chapters for the international community.

Part III takes on in-service teachers' knowledge improvement through their teaching practice and professional development. Teachers' knowledge acquisition and professional development have been a focus for the mathematics education community for many years (e.g., Li & Even, 2011; Ponte & Chapman, 2016; Wilson & Berne, 1999; Zaslavsky, Chapman, & Leikin, 2003). It is likewise in China that how to help teachers learn and improve through their teaching and professional development has been a tradition with some important and unique features (e.g., Huang, Ye, & Prince, 2016; Li & Huang, 2013; Li, Huang, Bao, & Fan, 2011). The first five chapters, contributed by Chinese mathematics educators as insiders, share five different in-service teachers' learning and professional development opportunities spanning daily school-based activity to special programs. In Chapter 9, Pu, Sun and Li explore teachers' knowledge improvement through intensive textbook studies in a teaching research group. With the critical role of textbooks in Chinese school education, teachers use textbooks not only for classroom instruction planning, but also for helping improve their own knowledge about mathematics, curriculum and pedagogy. Huang and Huang (Chapter 10) present a case study of an experienced teachers' learning through his participation in a master teacher workstation program. The teacher's learning and knowledge improvement are exemplified and examined through his development and implementation of a new teaching approach. In Chapter 11, Yang and Zhang examine teachers' learning and knowledge development through the Teaching Research Group (TRG), a formal component of school structure that has been established in every school in China since 1952. Through a case study, they argue that Chinese teachers developed their Pedagogical Content Knowledge (PCK) mainly through discussing about and reflecting on issues related to classroom teaching in TRG activities. In Chapter 12, Liang focuses on teachers' development and observation of public lessons. In particular, she describes different forms of public lessons and how teachers' engagement in such activities is a productive way to help them acquire and accumulate necessary mathematical content knowledge for teaching. Li and Huang (Chapter 13) examine apprenticeship practices and how such practices provide opportunities for developing teachers' expertise and teaching practice through case studies. In the commentary Chapter (Chapter 14) for this part, Stillman provides not only a summary of these five chapters, but also a detailed recount and reflection of each chapter. She further indicates that nearly all different aspects of MKT, as conceptualized by Ball and her colleagues, have been developed in one way or another through different professional learning and development presented in these five chapters.

Part IV devotes to general reflection and conclusion for the book. It contains two chapters that help provide readers different lens in viewing teachers' knowledge acquisition and professional development in different systems and cultural contexts. Rowland in Chapter 15 reflects on recent initiatives by the UK government that emphasize the learning from several high-performing education systems in East Asia, especially mainland China. He discusses several important aspects, as associated with the situation in the UK and some related studies in the West, from reading the book including, textbooks, pre-service mathematics teacher knowledge, mathematics teachers' learning-in-community, mastery and morality. He concludes with an emphasis on the importance of learning from each other. In Chapter 16, Wu makes close connections between what can be learned from reading the book to the situation in the US. Based on the features of different programs and practices illustrated through different chapters, he makes a call for emphasizing on (1) teachers' acquisition of content knowledge that they will teach and (2) making teachers' knowledge acquisition and improvement as a lifelong learning and professional development process in the US.

REFERENCES

An, S., Kulm, G., & Wu, Z. (2004). The pedagogical content knowledge of middle school mathematics teachers in China and the US. *Journal of Mathematics Teacher Education, 7*, 145–172.

Ball, D. L., Thames, M. H., & Phelps, G. (2008). Content knowledge for teaching: What makes it special? *Journal of Teacher Education, 59*, 389–407.

Blömeke, S., Hsieh, F.-J., Kaiser, G., & Schmidt, W. H. (Eds.). (2014). *International perspectives on teacher knowledge, beliefs and opportunities to learn.* Dordrecht: Springer.

Depaepe, F., Verschaffel, L., & Kelchtermans, G. (2013). Pedagogical content knowledge: A systematic review of the ways in which the concept has pervaded mathematics education research. *Teaching and Teacher Education, 34*, 12–25.

Döhrmann, M., Kaiser, G., & Blömeke, S. (2012). The conceptualization of mathematics competencies in the international teacher education study TEDS-M. *ZDM: The International Journal on Mathematics Education, 44*(3), 325–340.

Hill, H., Ball, D. L., & Schilling, S. (2008). Unpacking "pedagogical content knowledge": Conceptualizing and measuring teachers' topic-specific knowledge of students. *Journal for Research in Mathematics Education, 39*(4), 372–400.

Hsieh, F.-J., Lin, P.-J., & Wang, T.-Y. (2012). Mathematics-related teaching competence of Taiwanese primary future teachers: Evidence from TEDS-M. *ZDM: The International Journal on Mathematics Education, 44*(3), 277–292.

Huang, R., & Li, Y. (Eds.). (2017). *Teaching and learning mathematics through variation.* Rotterdam, The Netherlands: Sense Publishers.

Huang, R., Ye, L., & Prince, K. (2016). Professional development system and practices of mathematics teachers in Mainland China. In B. Kaur, K. O. Nam, & Y. H. Leong (Eds.), *Professional development of mathematics teachers: An Asian perspective* (pp. 17–32). New York, NY: Springer.

Li, Y. (2014). Learning about and improving teacher preparation for teaching mathematics from an international perspective. In S. Blömeke, F.-J. Hsieh, G. Kaiser, & W. H. Schmidt (Eds.), *International perspectives on teacher knowledge, beliefs and opportunities to learn* (pp. 49–57). Dordrecht: Springer.

Li, Y., & Huang, R. (2008). Chinese elementary mathematics teachers' knowledge in mathematics and pedagogy for teaching: The case of fraction division. *ZDM: The International Journal on Mathematics Education, 40*, 845–859.

Li, Y., & Huang, R. (Eds.). (2013). *How Chinese teach mathematics and improve teaching.* New York, NY: Routledge.

Li, Y., Huang, R., Bao, J., & Fan, Y. (2011). Facilitating mathematics teachers' professional development through ranking and promotion in Mainland China. In N. Bednarz, D. Fiorentini, & R. Huang (Eds.), *International approaches to professional development of mathematics teachers* (pp. 72–85). Ottawa: University of Ottawa Press.

Li, Y., Ma, Y., & Pang, J. (2008). Mathematical preparation of prospective elementary teachers. In P. Sullivan & T. Wood (Eds.), *International handbook of mathematics teacher education: Knowledge and beliefs in mathematics teaching and teaching development* (pp. 37–62). Rotterdam, The Netherlands: Sense Publishers.

Ma, L. (1999). *Knowing and teaching elementary mathematics.* Mahwah, NJ: Lawrence Erlbaum Associates.

Ponte, J. P., & Chapman, O. (2016). Prospective mathematics teachers' learning and knowledge for teaching. In L. English & D. Kirshner (Eds.), *Handbook of international research in mathematics education* (3rd ed.). New York, NY: Taylor & Francis.

Schmidt, W. H., Blömeke, S., & Tatto, M. T. (Eds.). (2011). *Teacher education matters: A study of middle school mathematics teacher preparation in six countries.* New York, NY: Teachers College Press.

Yeping Li
College of Education and Human Development
Texas A&M University
USA
and
Shanghai Normal University
Shanghai, China

Rongjin Huang
Department of Mathematical Sciences
Middle Tennessee State University
USA

HUK-YUEN LAW, HAK PING TAM,
XIANHAN HUANG AND YU-JEN LU

2. EDUCATIONAL SYSTEMS AND MATHEMATICS TEACHER EDUCATION SYSTEMS IN MAINLAND CHINA, HONG KONG AND TAIWAN

INTRODUCTION

Under the umbrella of education reform, the mathematics curricula in Mainland China, Hong Kong and Taiwan underwent critical changes by the turn of the millennium, not just to cope with the various political, social and educational needs of these regions but also to align globally their own educational systems with those around the world. Consequently, the issues of how mathematics teacher education systems in these regions would have undergone their own transformations deserve our special attention. Such changes entail complex negotiations among various stakeholders, including the policy makers, the university teacher educators, the prospective and in-service teachers, the school administrators, and even the parents. In this chapter, we shall look into the mathematics teacher training programmes in the contexts of these three educational systems from three different perspectives: How have different regions made an attempt to address the issue of anchoring the linkage between educational theory and classroom practice in reforming their own mathematics education systems? How would the criteria for qualifying mathematics teachers have been established in these regions? And how have these regions developed their own mathematics teacher education programmes (including details about programme structures in these regions) to enhance the professional growth of both prospective and in-service mathematics teachers?

The educational systems in these three regions will first be briefed. Grounded in these different contexts, the mathematics teacher training (preparation and professional development) programmes in these three regions will be delineated and the issues related to the on-going innovations of their own mathematics education programmes will be discussed.

After briefly introducing teacher preparation and/or teacher professional development systems in these regions, this chapter will focus on how different education systems provide programmes for preparing in-service teachers; pathways to teacher certification. Additionally, the mechanism each system utilizes to continuously develop practicing teachers' profession knowledge will be discussed. Before drawing the conclusion of what lessons might be learned from the development of these three regions in education in general and in mathematics teacher education in particular, we will also discuss the commonly-shared concerns over the theory-practice divide and how it would be addressed if not be resolved in the three educational contexts.

EDUCATIONAL SYSTEMS IN MAINLAND CHINA, HONG KONG AND TAIWAN

Teacher education systems across the world evolve within and interact with social, economic, and political contexts and characterize themselves by three major components, namely, (1) the entry to the profession; (2) the processes of learning to teach (including the teacher education programmes' structure and curricular sequence); (3) the outcome of such learning experiences (Tatto, Lerman, & Novotná, 2009, p. 15). In the following section we explore how mathematics teacher education in the three regions is enacted in institutions offering teacher education programmes as well as their influences on teacher knowledge and practice.

Mainland China

In Mainland China, the eighth round of curriculum reform was introduced at the beginning of the 21st century. Compared with previous reforms, this round of reform contains fundamental changes and faces dramatic challenges in design and implementation (MOE, China, 2004). The curriculum reform covers two stages (nine years of compulsory education and high school). In September 2005, the Ministry of Education stated that all primary and junior secondary schools in Mainland China had implemented the newly-reformed curriculum (see Figure 1).

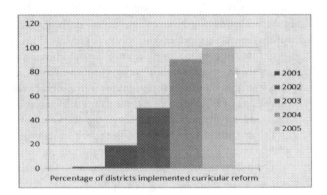

Figure 1. The percentage of province or city that implemented curriculum reform in Mainland China

The core of this curricular reform is to implement student-centred teaching. In 2010, the Outline of the National Plan for Medium- and Long-term Education

Reform and Development (2010–2020) (hereinafter referred to as the Education Plan Outline) was issued to guide the development of education for the next 10 years and also to emphasize the necessity and urgency of curricular reform implementation (MOE, China, 2010b). Based on the review on the decade-long implementation and the requirements of the Education Plan Outline, a total of 19 revised curriculum standards of all subjects in the compulsory stage were issued to correspond to the requirements of quality education, broaden and deepen the curriculum reform, and improve educational quality in Mainland China (MOE, China, 2011). For the mathematics curricular reform that began in 1999, the content standards include four fields: number and algebra, graphs and geometry, statistics and probability, and comprehensive practice.

How to train teachers to meet the requirement of curricular reform is a crucial issue during the reform. According to Education Minister Mr. Yuan (see Yuan, 2005), curriculum innovation would not be implemented without the comprehensive understanding and full support of frontline teachers. Therefore, several official documents and many workshops at various levels were introduced to direct teachers' training and intensify teaching quality improvement. In year 2011, the *Curriculum Standard for Teacher Education Programme (Experimental)* was issued to regulate the educational curriculum for pre-service teachers. In this document, three orientations were highlighted, namely, whole-person development-based, practice-oriented, and life-long learning. As for the content domain, this document emphasizes six aspects: adolescent development and learning, educational base, subject pedagogy and direction on student activity, psychological health and moral education, career ethic and professional development, and teaching practice (MOE, China, 2011b). A year later, three professional standards (experimental) for in-service teachers of kindergartens, primary schools, and secondary schools were issued consecutively (MOE, China, 2012) with four basic principles: emphasis on teachers' ethics, student-centered teaching, ability focused, and life-long learning. Clearly, these standards were designed to enhance the professional level of teachers and further promote the implementation of curricular reform.

Hong Kong

Since 1997 there have been a series of changes in the Hong Kong education system after the transition of its sovereignty from the United Kingdom to China. These include the changes to the language of instruction policies (advocating the use of Chinese as the main language in the classrooms) and to the senior secondary curriculum in line with those adopted in China and even the USA. As a result of the adoption of the new senior secondary curriculum, a completely different academic structure has been introduced in 2009. Under the re-structured education system (the so-called "3–3–4 scheme"), the students will receive a new Hong Kong Diploma of Secondary Education (HKDSE) after completing the six-year secondary schooling

(3 years for junior and 3 years for senior secondary) and the qualification of HKDSE will equip them to receive a further 4 years of studies of undergraduate university programmes or to enter a range of post-secondary, vocational and tertiary courses offered by a variety of institutions.

Under the reformed senior secondary mathematics curriculum (comprising of a Compulsory Part and an Extended Part), students may choose between three options: Option 1, take the Compulsory Part only; Option 2, take the Compulsory Part with Module 1 (Calculus and Statistics); Option 3, take the Compulsory Part with Module 2 (Algebra and Calculus). Lesson time for Option 1 involves about 10% to 12.5% (215 hours to 313 hours) while that for Option 2 or 3 entails up to 15% (375 hours) of the total lesson time available. All students will be expected to learn at least the Foundation Topics of the Compulsory Part whilst the mathematically more able students will be encouraged to take up at most one of the 'add-on' modules. The design of the new curriculum framework has highlighted the features of flexibility and diversity in order to cope with the diverse needs of students in the learning of mathematics for their future developments (Education Bureau, 2014). These features are reflected by the choices of the three options as offered in terms of their coverage of mathematics contents (amount of learning) as well as the abstractness of the mathematical concepts as embedded in the to-be-learnt topics (depth of learning). The rationales behind the curriculum design hint to target itself at the advocacy of the notions of 'effectiveness' and 'meaningfulness' in the learning and teaching of mathematics. These require teachers' extremely great efforts at using various pedagogies for delivering their classroom teaching in order to cope with the demands as derived from such a design. For example, the teachers can employ more sophisticated information technology tools or teaching aids to help students develop and apply abstract mathematical concepts. In addition, teachers need to strive at developing the generic skills and interest for their students through connections to daily life experiences in order to make the learning of mathematics more relevant and meaningful to the learners. Nonetheless, the teachers find it a great challenge to fulfil the newly-reformed curriculum aims of the Mathematics Education in two aspects. The first one is the challenge of the students' diversification in terms of their diversified learning needs for satisfying their different career pathways as well as their differentiated academic ability for the learning of mathematics. The other one is the challenge of creating time and space for fostering students' cognitive development leading to the accrual of the abstract thinking techniques for the solving of more complex mathematical problems. Such a challenge seems to be especially serious for the teaching of the Extended Part (Module 1 or Module 2) of the Mathematics Curriculum as the lesson time as allocated to it is much less than the time allocated to other independent subjects. Facing this time pressure, the teachers will find it hard to make use of even the normal lesson time (if not arranging extra lessons) for a more in-depth discussion of the abstract concepts as inherent in the contents of the Modules.

Taiwan

The pre-tertiary educational system of Taiwan is divided into 6 years of primary school education, three years of junior high and three years of senior high school education. Taiwan first implemented nine years free and universal education for its primary and junior high school students in 1968. School attendance, however, was not yet compulsory at that time. In 1979, the National Education Act was passed which specified that for those who reached school age but had missed the chance of receiving basic education should make it up by taking remedial education. It was not until 1982 when the Compulsory School Attendance Act was released, which required that all children between the age of six and fifteen must attend school (Government Information Office, 2006). Graduates from the junior high school can either pursue senior high school education or receive training in vocational schools. There had been time when ability grouping was practiced, especially in some secondary school classrooms. Yet in an effort to ensure equity of opportunity in learning for all students, the current official status stipulates that no ability grouping is allowed at any grade level (Tam, 2010). Furthermore, there is no streaming until the senior high school years.

Traditionally, Taiwan adopted a separate national curriculum at the primary, junior high and senior high school level of education. In 2001, the curricula for the primary and junior high level were unified under the official Grade 1–9 Curriculum Guidelines, which is the current curriculum at the time of writing. School mathematics learning is divided into four stages. Stage One spans from grades 1 to 2, Stage Two from grades 3 to 4, Stage Three from grades 5 to 6, and Stage Four from grades 7 to 9 (Ministry of Education, Taiwan, 2013). This system is going to change starting from the 2014 school year, when effort was invested to unify the Grade 1–9 Curriculum Guideline with the senior high school curriculum. As a result of this effort, the 12 Year Basic Education Curricula were compiled and will be implemented in 2019. Even though senior high level of education is not compulsory under the new policy, students will be subsidized by the government to attend schools at this level.

1994 marked an important year of change in the mathematics education in Taiwan when the Teacher Education Act was launched and was supplemented by the Teacher Education Law Enforcement Rules in 1995 (Li, 2001). In the past, teacher preparation in Taiwan was under the sole responsibilities of normal universities and teacher colleges, with the former training high school teachers and the latter training primary teachers. Starting in 1994, however, the new law allowed all universities to set up teachers training programmes. In order to maintain their enrolment, all the original teacher training institutes gradually changed their status towards comprehensive institutes. Meanwhile, the teacher colleges changed their names to universities of education to signify that they are now regular universities rather than special purpose universities. While this opens up more opportunities for those who are interested in becoming teachers to pursue their dreams, not all universities that attempt to set up teacher training programmes have adequate facilities or faculty to fulfil their responsibilities, thereby resulting in programmes of various levels of qualities (Lin, Wang, & Teng, 2007). Moreover, this change renders the supply of teachers into a big surplus later on.

There are two major issues regarding the mathematics teacher education system in Taiwan. A major concern is about the setting up of training programmes for pre-service teachers. The issues here centre on the identification of those who are suitable to become mathematics teachers as well as on providing them with appropriate training afterwards. Another matter of equally high concern is on how to deliver continuous training to in-service teachers. Since the socio-cultural environment changes all the time, curricula have to be reformed and implemented from time to time. How one should update the in-service teachers with details of the curriculum changes and equip them with new repertoire to teach are issues that policy makers need to address. For a while, in-service mathematics teachers received such information through a three-tier curriculum and instruction consulting mechanism that delineates the concept of reform agents and is done by way of promotion through the central government, local government and school subject leaders. They form a concerted effort to see through changes in the instructional practices of in-service teachers according to the spirit of the reformed curriculum (Chung, 2007).

MATHEMATICS TEACHER PREPARATION IN THE THREE EDUCATIONAL SYSTEMS

Mainland China

Teacher Qualification Certification (TQC) has been mandatory for those who want to become a teacher in Mainland China since 1995 (MOE, China, 1995). TQC requires candidates to have three aspects of knowledge: education, educational psychology, and proficient usage of Putonghua (the official national spoken language for the Mainland). Graduates from any normal university only need to take the Putonghua Proficiency Test because they already finished the former two aspects during their stay in the university. For others, they can only apply for TQC after passing the examinations on these three aspects. On the whole, most mathematics teachers (at middle and secondary school levels) in China usually graduate with a major in mathematics.

However, with the increasing requirements and social expectations in the curricular reform, the old mechanism could hardly guarantee the quality of teachers. Thus, two new mechanisms were introduced to identify teachers' qualification in recent years. The first mechanism is the new examination for teachers' qualification (MoE, 2011). It prescribes that all graduates, including students from any normal university, who want to become teachers will have to take the new examination. The methods and subjects of the exam differ depending on the stage (Table 1).

Table 1. Methods and subjects of the new examination for teachers' qualification

	Methods	Subjects of written examinations
Kindergarten	Written examinations and interview	Synthesis quality, knowledge and skills in child caring
Primary school	Written examinations and interview	Synthesis quality, educational knowledge and skills
Secondary school	Written examinations and interview	Synthesis quality, educational knowledge and skills, subject knowledge, and instructional abilities
Middle level Professional school	Written examinations and interview	Synthesis quality, educational knowledge and skills, professional knowledge, and instructional abilities

The content tested in the new examination was extended to five aspects:

- Basic teacher abilities, such as educational belief, career ethics, scientific literacy, reading and understanding, language expression, logical reasoning, and processing information
- Basic knowledge of the curriculum, student mentoring, and classroom management
- Basic knowledge of the subject
- Knowledge and methods of instructional design and implementing evaluations
- Ability to analyse and solve practical problems with learned knowledge

The second mechanism is the regular registration and validation of the certification for teachers. That is, all teachers who have had the TQC will be reassessed and revaluated every five years. The teacher may lose his/her certification if he/she cannot pass the regular assessment. The inspection covers five domains: career ethics, yearly assessment on performance, professional training, healthy body and spirit, and some criteria set by the educational department at the provincial level. To date, 13 provinces have become experimental districts and have implemented these two mechanisms. Through these mechanisms and by strengthening the management of teacher qualification, the central government hopes to improve the quality of teachers.

New mathematics teachers in the experimental regions have educational backgrounds similar to the previous teachers. However, the new teachers need to pass a strict and high demand examination before becoming a school teacher. Moreover, all teachers in these districts will have to experience regular teacher qualification validation every five years.

Hong Kong

All school teachers must be registered under the status either as 'permitted teacher' or 'registered teacher' under the Education Ordinance. A permitted teacher (PT) refers to the one who meets academic qualifications but has not yet received the teacher training that leads to the teacher qualification as required whereas a registered teacher (RT) holds teacher certifications such as post-graduate Diploma/ Certificate in Education (Education Bureau, 2014). The former after completing in-service teacher education will also be eligible to gain the qualified-teacher (RT) status.

There are five teacher education providers in Hong Kong. They offer a range of sub-degree, degree and postgraduate programmes for pre-service and in-service teachers. Among these, the Education University of Hong Kong (EdUHK)[1] is the largest teacher education provider. In the 2013/2014 academic year, there were about 4,500 full-time and 4,200 part-time students enrolling for the University Grants Committee (UGC)-funded programmes. The Hong Kong Baptist University (HKBU), the Chinese University of Hong Kong (CUHK), and the University of Hong Kong (HKU) also offer UGC-funded degree programmes for pre-service teachers and postgraduate programmes for pre-service and in-service teachers. The Open University of Hong Kong (OUHK), which is self-funded, offers degrees and postgraduate programmes for pre-service and in-service teachers (Hong Kong Government, 2014). Graduates of teacher education programmes from the institutes are eligible to become registered teachers. Up to the present, most mathematics teachers in Hong Kong are degree holders with teacher qualifications echoing what the Government has advocated.

Normally there are two pathways to becoming mathematics teachers in primary or secondary schools. First, students who have gone through HKDSE would apply for the five-year bachelor's degree full-time programme in education. In 2012, CUHK initiated a new Bachelor of Education (B.Ed.) programme of Mathematics and Mathematics Education with which the graduates are eligible to become a qualified mathematics teacher serving in both primary and secondary schools. Second, students who have a general bachelor's degree would apply for a post-graduate Diploma/Certificate in Education with Mathematics Major in either Primary or Secondary Subject Curriculum and Teaching (SCT). It normally takes one year for full-time study and two years part-time study for in-service teacher.

Other than the new B.Ed. programme as offered in CUHK, Table 1 delineates the mathematics teacher education programmes being provided by other institutes in Hong Kong. As shown in Table 2, the CUHK programme distinguishes itself by its *double major* (Mathematics and Mathematics Education) mode of study that equips the students to teach mathematics from Primary 1 up to Secondary 6 level[2] (the specifications and features of the programme will be delineated in a later section about Hong Kong).

Table 2. Mathematics teacher education programme in Hong Kong

Teacher education providers	Bachelor of education with mathematics major	Post-graduate diploma in education with mathematics major[#]
EDUHK	Primary	Primary/Secondary
HKBU	Nil	Primary/Secondary
CUHK	Primary and Secondary*	Primary/Secondary
HKU	Nil	Secondary
OUHK	Nil	Nil

[#] Separate programmes with Mathematics Major for either primary or secondary education
* Integrated Mathematics Education programme for both primary and secondary education

Taiwan

All universities that housed a teacher training programme will organize an annual qualification examination for interested undergraduate and graduate students from practically all majors to participate on the condition that they satisfied a certain set of criteria. The content matter being tested does not depend on majors. Instead, it usually surrounds such areas as educational theory, psychology, laws and assessment in relation to education. The set of criteria may include consideration pertaining to language abilities, psychological tests and familiarity of educational policies and news. Those who pass the examination will become pre-service teacher candidates. They must then undergo complete course training before they can participate in a half-a-year long teaching practicum. Successful candidates who complete all training course requirement will obtain diplomas from their programmes. They can then apply for the official teacher certificate examination (Teacher Education Act, 2014). Table 3 below shows the current subject requirement for the teacher certificate examination in Taiwan.

Table 3. Subject requirement for the teacher certificate examination for primary and secondary pre-service teachers

Exam subjects	Category 1	Category 2	Category 3	Category 4	Category 5
Primary level	Chinese language ability test	Educational principles and policy	Children development and counselling	Primary curriculum and instruction	Mathematics ability test (general mathematics and teaching method)
Secondary level	Chinese language ability test	Educational principles and policy	Adolescent development and counselling	Secondary curriculum and instruction	

Source: Ministry of Education, Taiwan (2014)

PRE-SERVICE TEACHER EDUCATION: ACQUIRING MATHEMATICS KNOWLEDGE FOR TEACHING

Mainland China

Teacher training programmes should focus on developing teachers' knowledge and reason towards their practice (Robertson, 2000; Tatto, 2007). Thus, this section of the paper addresses this issue by describing the pre-service teacher education in Mainland China.[3] Prior to the expansion of higher education in Mainland China, a bachelor's degree is the highest attained by most mathematics teachers.[4] With the increasing expectations on student teachers, certain normal universities have developed a new programme called the 4+2 model to cultivate high-quality pre-service teachers for schools. Thus, two models have been developed to cultivate future teachers. In this section, both the bachelor programme in normal universities and the 4+2 model will be introduced to illustrate the knowledge included for pre-service teachers of mathematics in Mainland China.

Normal universities equip their students by providing two approaches of courses and teaching practices. The first approach encompasses two aspects, namely, subject knowledge and education-related knowledge (Table 4).[5]

Given the two examples, a considerable number of mathematics courses and physical courses are expected to provide student teachers with solid knowledge on the subject. By contrast, the disparity on the number of educational courses in these two universities is obvious. Compared with University B, University A provides more educational courses to fulfil students' needs to become teachers.

As previously described, given the increasing requirements of curriculum reform, the existing programme remains inadequate. To develop high quality and research-oriented teachers and instructional administrators, some normal universities develop the 4+2 model providing a two-year degree of Master of Education to students with bachelor's degree. This model intends to consolidate the student teachers' educational knowledge and abilities. In University C, which is another key normal university directly administered by the Ministry of Education (MOE) (national level), a student teacher majoring in mathematics is required to take many courses on curriculum and instruction (Table 5) (2010). The 4+2 programme also provides students with an extended period of practice teaching (from 2 to 3 months). This opportunity allows student teachers to experience authentic instructional problems in actual school life.

Hong Kong

To cope with children's needs to take mathematics up to 12 years under the new school curriculum framework in Hong Kong, the B.Ed. double major programme in Mathematics and Mathematics Education as offered by CUHK is innovative by itself in terms of the design of its curriculum structure that embodies strands of study enabling students to acquire strong subject knowledge as well as the knowledge

Table 4. Programme designed for mathematic novice teachers of two normal universities at different levels

	University A (National level, under the direct administration of the Ministry of Education)	University B (Provincial level)
Goals on subject knowledge	1. Master extended mathematical knowledge and skills 2. Know the update status and achievements of mathematical science as well as common knowledge related to mathematics.	1. Solid and broad mathematics knowledge must be mastered, and the trend of mathematics development must be known. Good thinking abilities, such as space imagination, logical thinking, abstract thinking, and sensitive and divergent thinking, must be obtained to enhance the potential learning ability for further development. 2. Broad cultural literacy, good mentality, scientific thinking, and accurate expression of mathematics language must be achieved to ensure the competent teaching of mathematics in secondary school and the curriculum development of more than three elective mathematics subjects. 3. Modern educational technology and basic mathematics skills must be mastered to solve real-world problems. 4. The basic principles of education and psychology and the basic skills for conducting independent educational research must be mastered. The basic skills related to mentoring and classroom management must be honed. 5. The skills for accessing, organizing, and analyzing online information must be mastered. The skills for reading and translating common mathematics literature must also be obtained. The basic skills for writing papers on mathematics and mathematics education should likewise be honed.

(Continued)

Table 4. (Continued)

	University A (National level, under the direct administration of the Ministry of Education)	University B (Provincial level)
Subject knowledge (compulsory courses)	Mathematical Analysis 1&2&3/Advanced Algebra 1&2/Analytic Geometry/Modern Algebra/Ordinary Differential Equation/Function of Complex Variable/Function of Real Variable/Differential Geometry/Foundations of Probability Theory/Functional Analysis/Digital Approximation/Topology/Statistics 60 credits Basic Physics 1&2/Experiment of General Physics 6 credits	Mathematical Analysis 1&2&3/Advanced Algebra 1&2/Analytic Geometry/Foundation of computer culture/C++/Ordinary Differential Equation/Functions of Real Variable/Probability Theory and Mathematical Statistics/Function of Complex Variable/Introduction of Mathematics 54 credits Basic Physics, 4 credits
Education-related (compulsory courses)	Teacher and Instruction/Educational Psychology in School/Adolescent Psychology/Mathematic Curriculum and Instruction in Secondary School/Research on Curriculum Standard and Textbook in Secondary School/Educational Research Method/Modern Educational Technique/Professional Skills Training for Teachers/Mathematics Micro-Teaching in Secondary School/Application of Informational Technique in Mathematical Instruction 14 credits	Psychology of Mathematics Education Education Mathematics Pedagogy Modern Educational Technology Educational Research Methodology Training of Teaching Skills 15 credits
Teaching practices	Class observation, teaching practice, and educational survey (no more than 1 month) 7 credits	Class observation and teaching practice (twelve weeks) 11 credits

Table 5. Compulsory and core courses provided by 4+2 model (Master stage at University C)

Education-oriented (6 courses)	Educational theory/Educational research method/curriculum and instruction/Teacher ethic/Modern educational techniques/Eastern and western educational history
Mathematics education-focused (select 2 from 4)	Mathematical educational psychology/Theory of mathematical instruction/Research of basic mathematics/Modern mathematics and mathematics in secondary school
Mathematics-related (3 courses)	Functional analysis Abstract algebra (Algebraic topology or Differential geometry) Basic computer theory

as entailed within the domains of mathematics teacher education and pedagogy (see Table 6). To ensure that the student teachers of the programme are to have a rigorous training in mathematics, they are required to take a number of advanced mathematics courses (apart from a variety of elective courses) including Linear Algebra, Advanced Calculus, Mathematics Laboratory, Mathematical Analysis, and Complex Variables with Applications. The emphasis on a strong foundation in mathematical training equips the pre-service teachers with sound mathematical knowledge for visualizing the future need of understanding how teaching school mathematics can be done through the classroom communication (Sfard, 2007) from a mathematically advanced standpoint.

Table 6. Curriculum structure of B.Ed. double major programme for pre-service mathematics teachers

Courses	Credits
Mathematics related courses	47
Mathematics Education and Education	55
University core requirements	45
Total	147

Taiwan

In the past, pre-service primary teachers learned their mathematics via two routes depending on their background. Those who enrolled in mathematics related majors, they were required to take about twelve credit hours of mathematics courses. Those who enrolled in non-mathematics related majors needed to take two credit hours of general mathematics and two credit hours of a method course on mathematics. Only graduates whose majors were in mathematics could serve as mathematics teachers in secondary schools. However, under the Teacher Education Law of 1994, pre-service teachers at all levels must first be qualified to enroll in teacher training programmes. Whether pre-service teachers at the primary level need to take general mathematics and method course on mathematics is left to the discretion of individual education programmes. At the secondary level, pre-service teachers from all majors can prepare themselves by taking thirty credit hours of mathematics to become junior high school mathematics teachers and thirty-five credit hours to become senior high school mathematics teachers (Teacher Education Law Enforcement Rules, 1995).

Table 7 showed the official curriculum requirement for pre-service teacher education in Taiwan. A typical pre-service teacher education programme is composed of three main sections, namely, a specific subject curriculum, a professional curriculum and a general curriculum. Since all teachers are usually required to teach various subjects in primary schools that are beyond their majors, pre-service primary teachers are hence required to take ten credit hours for basic subject content courses. This is not required for pre-service secondary teachers because in Taiwan, only mathematics majors can teach mathematics in secondary school as required by law (Bill TaiChung-0920141412, 2003). The professional curriculum includes education courses plus instructional method and practices courses. Finally, the general curriculum refers to course requirement, such as general studies, that must be abided by all undergraduate students.

Table 7. The official curriculum requirement for pre-service teacher education

Course categories		Primary	Secondary
Required courses	Primary school teacher basic subject content courses	10	/
	Secondary and primary school teacher basic education courses	4	4
	Secondary and primary school teacher pedagogical knowledge courses	6	6
	Primary school teaching practicum and instructional method and practices courses	10	/
	Secondary school teacher teaching instructional method and practices courses	/	4
Electives		10	12
Total credits		40	26

Source: Bill TaiChung-0920141412: Course Requirement in Secondary and Primary School Pre-service Teacher Education Programmes (2003).

There is an emerging issue for which all mathematics teacher education programmes in Taiwan must prepare for themselves. Starting from the 1990's, there is an obvious drop in birth rate, the impact of which is now taking shape in our classrooms. Currently, the average class size for primary and secondary schools is dwindling. As a result, not as many teachers are needed as before on the job markets. In some schools, after some senior teachers have retired, their posts are not filled by recruiting formal teachers. Instead, substitute teachers are hired to fulfil the teaching duties. The effect of low birth rate is now directly affecting many teacher training programmes, many of which are encountering low enrolment problems (Lin, Wang, & Teng, 2007). A recent example is that the mathematics education programme of one university is merged into the Science Communication department. There is a possibility that various resources for mathematics teacher training may dwindle or spread thin as a result of this merge. This trend, perhaps being inevitable, will certainly throw doubt on the capacity of the departments if they can prepare future teachers with enough mathematics background for teaching. To meet such a challenge, new policies need to be made to delineate the mathematics and pedagogical requirement for future teachers.

CONTINUING PROFESSIONAL DEVELOPMENT: IMPROVING MATHEMATICS KNOWLEGE FOR TEACHING

Mainland China

Promoting teachers' professional development and life-long learning has been a crucial issue during curriculum reform. In Mainland China, various approaches have been developed for teachers' continuous learning. The first one is voluntarily pursuing Master of Education (professional degree), and the second is the teachers' continuous educational training.

The majority of normal universities provide Master of Education (professional degree) for in-service teachers who wish to broaden their educational knowledge and skills. University C, for example (Table 8), provides a programme that primarily aims to introduce extensive related knowledge related to daily instruction.

Table 8. Programme design of master of education (professional degree) at university C

Core courses (18 credits)	English/Politics (including moral education on teacher career)/ Principles of education/Educational technology/Educational psychology/Educational research methodology
Compulsory courses (12 credits)	Mathematics education/Assessment and evaluation of mathematics education/The history and philosophy of mathematics education/ Modern mathematics and secondary mathematics
Selective courses (select 2 courses) (at least 4 credits)	Psychology of mathematics education/Power point for mathematics teaching/Research cases of mathematics education/The principles of solving mathematics problems and mathematics competition/ Research on thinking skills of mathematics

Continuous professional training has become an urgent need among frontline teachers since the beginning of the curriculum reform. At the initial stage of the reform, teachers should attend several workshops provided at various levels to learn the core values and the recommended instructional model of the curriculum reform. In-service teachers are then required to assimilate new knowledge and develop an understanding of various educational theories by participating in different educational training programs. There were three types (levels?) of teacher professional development programmers. The first type is the National Teacher Training Program, which was introduced in 2009 (MoE, 2009). The programme aims to cultivate excellent teachers and to develop fruitful educational resources. Most of the projects conducted under this programme cover teachers handling a variety of subjects; hence, they may not cater specifically to the needs of mathematics teachers. The second type of professional development programme is conducted at the provincial and municipal level. Take city C as an example. The professional courses for in-service teachers to take include six aspects:

- Moral education and ethics of teachers
- Update or extension of professional knowledge
- Modern educational theory and practice
- Scientific research in education
- Instruction on skills training and modern educational techniques
- Modern technology and knowledge on social science

Theoretically, teachers of City C can select interested courses and study them either online or by attending a local university. However, the provided courses are limited, and teachers do not have much freedom to select appropriate courses. In 2014, mathematical modelling is the only course offered that is related to mathematical instruction.

The third type is a school-based teaching research system, which was initiated by the Minister of Education in 2010 (MoE, 2010). The system suggests the selection of problems that evolve into research questions during the teaching process and encourage teachers to take the lead during research with the support of academic scholars. This approach is designed for both the improvement of frontline teaching and the professional development of teachers.

Hong Kong

Any recommendations or teacher education programmes which are targeted at enhancing the effectiveness of teachers' continuing professional development (CPD) should put their emphases on promoting the improvement of teachers' subject content knowledge, pedagogical content knowledge, self-efficacy in teaching, as well as teaching and learning activities in the classroom (Whitehouse, 2011). Since 2006, the in-service mathematics teachers, like the teachers of any other subjects, are required to engage themselves in continuing professional development activities of not less than 150 hours in a three-year cycle (Pattie, 2009; see also ACTEQ, 2003). To further enhance professional development through post-graduate studies, they can enrol in some of the specialized mathematics education Master's degree programmes such as those offered by EdUHK (MA in Mathematics and Pedagogy) and CUHK (MSc in Mathematics Education). The programme curriculum of the former includes four core courses in Mathematical Studies (12 credits), and four core courses in Pedagogy in Mathematics (12 credits) whereas the study scheme of the latter entails the courses covering 9 credits of Mathematics courses (offered by the Department of Mathematics), 9 credits of Mathematics Education courses and 6 credits of Education courses (offered by the Faculty of Education). The design of these programmes has highlighted the changes in teachers' subject content knowledge and pedagogical content knowledge.

Taiwan

Starting from 2006, the Ministry of Education of Taiwan has set up a curriculum and instructional consultation mechanism. Within this system, invited mathematics education professors will guide and interact with expert teachers who are members of the MOE curriculum and instruction consulting team and local subject lead teachers, who in turn will guide and interact with classroom mathematics teachers. The idea here is to nurture the members of the consulting team and local subject lead teachers in their disposition of leadership in mathematics teaching and building up their roles as reform agents, thereby facilitating professional development of in-service mathematics teachers and the induction of curriculum reform in the actual classrooms.

Members of the MOE curriculum and instruction consulting team are selected from outstanding local subject lead teachers. All subject lead teachers will need to undergo a series of preparation courses and theme-based symposiums (Lu & Chung, 2010). It is expected that these members of the consulting team, upon completion of their training, can organize training sessions or workshops for mathematics teachers from various cities and counties, through which they can introduce what they have learned and facilitate the promotion of important skills and ideas to a broader circle.

Another approach that institutes continued education is by organizing various workshops for in-service teachers to attend. There are many workshops offered in relation to pedagogical knowledge, pedagogical content knowledge and instructional technology. For example, there is an 18 hours workshop attendance requirement set by the Ministry of Education for teachers to prepare themselves for the new 12 Year Basic Education Curricula. There are, however, relatively few workshops directly on mathematics as a subject by itself, such as those that introduce newer development of mathematics as well as their potential applications to school mathematics. This kind of enhancement in content knowledge is left to the responsibility of in-service teachers. They are expected to exercise self-regulated learning by reading relevant journal articles for primary to secondary school instructors. Some of these journals are organized by mathematics associations and the others by different publishers of mathematics textbook. Lastly, in-service teachers can pursue graduate degrees in mathematics education. Most masters and doctorate programmes in mathematics education require their students to take a certain amount of credit hours from mathematics departments to strengthen their mathematical background.

THE THEORY-PRACTICE DIVIDE: DEVELOPING STRATEGIES IN THE THREE REGIONS TO OVERCOME THE GAP

Mainland China

Based on the different programmes for mathematics practitioners in university A, B, and C, an extensive number of subject courses and limited educational courses are arranged in normal universities in Mainland China. Furthermore, there is a dearth of courses that can integrate theory and practice. Tchoshanov (2011) stated that teachers with more knowledge of the concepts and applications can significantly promote student achievement. Those with higher ability to connect mathematics and teaching are equally vital in enhancing student success. Therefore, the programme designers are suggested to begin by building linkage between abstract concepts and lively classroom instruction.

In addition, the problems arose from the implementation of the 10-year curricular reform, thereby also providing some valuable implications for the improvement of our teacher professional system. Many mathematics teachers still perceive the reform goals as ideal and unrealistic (Li, 2012). They find it difficult to implement innovative ideas in their classrooms (Liu, 2001; Ma, 2012; Yin, 2008). Compared with the limited educational courses provided at normal universities in Mainland China, it is suggested that the courses or workshops on how to understand the underpinning values of curriculum reform and how to effectively implement reform principles and curriculum changes should be included in the professional programme. Third, based on surveys, Han and her colleagues (2011) found that a mathematics teacher's pedagogical content knowledge of mathematics is a distinguishing feature that differentiates experts from mediocre teachers. Therefore, related course objectives to develop practical skills and instructional methods of teachers should also be included in the training systems for student teachers (Xu & Gu, 2014). Fourth, the knowledge of mathematics teachers, such as knowledge about the subject's historical and cultural background (Li, 2013) and its assessment and evaluation (Huang, 2011), remains deficient. Fifth, Chinese students performed well in international mathematical competitions and assessments (e.g., IMO, IAEP, PISA), which indicates that traditional mathematics education in China has some advantages (Han et al., 2011). Mathematics teachers in Mainland China find that integrating traditional instructional methods with current Western theories is a challenging issue (Zheng, 2005; Ma, 2012); this is another critical topic that should be addressed in designing and implementing teacher education programmes in China.

To supplement the knowledge described earlier about the current teacher training systems in Mainland China, the strategies of two approaches, namely, bottom-up and top-down, should be considered. For the top-down approach, the programme designers of universities must enhance the collaboration between tertiary institutions and frontline schools to access and organize practical and crucial subject knowledge for effective teaching and to integrate the knowledge with the current system. For the bottom-up approach, the school-based research system, which has already been introduced in Mainland China, must be further strengthened by providing additional resource support. The quality of classroom instruction and the professional abilities of teachers can both be deeply improved through research on daily instructional issues.

Hong Kong

All teacher educators have to face the challenge of addressing the theory-practice divide. Efforts have to be made to develop strategies to help the pre-service teachers develop their own theories grounded in practice (Holland, Evans, & Hawksley, 2011). Based on the interpretative framework, Everington (2013) advocates the use of action research as an effective way of bridging the theory-practice gap by fostering the pre-service teachers' capability to develop their own awareness from both the theoretical and practical aspects. Such capability is vital for all teaching practitioners as they need to make practical judgments in their own practice. Through their own participant observation, reflection, and theorizing, they create new ways of understanding the theory of developing the teaching strategies to improve their ongoing practice. In the case of Hong Kong, the newly-developed B.Ed. double major programme in Mathematics and Mathematics Education offered by CUHK included two courses, namely, the Action Research in Mathematics Classroom and the Mathematics Teaching Project Report. The purpose is to help the pre-service teachers develop through research experiences the ever important 'mind tool' for enhancing their own awareness whilst doing the teaching practice. In the course of the former, the key ideas and concepts related to lesson study, learning study (see Lo & Marton, 2012), as well as design experiment would also be included.

Under the paradigm of technical rationality, the traditional professional practices, such as in the fields of education and nursing, rely on the application of generalizable, university research-based knowledge. Yet, as argued strongly by Rolfe (1998), such practice would lead to the existence of the theory-practice gap unresolved. To overcome the theory-practice divide, the teachers, if they desire to have further professional development, would have to *reinvent* their teaching (Law, 2013) by reflecting *philosophically* about the values as expressed by their teaching practices (Laverty, 2006). As a way of advocating the in-service mathematics teachers to make a difference in practice, a course entitled "Researching Action in

Mathematics Teaching", grounded in the context of action research (Cain, 2011) and also in the adoption of lesson study (Kieran, Krainer, & Shaughnessy, 2013) as means of building connections between theories and practices, has been offered in the CUHK's MSc in Mathematics Education programme. In the context of a university-based course, the teachers would have chances to conduct inquiry into their own practice and thereafter to draw the concerted efforts at making the quality professional judgments through the practitioner-based research in their own settings.

Taiwan

It is well known that mathematics content knowledge and mathematics pedagogical content knowledge (MPCK) are two key components that should be part of the knowledge structure possessed by any mathematics teacher (Delaney, Ball, Hill, Schilling, & Zopf, 2008; Hill, Ball, & Schilling, 2008). Specific content knowledge is important in order for teachers to function as effective communicators by "decompressing" mathematical knowledge to the extent accessible by students. According to the data presented in Figures 2 and 3, Taiwan's pre-service teachers performed relatively well in terms of both mathematics content knowledge and

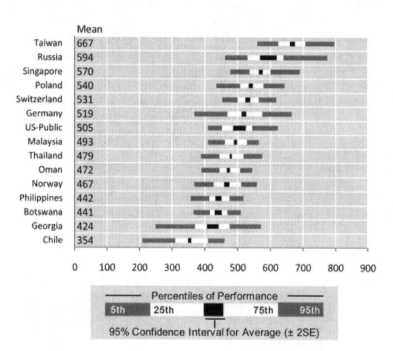

Figure 2. Countries' performance in mathematics content knowledge in TEDS-M at the lower secondary level.
Source: Hsieh, Wang, Hsieh, Tang, Chao, Law, Lin, Yiu, & Shy (2010)

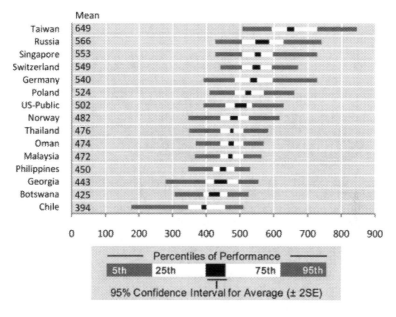

*Figure 3. Countries' performance in mathematics pedagogical content knowledge in TEDS-M at the lower secondary level.
Source: Hsieh, Wang, Hsieh, Tang, Chao, Law, Lin, Yiu, & Shy (2010)*

MPCK when compared to their peers from other countries that participated in the Teacher Education and Development Study in Mathematics (TEDS-M) (Hsieh et al., 2010; Tatto et al., 2012). Similar performances were also observed from the pre-service primary teachers participating in the same study. Hence based on this international comparative study alone, one may be tempted to assume that pre-service teachers in Taiwan have relatively strong foundations in specific content knowledge to teach mathematics.

In addition, a local report based on the TEDS-M study (Hsieh, 2012) indicated that of the 365 stratified sampled pre-service secondary school mathematics teachers, 25.8% of them reported that they have either Master's or Doctorate degrees in mathematics. Meanwhile, of the 923 stratified sampled pre-service primary teachers, 5.7% of them reported that they have graduate level degrees in mathematics. As for the in-service sector, data from the Ministry of Education (MOE, Taiwan, 2013) indicates that about 42%, 41% and 61% of the primary, the junior high and the senior high school teachers, respectively, hold Master's degrees from various disciplines. This certainly reflects that there is in Taiwan a growing trend for in-service teachers to pursue graduate degrees. One would probably assume that teachers with higher degrees would be better able to teach mathematics. A corollary to this assumption is that students studying under teachers with higher degrees will learn better than those studying under teachers without higher degrees.

There is, unfortunately, not much research done locally to confirm these assumptions. Rather, there are grounds to worry about pre-service teachers' knowledge in mathematics. For example, one should be aware that Figures 2 and 3 reflected the overall performance of Taiwan's pre-service teachers in the TEDS-M study. There are actually reports which pointed out that there were significant differences in mathematics content knowledge and MPCK among pre-service mathematics teachers trained by different teacher training programmes (e.g., Lin & Hsieh, 2011). This reflects that some pre-service teachers trained under some programmes may not be as qualified as one would prefer them to be. Despite the prevalence of attaining advanced degrees by teachers, there also lurks a hidden worry in terms of the in-service training for mathematics teachers. Most teachers in Taiwan regard attaining a Master's degree as the final stage of their academic growth. But as observed above, not many teachers pursued graduate training directly from the mathematics departments. Besides, most in-service training sessions being offered are not directly on mathematical topics. As a result, most teachers' knowledge of mathematics might merely maintain what they had learned from their university days, unless they have pursued further reading and personal enhancement on their own. Even if they had enrolled in graduate studies, most graduate courses offered in the mathematics departments were specialized courses and not catered for school teachers nor directly related to school mathematics. For example, there are studies that indicated some in-service teachers might not have enough specific content knowledge on concepts as general as sampling in the mathematical sense (Wang, 2013). Furthermore, Wang's study specifically followed a teacher with ten years of teaching experience and good content knowledge, but with deficiency in terms of pedagogical knowledge. The author pointed out that merely following the order of presentation in the textbook is not enough to deliver the lessons successfully, thereby affecting students in learning more in-depth knowledge about the mathematical concept of mean. This kind of discrepancy between theory and practice hampered the teachers in their delivery of instructional practices in classrooms.

This kind of discrepancy may not be efficiently handled by involving the teachers in doing action research alone. Rather, a better remedy is to organize more courses or workshops on school level mathematics from an advanced viewpoint. Such courses will help in-service teachers to see more structures in mathematics as well as the connection between advanced mathematics and school mathematics. But who can offer such courses for in-service teachers? Perhaps some mathematics education professors can handle this kind of courses. Still better, it will be more efficient to involve some mathematicians to join hands with mathematics education professors to organize high quality workshops that can enhance in-service teachers' view of school mathematics from an advanced standpoint. They can then be in a better position to teach mathematization and mathematical reasoning rather than merely teaching problem solving techniques according to various problem types. On the other hand, technical workshop on instructional technology, including those on the use of GSP and GeoGebra, can also be arranged for in-service teachers since these tools allow more chances for students to explore the properties of mathematical objects rather than passively observing demonstrations from their classroom teachers. There is definitely a place for mathematicians to participate in professional development programmes for in-service teachers.

DISCUSSION AND CONCLUSION

Within the limited scope of this chapter, we have attempted to depict, though not exhaustively the current scenarios of the three regions in terms of their latest developments in mathematics teacher education. In the context of the ongoing changes and developments of our own educational practices, the challenges ahead are inevitable as efforts have been drawn to continuously improve the educational system in general and the mathematics teacher education system in particular across the regions. In the case of the Mainland China, the educational system is still undergoing its own transformation that will bring in, to a certain extent the complexity of reshaping its teacher education system. The challenges ahead to the teacher educators entail the issues such as the unification of the teacher education curriculum framework and the standardization of teacher qualifications. In Hong Kong, the teacher education distinguishes itself by its own features of flexibility and diversity in terms of the design of curriculum framework and the pathways to the qualifications of teaching practitioners. The challenges ahead to teacher educators will be enormous as they need to keep a close eye on the newly-developed mathematics teacher education programmes in response to the recently established educational system (the "3–3–4 scheme") and the foreseeable review of the new senior secondary mathematics curriculum. In Taiwan, efforts have been made to further strengthen the rigor of teachers' mathematical knowledge through the framing of the curriculum requirement for pre-service teacher education, particularly when there is a down turning of the teacher job market due to the low birth rate. Now that fewer teachers are in demand, higher requirements can be imposed on the selection of new teachers, thereby resulting in a more elite group of mathematics teachers at the entrance level. Efforts are now directed towards maintaining Taiwan's advantage as reflected from the TEDS-M study.

In examining the future directions for research in teacher education, Grossman and McDonald (2008) have proposed a framework including mainly four strands: the pedagogy of classroom interaction (instructional explanations), the pedagogy of enactment (sustained practice), the contexts of teacher education (the national and state policies, institutional contexts, and local districts and labour markets), and the pathways toward teacher certification. This chapter has discussed some issues around these strands such as the diverse teacher education systems, which evolved within and interacting with different socio-cultural as well as socio-historical contexts of the three regions, and how the pathways toward teacher certification differ from each other within those systems. The discussion on the advocacy of adopting action research and lesson study such as in the context of Hong Kong is hoped to provide insights into improving mathematics teacher education systems in terms of sustained professional practice and instructional enhancement of mathematical explanations in the classrooms. As argued by Ulvik (2014), however, it remains as ever a serious challenge to all teacher educators in adopting action research to bridge the gap between theory and practice as reflected from its complex and time-consuming nature. It deserves, perhaps, to have our efforts at undertaking further investigation into how the strategies to overcome the theory-practice divide can be developed and implemented in the three regions in terms of how action research as well as lesson or learning study can be facilitated within and beyond the undergraduate mathematics teacher education programmes. Another suggestion for further research (see Tatto, Lerman, & Novotná, 2009) may also come from the development of comparative framework through which we may have a better understanding of how the systemic conditions in the three different regions' education systems would contribute to the conceptualization of viable policy for enhancing the efficiency and effectiveness of mathematics teacher education.

NOTES

[1] The Hong Kong Institute of Education (HKIEd) was officially retitled The Education University of Hong Kong (EdUHK), effective 27 May 2016.
[2] Starting from 2016 onwards, the intakes of the Bachelor of Education (B.Ed.) programme of Mathematics and Mathematics Education are eligible to become a qualified mathematics teacher serving in secondary schools only upon their graduation from it.
[3] For a large portion of pre-service teacher education programmes, they are designed for both primary teachers and secondary teachers.
[4] Given the likelihood of qualification inflation, certain frontline schools, especially key institutions, indicate that they require their teachers to have higher degrees, such as Master of Education (academic degree) and Ph.D. Such a phenomenon is questioned by the public. On the one hand, the graduates of master's or doctorate degrees focus only on the research of theoretical mathematics. On the other hand, their bachelor counterparts possess more subject knowledge that may be too extreme for students. Instead, educational knowledge and skills are crucial for their instruction.
[5] This table only lists compulsory courses.

REFERENCES

Advisory Committee on Teacher Education and Qualifications (ACTEQ). (2003). *Towards a learning profession: The teacher competencies framework and the continuing professional development of teachers*. Hong Kong: Government Printer.
Bill TaiChung-0920141412. (2003). *Course requirement in secondary and primary school pre-service teacher education programs* [in Chinese].
Cain, T. (2011). Teachers' classroom-based action research. *International Journal of Research & Method in Education, 34*(1), 3–16.
Delaney, S., Ball, D. L., Hill, H. C., Schilling, S. G., & Zopf, D. (2008). Mathematical knowledge for teaching: Adapting U.S. measures for use in Ireland. *Journal of Mathematics Teacher Education, 11*, 171–197.
Education Bureau. (2014, July 27). *Teacher registration*. Retrieved from http://www.edb.gov.hk/en/teacher/list-page.html

Everington, J. (2013). The interpretive approach and bridging the "theory-practice gap": Action research with student teachers of religious education in England. *Religion & Education, 40,* 90–106.

Government Information Office. (2006). *Taiwan yearbook 2006.* Taipei: Author. Retrieved from http://www.gio.gov.tw/taiwan-website/5-gp/yearbook/18Education.htm

Grossman, P., & McDonald, M. (2008). Back to the future: Directions for research in teaching and teacher education. *American Educational Research Journal, 45*(1), 184–205.

Han, J. W., Wong, N. Y., Ma, Y. P., & Lu, N. G. (2011). Research on teacher knowledge of middle school teachers: Based on maths teachers in municipalities in northeast China. *Educational Research, 4,* 91–95.

Hill, H. C., Ball, D. L., & Schilling, S. G. (2008). Unpacking pedagogical content knowledge: Conceptualizing and measuring teachers' topic-specific knowledge of students. *Journal for Research in Mathematics Education, 39,* 372–400.

Holland, M., Evans, A., & Hawksley, F. (2011, August). *International perspectives on the theory-practice divide in secondary initial teacher education.* Paper presented at the Annual Meeting of the Association of Teacher Education in Europe, University of Latvia, Latvia.

Hong Kong Government. (2014, April). *Hong Kong: The fact education.* Retrieved from http://www.gov.hk/en/about/abouthk/factsheets/docs/education.pdf

Hsieh, F. J. (Ed.). (2012). *Teacher Education and Development Study in Mathematics (TEDS-M)* [in Chinese]. Taipei: Department of Mathematics, NTNU.

Hsieh, F.-J., Wang, T.-Y., Hsieh, C.-J., Tang, S.-J., Chao, G., Law, C.-K., Lin, P.-J., Yiu, T.-T., & Shy, H. Y. (2010). *A milestone of an international study in Taiwan teacher education: An international comparison of Taiwan mathematics teacher preparation.* Taipei: Taiwan TEDS-M 2008. Retrieved August 4, 2014, from http://www.dorise.info/DER/05_TEDS-M/TEDS-M_2008.pdf

Huang, Q. A. (2011). Whither mathematics curriculum reform: A survey of mathematics teaching and curriculum reform of basic education. *Journal of Mathematics Education, 20*(3), 12–16.

Kieran, C., Krainer, K., & Shaughnessy, J. M. (2013). Linking research to practice: Teachers as key stakeholders in mathematics education research. In K. Clements, A. Bishop, C. Keitel, J. Kilpatrick, & F. Leung (Eds.), *Third international handbook of research in mathematics education* (pp. 361–392). New York, NY: Springer.

Laverty, M. (2006). Philosophy of education: Overcoming the theory-practice divide. *Paideusis, 15*(1), 31–44.

Law, H. Y. (2013). Reinventing teaching in mathematics classrooms: Lesson study after a pragmatic perspective. *International Journal for Lesson and Learning Studies, 2*(2), 101–114.

Li, Y. H. (2001). *History of teacher education in Taiwan* [in Chinese]. Taipei: SMC Publishing, Inc.

Lin, H. F., Wang, H. L., & Teng, P. H. (2007). The current situation, policy and prospect of elementary and secondary teacher education in Taiwan. *Journal of Educational Research and Development, 3*(1), 57–80.

Lin, P. J., & Hsieh, F. J. (2011). *Teacher education and development study in mathematics 2008* [in Chinese]. Retrieved from http://www.dorise.info/DER/download_TEDS-M/1000601bookpublication.pdf

Liu, C. H. (2001). Several issues needed to be focused in contemporary mathematical instructional reform. *Curriculum, Teacher Material and Method, 9,* 48–49.

Lo, M. L., & Marton, F. (2012). Towards a science of the art of teaching: Using variation theory as a guiding principle of pedagogical design. *International Journal of Lesson and Learning Studies, 1*(1), 7–22.

Lu, Y. J., & Chung, J. (2010). Essentials of developing a mathematics teacher leader project. In M. M. F. Pinto & T. F. Kawasaki (Eds.), *Proceedings 34th conference of the international group for the psychology of mathematics education* (Vol. 3, pp. 233–240). Belo Horizonte: PME.

Ma, Y. P. (2012). The ten-year curriculum reform of mathematics in China. *Educational Journal, 40*(1–2), 79–94.

Ministry of Education. (1995, November 12). *The regulation of teacher qualification.* Retrieved from http://www.moe.edu.cn/publicfiles/business/htmlfiles/moe/moe_620/200409/3178.html

Ministry of Education. (2004, November 10). *Instruction and curriculum reform.* Retrieved from http://www.moe.gov.cn/publicfiles/business/htmlfiles/moe/moe_368/200411/4404.html

Ministry of Education. (2009, October 26). *The introduction of "national teacher training programme."* Retrieved from http://www.gpjh.cn/cms/sfxmtingwen/635.htm

Ministry of Education. (2010, April 27). *The opinions of Ministry of Education on deepening compulsory educational curriculum reform and quality education.* Retrieved from http://www.moe.edu.cn/publicfiles/business/htmlfiles/moe/moe_711/201007/92800.html

Ministry of Education. (2010, July 29). *The outline of the national plan for medium- and long-term education reform and development (2010–2020).* Retrieved from http://www.moe.gov.cn/publicfiles/business/htmlfiles/moe/s4693/201008/xxgk_93785.html

Ministry of Education. (2011, October 08). *The opinion of Ministry of Education on promoting curriculum reform on education programme for teachers.* Retrieved from http://www.moe.gov.cn/publicfiles/business/htmlfiles/moe/s6049/201110/xxgk_125722.html\

Ministry of Education. (2011, November 25). *The first national examination for teacher certification will conduct.* Retrieved from http://www.moe.gov.cn/publicfiles/business/htmlfiles/moe/s5147/201111/127126.html

Ministry of Education. (2011, December 28). *The notice of printing curriculum standards of subjects of compulsory education.* Retrieved from http://www.moe.gov.cn/publicfiles/business/htmlfiles/moe/s8001/201404/xxgk_167340.html

Ministry of Education. (2012, September 13). *The notice of Ministry of Education on printing professional standards for teachers of kindergarten, primary, and secondary schools.* Retrieved from http://www.moe.edu.cn/publicfiles/business/htmlfiles/moe/s6991/201212/xxgk_145603.html

Ministry of Education. (2013). *Yearbook of teacher education statistics, the Republic of China 2012* [in Chinese]. Taipei: Author.

Ministry of Education. (2013, May). *Grade 1–9 curriculum guidelines: Mathematics.* Taipei: Author.

Ministry of Education. (2013, August 21). *The notice of Ministry of Education on printing temporary regulations on teacher qualification examination and regular registration of teacher qualification.* Retrieved from http://www.moe.gov.cn/publicfiles/business/htmlfiles/moe/s7085/201309/xxgk_156643.html

Ministry of Education. (2014). *Order is hereby given, for the revision on Paragraph 3, 4, and 7 of "directions regarding the categories and questions for the teacher assessment examination: Pre-school through grade 12"* (revised directions take into force from 3rd September, 2014). Retrieved from http://gazette.nat.gov.tw/EG_FileManager/eguploadpub/eg019116/ch05/type2/gov40/num11/Eg.htm

Pattie, L.-F. Y. Y. (2009). Teachers' stress and a teachers' development course in Hong Kong: Turning 'deficits' into 'opportunities.' *Professional Development in Education, 35*, 613–634.

Robertson, S. L. (2000). *A class act: Changing teachers' work, globalisation and the state.* London: Taylor & Francis.

Rolfe, G. (1998). The theory-practice gap in nursing: From research-based practice to practitioner-based research. *Journal of Advanced Nursing, 28*, 672–679.

Sfard, A. (2007). When the rules of discourse change, but nobody tells you: Making sense of mathematics learning from a cognitive standpoint. *Journal of the Learning Sciences, 16*, 565–613.

Tam, H. P. (2010). A brief introduction to the mathematics curricula of Taiwan. In F. K. S. Leung & Y. Li (Eds.), *Reforms and issues in school mathematics in East Asia* (pp. 109–128). Rotterdam, The Netherlands: Sense Publishers.

Tatto, M. T. (2007). *Reforming teaching globally.* Oxford: Symposium Books.

Tatto, M. T., Lerman, S., & Novotná, J. (2009). Overview of teacher education system across the world. In R. Even & D. L. Ball (Eds.), *The professional education and development of teachers of mathematics: The 15th ICMI study* (pp. 15–23). New York, NY: Springer.

Tatto, M. T., Schwille, J., Senk, S. L., Ingvarson, L., Rowley, G., Peck, R., Bankov, K., Rodriguez, M., & Reckase, M. (2012). *Policy, practice, and readiness to teach primary and secondary mathematics in 17 countries: Findings from the IEA Teacher Education and Development Study in Mathematics (TEDS-M).* Amsterdam: IEA.

Tchoshanov, M. A. (2011). Relationship between teacher knowledge of concepts and connections, teaching practice, and student achievement in middle grades mathematics. *Educational Studies in Mathematics, 76*(2), 141–164.

Teacher Education Act. (2014). *Laws & regulations.* Retrieved from http://edu.law.moe.gov.tw/LawContent.aspx?id=FL008769

Teacher Education Law Enforcement Rules. (1995). *Laws & regulations.* Retrieved from http://edu.law.moe.gov.tw/LawContent.aspx?id=FL008776

Ulvik, M. (2014). Student-teachers doing action research in their practicum: Why and how? *Educational Action Research, 22,* 518–533.

Wang, C. H. (2013). *A case study on the relationship between the knowledge of average and instructional practices by two junior high school mathematics teachers* (Unpublished master's thesis). National Taiwan Normal University, Taipei.

Whitehouse, C. (2011). *Effective continuing professional development for teachers.* Manchester: AQA Centre for Education Research and Policy.

Xu, Z. T., & Gu, L. Y. (2014). Investigation of knowledge of curriculum and content of pre-service teachers. *Journal of Mathematics Education, 23*(2), 1–5.

Yin, X. (2008). Expansion and extension: Development strategies of curriculum reform of "primary mathematics teaching theory" in normal universities and colleges. *Curriculum, Teacher Material and Method, 1,* 79–82.

Yuan, G. R. (2005, July 25). *Providing guarantee of quality teachers for curriculum reform.* Retrieved from http://www.moe.gov.cn/publicfiles/business/htmlfiles/moe/moe_233/200507/10716.html

Zheng, Y. X. (2005). Curriculum reform of mathematics in the year of 2005: Review and prospects. *Curriculum, Teacher Material and Method, 9,* 45–50.

Huk-Yuen Law
The Chinese University of Hong Kong
Shatin, Hong Kong

Hak Ping Tam
National Taiwan Normal University
Taipei, Taiwan

Xianhan Huang
The University of Hong Kong
Pokfulam, Hong Kong

Yu-Jen Lu
National Taiwan Normal University
Taipei, Taiwan

FENG-JUI HSIEH, SIN-SHENG LU, CHIA-JUI HSIEH,
SHU-ZHI TANG AND TING-YING WANG

3. THE CONCEPTION OF MATHEMATICS TEACHERS' LITERACY FOR TEACHING FROM A HISTORICAL PERSPECTIVE

INTRODUCTION

Countries' cultural traditions impact their educational conceptions and practices (Leung, 2006). Chinese culture is well known for its Confucian tradition. Confucius (551 B.C.–479 B.C.) was declared *zhi sheng xian shi* (the greatest sage and teacher) and *wan shi shi biao* (a model teacher of all ages). The Confucian educational conception is indubitably dominant, and influences Chinese education conception and practices; however, it is not the only influencing conception when it refers to mathematics education. The ancient philosopher MoZi (ca. 470 B.C.–ca. 391 B.C.) is also famous in Chinese societies and has been considered to have influenced the development of science education in Chinese society.

The origin of mathematics education can be traced as far back as to approximately 4000 B.C., when mathematics ideas and their transmission were connected closely with daily life and production. Evidence from archaeological excavations has revealed that, between 1066 B.C. and 221 B.C., mathematics education had been formally included in the early forms of schools, and the position of mathematics teachers included their respective responsibilities. Assuming "mathematics teaching" as an occupation first appeared ca. 1300 A.D. in China (Yan & Wen, 1988a, 1988b, 1988c). These events show that the concept of teachers has existed for centuries in Chinese society.

This chapter chiefly examines and enhances our understanding on how traditional views and practice regarding mathematics teachers' literacy, namely intention and competence (knowledge and ability) for teaching and preparation influence current mathematics teachers' literacy and their acquisition. The focus of the competence of mathematics teachers is on the subject of mathematics and mathematics pedagogical dimension. Within the pedagogical dimension, many conceptions influenced by ancient conceptions is common to all fields, not only mathematics. Therefore, a considerable amount of content in this chapter is related to the general concept of all academic teachers, with the inclusion of mathematics teachers. The specification of mathematics teachers is mentioned for further illustration.

Y. Li & R. Huang (Eds.), *How Chinese Acquire and Improve Mathematics Knowledge for Teaching*, 37–56.
© 2018 Koninklijke Brill NV. All rights reserved.

In the first section, we discuss the rise of current teacher education institutions and their special features that have a strong impact on current teacher education systems. In the second section, we describe the connotation of the traditional conceptions of teachers' literacy for teaching, in addition to their origins, which have been passed down to the present. We focus on the conceptions of the intention, learning and teaching, and knowledge requirement of teachers in all academic fields. In the third section, we discuss the special conceptions of mathematics teacher literacy. This section includes the interaction of the development of mathematics and mathematics education, and the conceptions of traditional mathematics teacher qualification practices, and how they influence and deviate from current practices. A brief comparison of a Western model of mathematics teacher knowledge with Chinese conceptions is provided in this section. In the concluding section, we summarize some points from the previous sections, and recommend areas that warrant future consideration.

Two ancient Confucianism books, among others, are cited throughout this chapter: *Analects* and *Xue Zi* (a chapter in *the Book of Rites*). *Xue Zi* is considered the earliest academic text (originating before 200 B.C.) in the world, with descriptions and discussions of education problems (Zhang & Jiang, 2009).

BECOMING A TEACHER: ITS RISE AND ORIGIN

The Routes of Becoming Teachers in Ancient China

Ancient Chinese was governed by imperial dynasties starting from about 2100 B.C. As mentioned above, as early as 1066 B.C. to 221 B.C., ancient China had early forms of schools (Yan & Wen, 1988a, 1988b, 1988c). There were two types of teachers: (a) those in the imperial government, and (b) those in old-style private schools. The routes for becoming teachers were not through group training as in the present. In ancient China, one of the most crucial reasons for academic studying was the pursuit of official positions in the imperial government. These positions required knowledge in diverse fields such as politics, military, water engineering, and astronomy affairs. The fields of water engineering and astronomy affairs were related to mathematics knowledge. The descendants of officials in the imperial government were allowed to study in the imperial schools, whereas the commoners only had the opportunity to study in private schools.

Teachers in private schools were former learners. To become a teacher outside the imperial government, a man[1] usually had to study hard on his own, and perhaps also followed other knowledgeable people. When he had acquired a substantial amount of knowledge and other learners started to query (learn from) him, he could then begin his teaching career. If his disciples considered him a good teacher, they would tell others, and increasingly more people would begin to learn from him. Some famous teachers even had thousands of students. In such cases, the older disciples would teach their younger counterparts, from one generation to the next. The teachers in imperial governments were appointed by the emperors; they were either imperial officials who passed the imperial examination or famous, knowledgeable, commoner teachers.

The appointing route of becoming teachers in imperial government was terminated as the dynastic system ended. However, the route of becoming old-style private teachers has a continuous influence on the qualification of teachers at present in Chinese society. This route revealed critical conceptions of teacher qualification. First, the evaluation of teacher qualifications was conducted by learners and local (or partially global) communities, rather than any other formal assessment agents. Second, the evaluation of teacher qualification was conducted in a practical field (i.e., the actual learning sites), rather than on a selected outside site. Third, the evaluation appeared under real teaching and learning situations, rather than under simulated conditions or by sitting-in on any paper-and-pencil examinations. Fourth, the evaluation examined integrated, holistic literacy for teaching, rather than literacy composed of different components such as subject matter or pedagogical competence. Fifth, teacher qualification was awarded directly based on the results of teaching, rather than certificates provided by any official capacity.

These conceptions of teacher qualification survived for thousands of years in Chinese society. Although in the past one hundred years modern teacher qualification systems, having adopted framework-based teacher qualification, have been established as official systems, the ancient practice of teacher qualification in private schools has not faded. The qualification of teachers in a *Bu-Xi-Ban*, a type of after-class school in Chinese society, is still determined using the same conceptions as those from ancient China. Because the majority of students have attended a *Bu-Xi-Ban*, and even for several years,[2] these practical conceptions of teacher qualification currently have a strong impact in Chinese society.

The Development of Teacher Schools and the Affiliated Student Schools

In China, the first teacher education institute, the normal school in Nanyang Gong Xue (a secondary school at that time, and Shanghai Jiao Tong University at present), was established in 1897, during the period of Emperor Guangxu of the Qing dynasty. From 1897 to 1904, more than 10 teacher education institutions were established. The reason for the establishment was China's loss of the First Sino-Japanese War, resulting in a claim made by Qi-Chao Liang and You-Wei Hang that China had to abandon the old convention of "talented education," which educated mainly people who sought to become scholar-officials in the imperial dynasty; instead, China had to educate all people like other countries, including Japan that had defeated China in the war. They clearly asserted that China had to "discard old convention, we must establish scholarly schools, and to achieve these, the foremost thing is to build normal schools."

As a response to this demand, the first official teacher education policy in mainland China was established in 1904, with the major goal of educating a sufficient number of professional teachers from China to reduce the number of aboard teachers[3] (Li, 2006). Another Chinese society, Taiwan, colonized by Japan, also initiated its teacher education institutions relatively concurrently. In 1899, the Japan government in Taiwan established three "normal schools" to educate Taiwanese teachers (Wu, 1983, p. 18). This marked the first time that the Taiwan people had the opportunity to be educated as teachers, and to be equipped to teach mathematics in the formal teacher education system (Hsieh et al., 2010).

Chinese teacher education did not have explicit standards in this early period; instead, a teacher's qualification was assessed by completing the courses required in the curriculum through the designated years (Lee, 2001, pp. 8–11). The curriculum covered educational courses; teaching methodology and general education were also typical requirements in many teacher education institutions. The history of education in different regions such as Europe was also included (Li, 2006). This initial concept of preparing teachers remains in practice currently; Chinese society still retains the feature of training teachers through accredited courses to attain a degree, rather than through standards-based criteria.

In the initial period of establishing teacher education schools, teacher preparation focused considerably on the practical experience a student teacher should gain. Certain teacher preparation institutions had even established primary schools for their student teachers to conduct a practicum (Lee, 2001, p. 9; Li, 2006). The concept of including a practicum was not merely borrowed from other countries, but was deeply rooted in ancient China. The most renowned teachers such as Confucius in ancient China did not graduate from any preparation institutes; rather, they began teaching in practical sites as knowledgeable people. They became famous because they relied on their teaching experience, improved their teaching skills, and educated many noted disciples in these practical sites.

The concept of establishing affiliated schools has a considerable influence on current teacher preparation practices in Chinese society; for example, both National Beijing Normal University and National Taiwan Normal University have affiliate high schools, and numerous primary teacher education institutes also have affiliate (experimental) schools. Although some of the affiliated schools have been established years ago, some of them are newly affiliated. In 2005, National Chengchi University (including numerous teacher education units) in Taiwan added an affiliate high school to implement their educational concepts. Even this year, 2014, National Chung Hsing University in Taiwan added an affiliate senior high school and an affiliate vocational high school as partners to fulfill numerous educational purposes.

THE CONTINUUM OF TEACHER LITERACY CONCEPTION

Teachers have enjoyed a great reputation from ancient China, and the standards of their literacy have been established through a practical operation that emerged naturally in Chinese society. Confucius discussed the features of becoming and qualifying a teacher on numerous occasions. One of his most famous dictums is as follows: "If a man[4] keeps reviewing his old knowledge and then acquiring new knowledge, he may be a teacher of others" (Analects, 2,11). This means that, if a man studied hard and had his own and new appreciation of the subject he had studied, then he may be considered a teacher of others. This concept showed a subject-matter competence (knowledge and ability)- and attitude (toward study)-based conception of the qualification of teachers, and it continues in the present.

Although this speech about the qualification of teachers in *Analects* did not include any pedagogical parts, descriptions in other Chinese ancient books have explicitly shown the pedagogical parts of teacher qualification in Chinese society. An example is shown in *Xue Zi*, which states the following: "When a noble man knows the causes that make instruction successful and those that make it fail, he can become a teacher of others." These extracts showed that the subject-matter, pedagogical, and attitude parts were all included in teacher literacy in the ancient times. They also showed that the ancient conception was rather abstract and general. Nevertheless, most of these general conceptions have deep and profound influence in Chinese societies from the past days to the present days through a direct pass down or a gradual transformation.

Recently, Chinese researchers Lin, Shen, and Xin (1996) proposed a framework of the structure of teacher literacy, and conducted their empirical research accordingly. They claimed that teacher literacy had to include the following components: ideal of the teacher's profession, concepts on education, knowledge level, mentoring ability in teaching, and teaching behaviors and strategies.[5] The five components can be re-structured into four categories, with the first two components as a whole and may be referred to as the *Ren Shi* conception (will be delineated in the next section). The remaining components may be referred to as teacher competence and re-divided into three categories: student learning, teaching method, and knowledge requirement. Each category has a unique conception, and all conceptions have firm historical roots that have been addressed substantially by educators and philosophers in ancient China. Although the "format" of the current conception of competence may look different from the ancient one, but when considering the content and substance, many of them are in line with the ancient conception. The following paragraphs illustrate the conception that originated from ancient China and remains active today (no attempt has been made to be inclusive).

The Continuum of the Concept of Ren Shi from Past to Present

Without a background on the concept of the teacher's role in Chinese society, a discussion of teacher literacy for teaching cannot reflect the genuine intensity of actual teaching by a teacher and the unique Chinese conception of teacher-student interaction. The teacher's and student's relation typically influences the amount of effort exerted by students into studying, and teachers into teaching and assisting extra to students.

Chinese society believed that teachers play critical roles in student life. The proverb, "a teacher a day, the father forever" provides a basic concept of the teacher's role. Confucius and his followers discussed the teacher's roles frequently. Yu Han (768–824), a famous Confucian follower during the Tang dynasty, indicated the teacher's role with a statement in the book *Shi Shou*[6]: "a teacher, to teach the fundamental principles [of how to behave correctly in different respective positions], to instruct the academic works, and to clear up confusion." This portrayed an image of being a *Ren Shi*,[7] a teacher for people development, rather than merely a *Jing Shi*,[8] a teacher for the development of academic knowledge. Confucius always discussed his thoughts with his disciples, and how he himself acted as a model for them to follow. Zengzi, a Confucian disciple, explained Confucius's words to other disciples: "The doctrine of our master [teacher] is to exercise faithfulness and be benevolent to others—this, and nothing more" (Analects, 4,15). Confucius also indicated that, "never refuse to instruct any person" (Analects, 7,7); "in teaching, there is no distinction of class" (Analects, 15,38); and "instruct others, and never feel wearied" (Analects, 7,2). Basically, Confucius set a high moral standard and positive view on education for teachers.

The conception of *Ren Shi* in teacher literacy has been adopted with great support for 2,500 years from the past to the present. It is still a current concept in Chinese educational systems, although it has diminished slightly because of the Westernization movement. This conception illustrated the requirement of teachers' personalities, responsibilities, spiritual condition and behaviors in Chinese societies from ancient to present times; that is, a teacher must (1) act as a student model, (2) instruct students academic knowledge, (3) clarify confusion in various parts of students' study, (4) teach fundamental principles of how to behave (5) educate students without a distinction of classes (6) prepare students with passion, and without refusal and weariness.

Thus, a teacher in general, or a mathematics teacher in particular, is supposed to be an instructor, supervisor, guardian, and model for students. Accordingly, teachers usually consider that they are obligated and have the right to require students to study hard, to suggest (or arrange) students' direction in career choice, or to help students after class and act as a guardian. Teachers also manage intensive pressure from students' performance in examinations, especially for large-scale entrance examinations. Certain teachers believe that their students' success in examination is their success, and the students' loss is their own. Consequently, mathematics teachers often instil in their students the importance of studying mathematics, being hardworking, and succeeding in learning, because they do believe it and regard it is their responsibility to tell their students and assist their students to achieve. The *Ren Shi* concept that is expected from mathematics teachers as they enter their classroom to teach may be exemplified by a story detailing the professional lives of Taiwan mathematics teachers in a book written by Schmidt et al. (2011, pp. 12–15).

The Influence of the Ren Shi Concept in Present Teacher Education

The ancient conceptions of teachers being required to possess high passion, high moral and positive views on education and act responsibly influence the teacher qualification conception. Numerous teacher preparation institutions set a minimal requirement on the moral grades for teacher qualification. Moreover, to pass the school-based practicum, prospective teachers must usually display these characteristics. Among others, the mottos of two normal universities in Chinese society demonstrate the high standards of passion for teachers. The first example is the motto of National Taiwan Normal University: "to be genuine, in accord with what is right, diligent, and without indulging in luxurious and expensive habits" (*cheng zheng qin pu*). The motto of Beijing Normal University is as follows: "learning to be a teacher for developing people, and behaving as a model of the world" (*xue wei ren shi, xing wei shi fan*).

The Continuum of the Conception of Learning from Past to Present

The conception of psychology for learning. Ancient Chinese teachers valued psychology for learning: the study of students' mind and behaviors in learning. *Xue Zi* claimed that teachers must know four defects in learners when learning: Some attempted to learn a massive amount, which exceeded their affordances; some studied only a few subjects; others overestimated the ease of learning; and others stopped learning inappropriately. *Xue Zi* futher argued that these four defects arose from differences in learners' minds. A teacher had to know the learners' mind to save them from these defects. This concept of emphasizing psychology has been explicitly raised for 1,000 years. It had a longer history rooted in *Analects*, which described how Confucius observed his disciples' defects and merits and what was in their minds in relation to learning. There is still a great emphasis on the importance of psychology for Chinese teachers as evidenced by requiring future teachers to take courses relating to psychology in the teacher preparation programs (Hsieh, Lin, Chao, & Wang, 2013).

The ancient Chinese conception on how teachers should direct learners' attitude or motivation toward learning can be found in *Analects* (*Analects*, 7,8), which stated that a teacher should require students to exhibit their real desire of learning before the teacher enlightened them. This conception demonstrated a focus on the learner's readiness and motivation, and has been passed down to the present. Compared with the system implemented in Western society, which allows students to take different levels of classes for the same courses (topics), the Chinese system of taking the same levels for all students seems to adjust less to meet learner readiness and require more on the learner's self readiness. In actuality, Chinese practice places the obligation concerning readiness on classroom teachers, rather than on school policies.

The conception of learning. The Confucian conception of the learning process entailed a sequence of concepts: learning (listening or reading), querying, thinking, judging, and using. In other words, teachers should develop in learners a habit of understanding what the teachers have described, and clarifying doubts before learners practice what they have learned. Specifically, Confucian concepts have emphasized a mimicking of previous sages such as Confucius himself through the learning process. By contrast, MoZi's learning concept emphasized innovation, logical reasoning, and practice. He argued that the sages who learners mimicked must have attempted novel ideas or ways of acting to become sages; therefore, learners would gain nothing by mimicking them. He focused on learners' understanding of the origins, the reasons, and classes by analogizing and inducing, rather than focusing on isolated concepts described in books or by teachers (Yang, 2002). These two strands of learning philosophies grew and declined in Chinese history, but had never been eliminated. In recent years, as Western teaching concepts were introduced into Chinese society and the modern scientific technology embedded a sense of innovation, MoZi's concepts on learning seemed to resurface because they coincided more with Western views and current technological approach than the Confucian ones did.

The description from the best-known books related to teaching, learning, and philosophy in ancient China show that the development of teaching competence occurred through the learning and reflection on the process of practical teaching in ancient China. *Xue Zi* indicated, "One learns, then realizes the deficiencies, teaches, then realizes the struggles." This concept considered teaching and learning to be tools for the development of each other's mastery. This concept motivated a learner's willingness of teaching, and initiated a potential future career as a teacher.

The Continuum of the Conception of Teaching Method from Past to Present

The Chinese have focused substantially on the teaching method. Ancient China emphasized the importance of telling, querying, and peer discussion in their teaching. The telling method is not the same as its superficial meaning and a clarification will be given in the next section.

The conception of heuristics-persuasion. The context of *Analects* explicitly showed the telling and quering methods. The telling method is crucial because only the elite had the opportunity to learn in ancient China, and thus, the taught content was extremely difficult, and would have been impossible to understand had no explanation been provided. These features remain, as shown in the Chinese curriculum, which usually includes difficult content, and must be explained before students can start any tasks. The Chinese telling method is different from rote learning or instrumental learning; it is more akin to the heuristic approach to teaching. One of Confucius's disciples described the way Confucius taught him as "gradually and skilfully leading." *Xue Zi* indicated that, as a teacher, "his words are brief, but far-reaching; unpretentious, but profound; with few metaphors, but illustrative." *Xue Zi* stated that teachers should be able to "make broad metaphors." These statements show the heuristic method of teaching in Confucianism

(Zhang & Song, 2005, 2006). The impression that Chinese teachers admire rote learning is incorrect; we, the authors, as insiders in Chinese society, realize that our mathematics teachers implement various heuristic means, including paving a way for further understanding, connecting between ideas, giving hints, explaining, providing feedbacks, and using metaphors to inspire student thought (Cao, Li, & Clarke, 2011).

Querying has been considered a critical facilitator for teaching and learning in ancient China, and occurred in two ways: (a) querying by the teachers, and (b) querying by the learners. The Confucian concept on transmitting knowledge emphasized "querying by learners." *Analects* always presented Confucius's concepts as answers to the questions asked by his disciples or others. A philosophy of "speaks when being asked, stops when not being asked" (a sentence spoke by a Confucian follower quoted in the book *MoZi, Gong Meng*) was implemented in Confucianism.

The coordination and balance of telling and querying was a dilemma in ancient Chinese societies. MoZi did not agree with the Confucian concept of querying; rather, he adopted a more positive attitude toward telling. He argued that, "without intense persuading, people will not understand" (*MoZi, Gong Meng*). However, MoZi's concept of persuading is not cramming. He indicated that, if a teacher was simply telling massive information without considering learners' conditions, the teaching went against the correct ways (Zhang & Wang, 2009). Because Mozi expressed a more positive attitude toward scientific knowledge, and this type of knowledge is not self-explanatory for most learners, a method of intense persuasion with a suitable approach can be considered a conception of transmitting scientific knowledge in ancient China. This conception is in agreement with the current implementation in Chinese mathematics classrooms regarding teacher-student interactions, which includes paving a way, explaining, leading, and evaluating (Cao, Li, & Clarke, 2011). It is not a rote learning method with only telling; instead, it is a heuristic method with intensive persuasion; this conception may be regarded as "heuristics-persuasion."

The conception of peer discussion. *Analects* explicitly noted the importance of peer discussion through a detailed description of a mutual discussion between Confucius's disciples. Moreover, it also showed that senior disciples usually tutored their junior counterparts. *Xue Zi* claimed, "Learning alone without friends together makes one isolated, ignorant, and ill-informed." Zhang (2008) presents an insightful view regarding the integration of collaborative learning and independent learning in Chinese societies in the following:

> Western educators emphasise individual learning while Chinese educators pay more attention to group learning. In China, teachers always assign high-achieving students to help low-achieving students. If low-achieving students made good progress in their exams, the supportive high-achieving students would be praised by their teachers. (p. 556)

The conception of doctrine of mean. The features of teaching methods as anticipated for implementation by teachers in Chinese society may be viewed around the central philosophical concept of *The Doctrine of Mean*. A sentence on how teachers should teach in *Xue Zi* best exemplifies this concept: "to lead the ways but not pull, to command but not to constrain, and to enlighten but not to complete." This philosophical concept has been practiced in Chinese society throughout history in numerous fields, including education. When teachers make decisions in various teaching scenarios, they tend to adopt the doctrine, and their actions are thus more likely to match the anticipation of their students, who had also been raised in the same society in which *The Doctrine of Mean* is pervasive.

The Continuum of the Conception of Knowledge Requirement from Past to Present

The conception of respecting theoretical-based knowledge. The belief that knowledge is valuable in its pure (theoretical-based) form has occupied a position in education in Chinese society throughout history. The intensive competition in examinations, the imperial examinations (*ke ju kaoshi*) for scholar-officials in ancient Imperial China, has continued to influence current examinations in the teacher education process and qualification tests. Although school-based teaching practice is also considered a major part of the teacher preparation process, the example that prospective teachers are required to complete in paper-and-pencil form at teacher qualification examination after they complete their school-based practicum in Taiwan demonstrates the emphasis on pure knowledge in Chinese society. The following table shows one such item in the subject, Development and Adolescent and Counseling, in the national teacher qualification assessment in Taiwan, required to be taken by all prospective teachers including those register to be mathematics teachers (TQC, 2014):

Which of the following counselling therapies focuses on the idea that everyone who makes a choice should be responsible for his/her behavior, thoughts, and feelings?
(A) A. Beck's cognitive therapy
(B) W. Glasser's reality therapy
(C) B. Skinner's behavioral therapy
(D) S. Freud's psychoanalysis therapy

This item shows the focus on memorizing theoretical-based knowledge, and the competence of thinking, judging, and using, which is in agreement with the focus of Confucian concepts of learning.

At present, it remains a dilemma for mathematics educators to determine whether their teacher preparation curriculum includes too many theoretical-based courses. The wave of introducing theorems from the West is still ongoing, but a voice for the inclusion of more training on practical teaching skills is relatively strong (Hu & Zhang, 2012).

The conception of massive and profound knowledge requirement. The aforementioned routes of becoming teachers in ancient China showed the conception of extensive and profound knowledge requirement. Confucius's disciple, Hui Yan (521 B. C.–481 B. C.) admired the doctrines (knowledge) of his teacher Confucius in the words: "When I looked up at them, they seemed to become more and more high; When I tried to dig into them, they seemed to become more and more firm."

Confucius also described, using himself as a model, a concept of diligence and earnestness in lifetime learning as a teacher. He depicted himself through the words, "Understanding knowledge silently; learning tirelessly" (*Analects*, 7,2); "I have been the whole day without eating, and the whole night without sleeping; occupied with thinking. It was of no use. The better plan is to learn" (*Analects*, 15,30).

THE CONCEPTION OF MATHEMATICS TEACHER LITERACY

The development or evolution of the conceptions of mathematics teachers' literacy was under the aforementioned broad conceptions for all teachers of acedemia. However, the development of mathematics and its teaching in ancient China shaped some specifications about mathematics teachers.

The *Ren Shi* conception for teaching may be demonstrated by the fact that Chinese mathematics teachers sincerely introduce massive and deep mathematics concepts to students and require students to do numerous exercises. They do this with the purpose of providing students with more chances for success. The teachers constantly strive to equip students with the knowledge that is in their best interest, for they regard student success as their own responsibility. This section will focus on the mathematics-related literacy of mathematics teachers.

The Development of Mathematics and Its Education

The development of mathematics and mathematics education were mutually interacted and influenced in Chinese history (Cai, 2013). As early as the Shang dynasty (1046 B.C. to 221 B.C.), mathematics was a required subject in early forms of schools, and the word "shi" (teacher) and "jiao xue" (teaching) were already being used (Yan & Wen, 1988a). This marked the period while mathematics was deviated from daily-life knowledge and became a required academic subject to study. In this period, fractions, ratios, angles, and the concept of negative numbers had been used (Yan, 1965a). In the period of 220 B.C. to 581 A.D., Chinese mathematicians wrote an array of famous mathematics books. The first one (that could be preserved to date) was *Zhou Bi Suan Jing*, and the most famous one that has had a far-reaching influence on Chinese mathematics development was *Jiu Zhang Suan Shu* (the nine chapters on the mathematics art), which can be considered as the first Chinese mathematics textbooks. The most prominent mathematicians of this time were Zhao Shuang, Liu Hui, and Zu-Chong Zhi. Furthermore, these mathematicians became the early mathematics teachers (Yan & Wen, 1988a).

In the Qing Dynasty, Western missionaries who were typically knowledgeable in mathematics and astronomy entered China and started to communicate and cooperate with Chinese mathematicians. Many Western modern mathematics skills, principles, concepts, and fields were brought into China and books were translated to Chinese. This wave was held even in imperial governments. The Kangxi emperor (1661–1722) was surprised by and admired the accuracy and usage of Western mathematics in astronomy. He extensively studied and learned algebra and geometry with these missionaries and promoted the integration and propagation of modern Western mathematics into Chinese mathematics (Bai, 2008). Mathematics became even more important than before. After the Second Opium War (1856–1860), the Qing imperial government noted the vulnerabilities of China in war and recognized mathematics as the foundation of the engineering industries that were necessary to succeed in war. The imperial government thus advocated learning both Chinese and Western mathematics, such as concepts in Euclid's Elements, trigonometry, and calculus (Cai, 2013; Yan, 1965a, 1965b), aspiring to address the manifest deficiencies of the Chinese military. The strength in Chinese mathematics—algorithm (procedural and instrumental) and calculation by using Abacus Calculation and Rod Calculation (Cai, 2013) and the strength in Western mathematics—Structure and theorem were all included in the instruction of mathematics (Fu, 2006).

In 1906, the Qing imperial government enacted an educational law which explicitly raised the significance of mathematics and science. In the same year, the government announced the first formal national curricular standards and a national aim of educating common people all over the nation rather than only educating elites, which was the original intent of compulsory education. With this aim, the government was forced to admit that China did not have enough mathematics textbooks to use and should borrow Western mathematics textbooks (MOE, 1934, pp. 3, 13). This flow of mathematics and mathematics education development showed that mathematics was for elite imperial officials or elite common people until about a hundred years ago. Mathematics teachers were the elites among elites. Many current conceptions on mathematics education, enacted by mathematics teachers, and on the requirements of mathematics teachers' literacy can be traced back to the developmental flow.

The Conception of Massive and Profound Knowledge Requirement

In the teacher education programs in Chinese societies, the mathematics teachers are required to take almost all courses offered for a regular mathematics major. If one wanted to become a mathematics teacher, educational courses are addendums (see examples from Taiwan; Hsieh, Lin, Chao, & Wang, 2013, p. 79; Schmidt et al.,

2011, pp. 37–38). Teacher education programs require considerable large number of course credits for future mathematics teachers. An international comparison study MT-21 about future mathematics teacher preparation systems and results at lower-secondary level in six countries, including Taiwan, South Korea, and the U.S., showed that Taiwanese future mathematics teachers were required to study about 30% more course hours than South Korea and the U. S. (Schmidt et al., 2011, p. 79). Another international comparison study TEDS-M, sponsored by IEA, showed that Taiwanese future mathematics teachers took the second most amounts of tertiary-level mathematics topics among participating countries (less than that of Russia). Taiwan had 15 core topics for future mathematics teachers, while the U. S. only had 2 core topics (Hsieh, Yang, & Shy, 2012, p. 149). With these heavy demands on academic works, Taiwanese future teachers performed the best on the mathematics content knowledge (Hsieh & Wang, 2012) and mathematics pedagogical content knowledge examinations (Hsieh, 2012) among all participating countries. These data exemplified the Chinese conception of requiring massive and profound mathematics knowledge.

The conception of massive and profound knowledge for mathematics teachers originates in ancient China, where all of the famous mathematics teachers were extremely knowledgeable both in breadth and profundity. Chinese scholar Shi-Jie Zhu (1249–1314), the first person to accept "mathematics teaching" as an occupation, wrote the well-known mathematics books *suan xue qi meng* and *si yuan yu jian*, which encompassed his own mathematics inventions. One of the most famous mathematics teachers, Hui Yang (1238–1298), wrote numerous mathematics textbooks, which have introduced various mathematics methods, among are some considered to be the principal methods in the world, such as the Pascal's Triangle. These examples have shown the requirement of massive and profound mathematics knowledge for mathematics teachers in ancient China.

The Conception of Teaching and Learning for Mathematics Teachers

The first mathematics teaching methods explicitly shown in a book was written by Hui Yang, in 1274. In the beginning of one of his mathematics textbooks, *cheng fa tong bian ben mo*, he included *xi suan gang mu*, which suggested how to teach mathematics based on his lengthy experience of teaching mathematics. He indicated the following: "(1) The teaching method should consider from ease to difficulty, and gradually introduce the topics in an appropriate order; (2) teaching must pay attention to requiring students to do exercises, and to positively cultivate their skills on mathematics operations. Require students to draw three other cases through an analogy when given one instance.[9] 'A diligent learner may research by himself when touching a class of topics; in this case, why should the teacher teach him?' (3) In the teaching process, pay attention to any details. Even annotations should be taught, and ask students to think them over in detail" (Yan & Wen, 1988a). All of these concepts have been passed down to the present.

For learning and teaching mathematics, rather than the Western learner-centred concept, Chinese mathematics teachers share some common conceptions that originated from ancient China. This chapter selected the following ten to illustrate a picture of these conceptions.

1. Requiring students to put efforts in their studies; it originated from the diligent tradition of the agricultural society in China (Zhang, & Yu, 2013, p. 20) and Confucian's concept about study earnestly (*Analects*, 1,14; 7,2; 7,19). This diligent tradition has far-reaching and broad influence on most learning and teaching conceptions, including those that will be mentioned below.
2. Asking students to do a lot of exercises; it originated from the situation that ancient Chinese textbooks were written as the set of problem- and solution-banks; to learn mathematics was to solve problems.
3. Encouraging students to challenge themselves with difficult problems; it originated from the situation that mathematics was only for elites in ancient China, and these learners had strong motivation to challenge themselves.
4. Suggesting students to construct multiple solutions; it also originated from the elite education.
5. Helping students perform well in examinations; it originated from the fact that succeeding in imperial examinations was followed by a position in imperial governments (Zhang & Yu, 2013, p. 24) and the *Rhi Shi* spirit.
6. Requiring students to describe solutions clearly and accurately; it originated from the Confucian's concept of expressing carefully and not repeating a fault (*Analects*, 1,14; 6,2) and the written format of imperial examinations (Zhang & Yu, 2013, p. 24).
7. Developing students' computation skills; it originated from the Chinese tradition of using Rod Calculation and Abacus Calculation.
8. Fostering students' logical reasoning and proving abilities; it originated from the reasoning nature of Confucius doctrines (Zhang & Song, 2006), and the synthesisation of Western mathematics.
9. Preparing students' habits of understanding the reasoning behind mathematics procedures or theorems rather than only knowing how to solve problems; it originated from the elite education tradition that all learners were prepared with the intention for them to become scholars.
10. Using heuristics-persuasion in mathematics instruction; it originated from the Confucius' and Mozi's conceptions.

It is a shared conception that mathematics teachers should be prepared with this common learning and teaching conceptions. One common root of these conceptions comes from the profound characteristics of mathematics that prohibits the easy ways of learning, such as only memorizing, self-discovering, or by story-telling. Although these conceptions were explicitly or implicitly utilized in mathematics teacher preparation, it remains unknown about how to combine these conceptions with practical teaching methods. It is worthy to mention that these conceptions do not encourage the use of modern technology such as calculators. Even in recent years, some mathematics teacher educators continue to advocate the reduction in the number of modern multimedia courses in the teacher preparation curriculum; instead, they recommend using more blackboard instruction and teacher-student interactions to encourage students, rather than using multimedia for motivating students in mathematics classes (Hu & Sun, 2013).

Comparison of Chinese Ancient Conception with Current Western Conception

To illustrate the literacy of Chinese mathematics teachers, we compare the Chinese conceptions rooted in traditional Chinese society with one of the frequently mentioned Western construct, proposed by Hill, Ball, and Schilling (2008). Hill, Ball and Schilling's structure of mathematical knowledge for teaching (MKT) includes two broad categories and each with three subdomains: (1) subject matter knowledge: specialized content knowledge (SCK), common content knowledge (CCK), and knowledge at the mathematical horizon; (2) pedagogical content knowledge (PCK): knowledge of content and students (KCS), knowledge of content and teaching (KCT), and knowledge of curriculum. More details about this structure will be mentioned when it applies.

For mathematics knowledge, Hill, Ball and Schilling considered CCK and SCK as types of subject knowledge teachers used in the work of teaching, where no knowledge of students or teaching was entailed. They did not specify the role of knowledge at the mathematical horizon in the work of teaching or in the knowledge set of teachers. In contradistinction to this western view, Chinese mathematics teachers are required to equip massive and profound knowledge at the mathematical horizon.

CCK is meant by original subject matter knowledge, which definitely is included in the literacy of Chinese mathematics teachers currently and historically. SCK is a newer conceptualization proposed by Hill, Ball and Schilling, which indicates "the mathematical knowledge that allows teachers to engage in particular teaching tasks, including how to accurately represent mathematical ideas, provide mathematical explanations for common rules and procedures, and examine and understand unusual solution methods to problems" (pp. 377–378). The authors claimed that no knowledge of students or teaching is entailed in this type of knowledge. Some of the concepts of SCK are rather new to Chinese societies because traditional mathematics education concerned only the elites. For example, Chinese mathematics teachers are limited to various ways of providing mathematical explanations, especially those with simple ways, but they were equipped with understanding unusual solution methods and representing mathematical ideas accurately.

For the category of PCK, Chinese mathematics teachers focused more on KCT than on KCS and on knowledge of curriculum. Many specific mathematical teaching principles or conceptions were raised in ancient China and passed down to the present, but for those relating to KCS, such as students' misconceptions,

Chinese mathematics teachers (also regarded as mathematicians) do not pay much interest to them. One of the reasons is that mathematics was only for elites and these people were supposed to resolve any difficulties through studying and investigating. Regarding to knowledge of curriculum, it previously was not and currently is not emphasized in teacher preparation programs. Chinese mathematics teachers usually have strong mathematics background, so most of them realize the trajectories of mathematics content. They also understand the implemented curriculum because Chinese societies have implemented national standards starting in 1906. They are not required to take courses for them to understand the mathematics curriculum.

CONCLUSION

The teacher is the lynchpin of education, economic, and social reformation (Furlong, Cochran-Smith, & Brennan, 2009, p. 1). Different philosophies in various cultures have influenced education theory, policy, and practice throughout the respective regions (Smith & Hu, 2013). This raises an imperative question regarding how the traditional conception of mathematics teacher literacy in ancient China has influenced the current practice of mathematics teachers' literacy and their acquisition.

This chapter noted several conceptions that originated in ancient China and have been passed down to the present. The conceptions cover different areas, including teachers' personality, competence for teaching, qualification, and preparation. The conceptions include the following: A (mathematics) teacher is an instructor, supervisor, guardian, and model for students; teachers are obligated to, and have the right to require, be involved in, and help students' study, life and future career; and teaching and learning practice is implemented in accordance with the philosophy of *The Doctrine of Mean*.

A critical point related to mathematics teacher's conceptions on teaching method is as follows: The traditional Chinese conception on mathematics teaching can be considered a heuristics-persuasion method that includes intense persuasions with suitable heuristics means, such as using metaphors and questions to inspire student thought. Combining the heuristics-persuasion method with a focus on cultivating operational skills and logical abilities through difficult exercises conveys the essence of mathematics teaching in Chinese society. Because the intense persuasion of difficult materials usually includes telling, the Chinese teaching method has often been misunderstood as a way of cramming or focusing on rote learning; we have attempted to clarify this point in this chapter.

At present, a profound feature of the qualification of mathematics teachers is related to the substantial amount of mathematics knowledge and theoretical-based pedagogical knowledge. The conceptions of massive and profound knowledge and theoretical-based knowledge have been passed down from ancient times. However, regarding the pedagogical issue, ancient teachers gained their competence mostly from the teaching and learning process on an actual teaching site. The current educational systems in Chinese societies provide too few opportunities for a practicum, even if certain universities attached affiliated schools for their student teachers to practice; this may result in a disconnect of the practical teaching circumstances in the qualification of teachers. It remains a dilemma for mathematics educators to determine whether their teacher preparation curriculum includes too many theoretical-based courses. The wave of introducing theorems from the West is still ongoing, but the voice demanding the inclusion of additional training for practical teaching skills is relatively strong (Hu & Zhang, 2012). The qualification of teachers in a *Bu-Xi-Ban* differs completely from that in official systems. Its conception of teacher qualification remains similar to its traditional counterpart, which has a strong feature of qualification being attained by learners without having to sit in on any paper-and-pencil examinations, in contradiction with the current dominant practice.

The currently implemented teacher education focused on competence-based conception, which emphasizes concrete, observable, and measurable competences (Huang, 2013, p. 8). By contrast, as demonstrated in this chapter, the traditional conception of mathematics teacher education in Chinese societies emphasized abstract and spiritual conceptions, wherein it combined with scrappily concrete ways of executing the conceptions. Each approach of addressing teacher education has its respective merits. How to synchronize the two types of conceptions is a task warranting further pursuit.

NOTES

[1] Females were not allowed to study in ancient China.
[2] The majority of high school students have attended a *Bu-Xi-Ban* in certain Chinese societies; for example, the news reported that, in Taiwan, more than 70% of ninth graders attended a *Bu-Xi-Ban* in many urban areas in 2009 (http://mag.udn.com/mag/edu/storypage.jsp?f_ART_ID=99320); and in Hong Kong, more than 70% of tenth graders attended a *Bu-Xi-Ban* in 2012 (http://the-sun.on.cc/cnt/news/20120519/00407_057.html).
[3] The aboard Christian missionaries entered China and became teachers.
[4] Women were forbidden from studying.
[5] The *ideal on the teacher profession* is related to passionate, responsible, and active views on the teacher profession. The *conception of education* refers to the belief of the relation between teaching and learning. The concept of *knowledge level* includes knowledge on academic subjects, on teaching practice, and on education and learning psychology.
[6] *Shi* means teacher, and *Shou* means discussing. *Shi Shou* is an essay discussing teachers' and students' respective roles, and persuading people to study with teachers to improve society.
[7] *Ren* means people and *Shi* means teacher.
[8] *Jin* refers to the traditional concept of knowledge.
[9] This is a case of using the heuristic teaching method.

REFERENCES

Bai, J. (2008). 老外眼中的康熙大帝 [*The Kangxi Emperor from the view of foreigners*]. Beijing: People's Daily Press.

Cai, T.-Q. (2013). 中國傳統文化與傳統數學、數學教育的演進 [Research on the evolution of traditional mathematics and mathematics education in the context of Chinese traditional culture]. 全球教育展望 [*Global Education*], *42*(8), 91–99.

53

Cao, Y.-M., Li, J.-Y., & Clarke, D. (2011). 數學課堂中啟發式教學行為分析—基於兩位數學教師的課堂教學錄像研究 [Analysis of heuristics method in mathematics classroom—based on a video analysis of mathematics instruction of two mathematics teachers]. 中國電化教育 [*China Educational Technology*]. 10,100–10,102.

Fu, H.-L. (2006). 李善蘭與中國數學教育近代化 [Li Shan Lan and the modernization of Chinese mathematics education]. 紀念《教育史研究》創刊二十周年論文集 (2)——中国教育思想史与人物研究 [*Proceedings of Memorial the 20th Anniversary of the Journal "Educational History Research" (2): History of Chinese Educational Thought and Character Study*]. 2191–2193.

Furlong, J., Cochran-Smith, M., & Brennan, M. (2009). Introduction. In J. Furlong, M. Cochran-Smith, & M. Brennan (Eds.), *Policy and politics in teacher education: International perspectives* (pp. 1–9). London: Routledge.

Hill, H. C., Ball, D. L., & Schilling, S. G. (2008). Unpacking pedagogical content knowledge: Conceptualizing and measuring teachers' topic-specific knowledge of students. *Journal for Research in Mathematics Education, 39*(4), 372–400.

Hsieh, F.-J. (2012). 中學數學職前教師之數學教學知能 [Mathematics pedagogical content knowledge of high school future mathematics teachers]. In F. J. Hsieh (Ed.), 臺灣數學師資培育跨國研究 [*Taiwan TEDS-M 2008*] (pp. 119–142). Taipei: Mathematics Department of National Taiwan Normal University.

Hsieh, F.-J., & Wang, T.-Y. (2012). 中學數學職前教師之數學知能 [Mathematics content knowledge of high school future mathematics teachers]. In F.-J. Hsieh (Ed.), 臺灣數學師資培育跨國研究 [*Taiwan TEDS-M 2008*] (pp. 93–117). Taipei: Mathematics Department of National Taiwan Normal University.

Hsieh, F.-J., Lin, P.-J., Chao, G., & Wang, T.-Y. (2013). Chinese Taipei. In J. Schwille, L. Ingvarson, & R. Holdgreve-Resendez (Eds.), *TEDS-M encyclopedia, a guide to teacher education context, structure, and quality assurance in 17 countries: Findings from the IEA Teacher Education and Development Study in Mathematics (TEDS-M)* (pp. 69–85). Amsterdam: IEA.

Hsieh, F.-J., Wang, T.-Y., Hsieh, C.-J., Tang, S.-J., Chao, G.-H., Law, C.-K., & Shy, H.-Y. (2010). *A milestone of an international study in Taiwan teacher education: An international comparison of Taiwan mathematics teacher (Taiwan TEDS-M 2008)*.

Hu, Q.-Z., & Sun, H. (2013). 基礎教育課改背景下高師數學教師教育教學方式改革研究 [Study on teaching methods reform of the mathematics education in normal colleges under the new mathematics curriculum in the elementary education]. *Journal of Jiangxi Institute of Education (Comprehensive), 34*(3), 3–5.

Hu, Q.-Z., & Zhang, Y.-X. (2012). 基礎教育課改背景下的高師數學專業"教育課程"改革研究 [Study on the reform of education courses of mathematics major in teachers' college in the background of reform in elementary education courses]. *Journal of Jiangxi Institute of Education (Comprehensive), 33*(3), 13–16.

Huang, J.-L. (2013). *The idea and practice of standards-based teacher education*. Taipei: National Taiwan Normal University Press.

Lee, Y.-H. (2001). 台灣師範教育史 [*The history of Taiwan normal education*]. Taipei: The National Institute for Compilation and Translation.

Leung, F. K. S. (2006). Mathematics education in East Asia and the West: Does culture matter? In F. K. S. Leung, K.-D. Graf, & F. J. Lopez-Real (Eds.), *Mathematics education in different cultural traditions: A comparative study of East Asia and the West* (pp. 21–46). New York, NY: Springer.

Li, H. (2006). China's teacher education before the kuimao education system [癸卯學制之前的中國師範教育]. *Journal of Xinzhou Teachers University* [忻州師範學院學報]. *22*(1), 54–57.

Lin, C.-D., Shen, J.-L., & Xin, Y. (1996). Composition of teacher quality and its routes of development [教師素質的構成及其培養途徑]. *Journal of the Chinese Society of Education* [中國教育學刊]. (6), 16–22.

Meng, G.-J. (2013). On school-running tradition and experience of the century—long HSNU [百年韓師辦學傳統與經驗]. *Journal of Hanshan Normal University* [韓山師範學院學報], *34*(4), 37–42.

MOE. (1934). 第一次中國教育年鑑 [*The first annals of Chinese education*]. Beijing: Author.

Schmidt, W. H., Blömeke, S., Tatto, M. T., Hsieh, F.-J., Cogan, L., Houang, R. T., Bankov, K., Santillan, M., Cedillo, T., Han, S.-I., Carnoy, M., Paine, L., & Schwille, J. (2011). *Teacher education matters: A study of middle school mathematics teacher preparation in six countries*. New York, NY: Teachers College Press.

Smith, J., & Hu, R. (2013). Rethinking teacher education: Synchronizing eastern and western views of teaching and learning to promote 21st century skills and global perspectives. *Education Research and Perspectives, 40*, 86–108.

TQC. (2014). *Teacher qualification certification from kindergarten to senior high school*. Retrieved July 9, 2014, from https://tqa.ntue.edu.tw/page_exam103/103hsdng.pdf

Wu, W.-X. (1983). 日據時期臺灣師範教育之研究 [*A study of the normal education in Taiwan during the Japanese colonial period*]. Taipei: Graduate Institute of History, National Taiwan Normal University.

Yan, B.-H., & Wen, X.-Y. (1988a). 中國數學教育史簡論 [A brief discussion on Chinese mathematics education]. *Shuxue Tongbao*, (6), 27–28.

Yan, B.-H., & Wen, X.-Y. (1988b). 中國數學教育史簡論(續) [A brief discussion on Chinese mathematics education(continued)]. *Shuxue Tongbao*, (7), 31–32.

Yan, B.-H., & Wen, X.-Y. (1988c). 中國數學教育史簡論(續) [A brief discussion on Chinese mathematics education (continued)]. *Shuxue Tongbao*, (8), 29–31.

Yan, D.-J. (1965a). 中國數學教育簡史 [Brief history of Chinese mathematics]. 數學通報 [*Shuxue Tongbao*], 8, 44–48.

Yan, D.-J. (1965b). 中國數學教育簡史(續) [Brief history of Chinese mathematics]. 數學通報 [*Shuxue Tongbao*], 9, 46–50.

Yang, Q.-L. (2002). 儒、墨、道教學傳統比較及其對現代教學的啟示 [The traditional Chinese teaching theories and their contribution to modern teaching]. *Journal of Nanjing Normal University (Social Science Edition), 7*(4), 87–94.

Zhang, C.-S. (2005). 教師專業化:傳統智慧與現代實踐 [Teacher professionalism: Traditional wisdom and modern practice]. *Teacher Education Research, 17*(1), 16–26.

Zhang, C.-S., & Jiang, F. (2009). 《學記》的教師思想與教師專業化 [Teacher thought and teacher professionalism in the book Xue Zi]. 紀念《教育史研究》創刊二十周年論文集——中國教育思想史與人物研究[C] [*Proceedings of the Twentieth Anniversary for the Memorial of the Initiation of <Jiao Yu shi Yan Jiu>—Chinese History of Thoughts and Characters Regarding Education[C]*], (2), 2245–2249.

Zhang, C.-S., & Wang, S.-L. (2009). 《學記》對注入式教學的病理分析及其現實價值 [Pathological analysis on lecture-based teaching and its practical value in the book Xue Zi]. 紀念《教育史研究》創刊二十周年論文集——中國教育思想史與人物研究[C] [*Proceedings of the Twentieth Anniversary for the Memorial of the Initiation of <Jiao Yu shi Yan Jiu>— Chinese History of Thoughts and Characters Regarding Education[C]*]. (2), 2047–2051.

Zhang, D.-Z., & Yu., P. (2013). 數學教育的"中國道路" [*"The Chinese road" of mathematics education*]. Shanghai: Shanghai Educational Publishing House.

Zhang, W. (2008). Conceptions of lifelong learning in Confucian culture: Their impact on adult learners. *International Journal of Lifelong Education, 27*(5), 551–557. doi:10.1080/02601370802051561

Zhang, Y.-D., & Song, X.-P. (2005). 儒家教學思想與當今若干教學觀念的辯証思考 [The dialectic reflection on the Confucian teaching thought and several of today's teaching concept]. *Journal of Jianggsu University (Higher Education Study Edition), 27*(4), 73–77.

Zhang, Y.-D., & Song, X.-P. (2006). 儒家教學觀與我國數學教學 [The Confucian teaching concept and Chinese mathematical teaching]. 全國高師會數學教育研究會2006年學術年會論文集[C] [*Proceedings of 2006 Conference on Annually Academic Meetings of National Mathematics Education Association[C]*], 1–7.

Feng-Jui Hsieh
Mathematics Department
National Taiwan Normal University
Taipei

Sin-Sheng Lu
Mathematics and Science College
Shanghai Normal University
Shanghai

Chia-Jui Hsieh
Department of Mathematics and Information Education
National Taipei University of Education
Taipei

Shu-Zhi Tang
Taipei Municipal Bailing High School
Taipei

Ting-Ying Wang
National Taiwan Normal University
Taipei

MARTINA DÖHRMANN, GABRIELE KAISER
AND SIGRID BLÖMEKE

4. THE CONCEPTION OF MATHEMATICS KNOWLEDGE FOR TEACHING FROM AN INTERNATIONAL PERSPECTIVE

The Case of the TEDS-M Study

THEORETICAL FRAMEWORKS FOR MATHEMATICS KNOWLEDGE FOR TEACHING

The knowledge of mathematics teachers and how it is accomplished during teacher education has been conceptualized differently over time as well as across research paradigms and countries. A first important model that characterized mathematics teachers' knowledge started from classroom practices and was focused on knowledge acquisition by observation in a kind of apprenticeship (Zeichner, 1980). During the 1990s, the cognitive basis of teachers' pedagogical practices started to emerge and first small-scale comparative studies were carried out (Kaiser, 1995; Pepin, 1999).

More recently, research has focused even more strongly on the knowledge base of mathematics teachers' classroom practice. Several large scale studies developed theoretical frameworks focusing on mathematics knowledge for teaching, as well as a few more qualitatively oriented studies. In the following, we first present selected theoretical frameworks and then describe the international comparative study TEDS as an example of these kinds of studies.

A milestone within the recent developments of conceptions of mathematics knowledge for teaching is the seminal work by Shulman (1986, 1987), in which he developed theoretical categories for the knowledge base of teachers through analyzing the specifics of these knowledge categories for the teaching profession. He distinguished the following categories of the knowledge base of teachers:

- content knowledge;
- general pedagogical knowledge, with special reference to those broad principles and strategies of classroom management and organization that appear to transcend subject matter;
- curriculum knowledge, with particular grasp of the materials and programs that serve as 'tools of the trade' for teachers;

- pedagogical content knowledge, that special amalgam of content and pedagogy that is uniquely the province of the teachers, their own special form of professional understanding;
- knowledge of learners and their characteristics;
- knowledge of educational contexts, ranging from the workings of the group or classroom, the governance and financing of school districts, to the character of communities and cultures; and
- knowledge of educational ends, purposes, and values, and their philosophical and historical grounds. (Shulman, 1987, p. 8)

Shulman (1987) emphasizes that among those categories "pedagogical content knowledge is of special interest because it identifies the distinctive bodies of knowledge for teaching." He describes pedagogical content knowledge as "blending of content and pedagogy" and as the "category most likely to distinguish the understanding of the content specialist from that of the pedagogue" (p. 8).

In further work, Shulman concentrates on three of these categories, all related to content, namely "subject matter content knowledge", "pedagogical content knowledge" and "curricular knowledge" (1986, p. 9). He describes subject matter content knowledge as "the amount and organization of knowledge per se in the mind of the teacher" and puts the understanding of the structures of the subject in the foreground in contrast to the pure knowledge of facts and concepts. Pedagogical content knowledge "is pedagogical knowledge, which goes beyond knowledge of subject matter per se to the dimension of subject matter knowledge *for teaching*." Pedagogical content knowledge is defined as "the particular form of content knowledge that embodies the aspects of content most germane to its teachability" (p. 9). The third category of content knowledge, namely curricular knowledge, refers to the curriculum and the programs for teaching specific subjects. Shulman (1986) describes curriculum and its associated materials as "the *materia medica* of pedagogy, the pharmacopeia from which the teacher draws those tools of teaching that present or exemplify particular content and remediate or evaluate the adequacy of student accomplishments" (p. 10).

The question about the specific features of the professional knowledge for teachers and how to distinguish it from other forms of professional knowledge has created many discussions. Shulman (1987) writes: "But the key to distinguishing the knowledge base of teaching lies at the intersection of content and pedagogy, in the capacity of a teacher to transform the content knowledge he or she possesses into forms that are pedagogically powerful and yet adaptive to the variations in ability and background presented by the students" (p. 15).

Although Shulman's work was ground-breaking and can be described as milestone in the development of the theory of teachers' professional knowledge, critique was developed emphasizing that the knowledge facets were not sufficiently defined in order to allow operationalized empirical research. Especially the distinction between Shulman's concept of subject matter content knowledge and pedagogical content

THE CONCEPTION OF MATHEMATICS KNOWLEDGE FOR TEACHING

knowledge remains unclear according to these critical positions (Ball et al., 2008). We will come back to this critique while describing the approach developed by Ball and others in order to overcome this weakness. Another researcher, Anne Meredith, criticizes that the pedagogical content knowledge as defined by Shulman (1986, 1987) "seems to imply one type of pedagogy rooted in particular representations of prior knowledge" (1995, p. 176). She continues that Shulman's concept of pedagogical content knowledge "is perfectly adequate if mathematical knowledge is seen as absolute, incontestable, unidimensional and static. On the other hand, teachers who conceive of subject knowledge as multidimensional, dynamic and generated through problem solving may require and develop very different knowledge for teaching" (p. 184).

The critique of Shulman's work led to other conceptualizations on teachers' knowledge. Fennema and Franke (1992) in their famous handbook chapter discuss that the critical word, *transform,* in Shulman's approach neglects the complexity of the interaction between teachers and students. "This transformation is not simple, nor does it occur at one point in time. Instead, it is continuous and must change as the students who are being taught change. In other words, teachers' use of their knowledge must change as the context in which they work change" (p. 162). Based on this critique they modify the model by Shulman by emphasizing that teachers' knowledge is characterized by its "interactive and dynamic nature" (p. 162). They distinguish the following components of teacher knowledge: knowledge of the content of mathematics, knowledge of pedagogy, knowledge of students' cognitions, and teachers' beliefs. "It also shows each component in context" (p. 162).

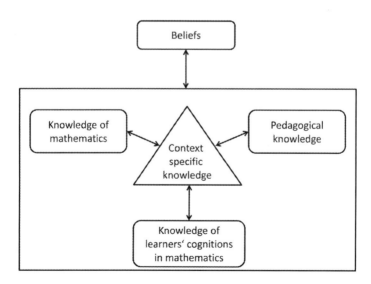

Figure 1. Teachers' knowledge developing in context (based on Fennema & Franke, 1992, p. 162)

Based on the description of teachers' knowledge as situated, which was a new perspective at this time, they explain the main characteristics of their model as follows: "The center triangle of our model indicates the teachers' knowledge and beliefs in context or as situated. The context is the structure that defines the components of knowledge and beliefs that come into play. Within a given context, teachers' knowledge of content interacts with knowledge of pedagogy and students' cognitions and combines with beliefs to create a unique set of knowledge that drives classroom behavior" (p. 162).

Departing from the critique that the two knowledge facets, subject matter knowledge and pedagogical content knowledge, are not distinguished precisely enough for measurement purposes, two US-American research projects based at the University of Michigan developed another modification of the Shulman model. The *Mathematics Teaching and Learning to Teach Project* (MTLT) and the *Learning Mathematics for Teaching Project* (LMT) define and distinguish between different knowledge facets functional for mathematics teaching. Widely discussed, especially in the US-American community, the MTLT project studies the interplay of mathematics and pedagogy in the teaching of elementary school mathematics. By looking closely at the mathematical and pedagogical work of teaching such as managing discussions, asking questions, interpreting students' thinking, the project aims to identify mathematical insight, appreciation, and knowledge that matters for teaching. In addition, the project aims to analyze and articulate ways in which it might be entailed in practice. The MTLT project developed the construct of Mathematical Knowledge for Teaching (MKT) as a part of quality mathematics instruction and defines MKT as "the mathematical knowledge used to carry out the work of teaching mathematics" (Hill, Rowan, & Ball, 2005, p. 373).

MKT embraces the two knowledge facets by Shulman, namely the subject matter knowledge and the pedagogical content knowledge, but differentiates them in various sub-facets (Figure 3). Subject matter knowledge includes both the mathematical knowledge that is common to individuals working in diverse professions and the mathematical knowledge that is specialized to teaching. It contains new strands that lie outside Shulman's conceptualization, namely common content knowledge (CCK) and specialized content knowledge (SCK). Common content knowledge (CCK) is, according to Hill et al. (2008), what Shulman likely meant by his original subject matter knowledge. It is this content knowledge that is used in the work of teaching in the same way as it is used in many other professions or occupations that also use mathematics. Specialized content knowledge (SCK) is a newer conceptualization and describes the mathematical knowledge that allows teachers to engage in particular teaching tasks, such as how to represent mathematical ideas or provide mathematical explanations. The third sub-facet, horizon content knowledge (HCK), is defined more as an awareness of the large mathematical landscape in which the present experience and instruction is situated than as practical knowledge. The second facet, pedagogical content knowledge, refers to Shulman's conceptualization. It contains knowledge of content and students (KCS) and is focused on teachers' understanding of how students learn a particular content. The second sub-facet, the knowledge of content and curriculum (KCC), refers to the arrangement of the mathematical topics within the curriculum and ways of using curriculum resources and materials. Knowledge of content and teaching (KCT) covers the knowledge about both mathematics and teaching such as the introduction of new concepts (for details see Hill et al., 2008; Ball et al., 2008).

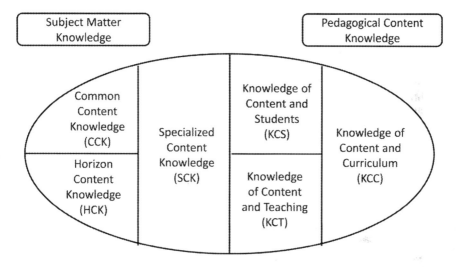

Figure 2. Mathematical knowledge for teaching (based on Ball, Thames, & Phelps, 2008, p. 403)

One central achievement of these two projects is the development of instruments to measure teachers' mathematical knowledge based on a series of multiple choice items. Although this measurement instrument originates from the U.S., it was applied in several other countries (Hill et al., 2007) and revealed severe cultural differences. For example, Ng (2012) studied Indonesian teachers' performances on geometry items and concluded that the initial measures may not be valid because of national differences between the United States and Indonesia in how shapes are classified. In contrast, Cole (2012) found that most items could be used validly in Ghana despite evidence of cultural incongruence in teaching practices between the U.S. and Ghana.

Another huge step forward achieved by these two projects is the identification of the relationship between teacher knowledge and students' achievements in mathematics and the evidence that teachers with weak knowledge transmit this to their students (Hill, Rowan, & Ball, 2005).

However, this new framework on mathematical knowledge for teaching has a few significant weaknesses such as not including teachers' beliefs, despite clear evidence from empirical research that teachers' beliefs about the nature of mathematics or about the genesis of mathematical knowledge strongly influences their teaching (Schoenfeld, 2011). Another problem is the proximity of various sub-facets such as the sub-facet specialized content knowledge and the sub-facets of knowledge of content and teaching and knowledge of content and students, which hardly can be differentiated (theoretically and empirically).

Looking back, Ball et al. (2008) reflect on the uncertainties and possible weaknesses of their model as follows:

"We are not yet sure, whether this may be a part of our category of knowledge of content and teaching or whether it may run across the several categories or be a category on its own right. We also provisionally include a third category within subject matter knowledge, what we call 'horizon' knowledge. ... Again we are not sure whether this category is part of subject matter knowledge or whether it may run across the other categories. We hope to explore these ideas theoretically, empirically, and also pragmatically as the ideas are used in teacher education or in the development of curriculum materials for use in professional developments" (p. 403).

The theoretical approach of the Knowledge Quartet – developed by research groups at the University of Cambridge – arose out of research into teachers' mathematical content knowledge. The approach refers to the theoretical conceptualization by Shulman, but takes up characteristics of the Fennema and Franke model by categorizing classroom situations where mathematical knowledge surfaces in teaching situations. "The purpose of the research from which the Knowledge Quartet emerged was to develop an empirically-based conceptual framework for lesson review discussions *with a focus on the mathematics content* of the lesson and the role of the trainee's mathematics subject matter knowledge (SMK) and pedagogical content knowledge (PCK). ... The focus of this particular research was therefore to identify ways that teachers' mathematics content knowledge – both SMK and PCK – can be observed to 'play out' in practical teaching" (Turner & Rowland, 2011, p. 197). Based on videotaped lessons of novice and trainee teachers the following four categories – foundation, transformation, connection and contingency – were distinguished in order to analyze the interplay of SMK and PCK. The first category, *foundation,* "is rooted in the foundation of the teacher's theoretical background and beliefs. It concerns their knowledge, understanding and ready recourse to what was learned at school, and at college/university ... It differs from the other three units in the sense that it is about knowledge 'possessed'" (Turner & Rowland, 2011, p. 200). The other three categories "focus on knowledge-in-action as demonstrated both in planning to teach and in the act of teaching itself" (Turner & Rowland, 2011, p. 200). The category, *transformation,* departs from Shulman's notion of the transformation of content knowledge into pedagogically powerful knowledge forms and mainly refers to the usage of instructional material, teacher demonstrations, and the choice of representations and examples. According to Turner and Rowland (2011), the category, *connection,* refers to "the coherence of the planning or teaching displayed across an episode, lesson or a series of lessons" (p. 201) and is characterized by making connections between procedures or concept, decisions about sequencing and so on. The last category, *contingency*, describes "the teachers' response to classroom events that were not anticipated in the planning" (Turner & Rowland, 2011, p. 202). This contingent action describes the adaptive and adequate response on children's ideas, the teachers' deviation from the agenda.

Although the Knowledge Quartet departs from the approach by Shulman with the strong and exclusive focus on SMK and PCK, it does not explicitly include curricular knowledge due to the traditionally minor role of curricular reflections in mathematics education in Britain.

The next theoretical approach developed by the German project, Cognitively Activating Instruction (COACTIV), also refers to the approach by Shulman, as it describes teaching as professional activity and knowledge as the core of professionalism. Departing from the theoretical approach of professional competence as defined by Weinert (2001), Baumert and Kunter (2013) describe competence as "the personal capacity to cope with specific situational demands" (p. 27). COACTIV uses a non-hierarchical model of professional competence as generic structural model, which is specified in Figure 3.

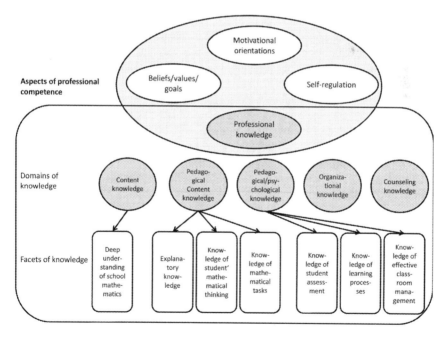

Figure 3. The COACTIV model of professional competence, with the aspect of professional knowledge specified for the context of teaching (based on Baumert & Kunter, 2013, p. 29)

The model distinguishes "between four aspects of competence (knowledge, beliefs, motivation, and self-regulation), each of which comprises more specific domains derived from the available research literature. These domains are further differentiated into facets, which are operationalized by concrete indicators" (Baumert & Kunter, 2013, p. 28). Concerning the mathematical content knowledge, COACTIV focuses on profound mathematical understanding of the mathematics taught at school, although theoretically four different levels of understanding of mathematics are distinguished starting with academic research knowledge as highest knowledge and ending with the mathematical everyday knowledge all adults should have. Pedagogical content knowledge is described by Baumert and Kunter (2013, p. 33) with three dimensions:

- Knowledge of the didactic and diagnostic potential of task, their cognitive demands and the prior knowledge they implicitly require, their effective orchestration in the classroom, and the long-term sequencing of learning content in the curriculum
- Knowledge of student cognitions (misconceptions, typical errors, strategies) and ways of assessing student knowledge and comprehension processes
- Knowledge of explanations and multiple representations

In addition, the model includes facets of general pedagogical knowledge such as pedagogical knowledge on effective classroom management and instructional planning. Moreover, the model covers various kinds of beliefs such as epistemological beliefs on the body of knowledge, beliefs about learning in a school subject area, and so on.

Like the LMT-study, the related empirical study provides insight into the strong relationship between teacher professional competency and students' achievements in mathematics.

However, the COACTIV-study has several weaknesses, namely the extended differentiation of the various competency facets, especially the general pedagogical knowledge, which were not covered in the main study, but only in the extension study, COACTIV-R.

From an international perspective the question arises whether the various facets of the professional competency and the professional knowledge of teachers can be distinguished empirically in different cultural context. Going back to the original conceptualization of PCK by Shulman as an amalgam of content and pedagogical knowledge, An, Kulm, and Wu (2004) compared the pedagogical content knowledge of mathematics teachers (PCK) between Chinese and U.S. groups and focused on fractions, ratio, and proportion. They found out that in contrast to the U.S. teachers, the Chinese had gained much of their knowledge through school-based in-service training led by expert teachers and continuous professional development activities, especially by observing each other's lessons and jointly discussing them (An, Kulm, & Wu, 2004; Paine, 1997; Paine & Ma, 1993).

These different studies point to the difficulty to reach international agreement on a definition of mathematical knowledge for teaching and how to acquire it. As Pepin (1999) pointed out, these differences also reflect differences in the meaning of mathematics didactics or pedagogy (for example called *Mathematikdidaktik* in German). Continental traditions are based on educational, philosophical, and theoretical reflections including normative descriptions of the teaching-and-learning-processes. In contrast, reflections on the knowledge transformation, its student-related simplification throughout the process to teaching knowledge, called *elementarization* in German, can hardly be found in English-speaking countries. From the beginning in English-speaking countries, research on mathematics knowledge and teacher education (Kaiser, 1999, 2002) was more outcome-based and thus, to a large extent, based on empirical studies in order to identify and determine influential factors as predictors of successful teaching and learning. As Westbury (2000) pointed out, the dominant features of the U.S. curriculum tradition was of organizational nature, referring to schools as institutions, where teachers were expected to be agents for an optimal school system.

Given these cultural differences an intriguing question is, whether it is possible to conceptualize the professional mathematics knowledge of teachers in a comparative study. The first enterprise in this respect was the international comparative study TEDS-M (Teacher Education and Development Study in Mathematics[1]), carried out in 2008 under the auspices of the International Association for the Evaluation of Educational Achievement (IEA) with 23,000 participants coming from 17 countries, the study was based on representative samples. Its aim was to understand how national policies and institutional practices influence the outcomes of mathematics teacher education. Referring to Shulman's model of the professional knowledge of teachers, the achievements in the knowledge facets, mathematics content knowledge (MCK) and mathematics pedagogical content knowledge (MPCK) were defined as outcomes and measures for the efficiency of teacher education.

Not surprisingly, the development of MCK and MPCK assessments was controversial in TEDS-M. Although international agreement was reached with respect to their core dimensions, national specifications had to be left out, as is common in comparative large-scale assessments. In the following, the nature of the TEDS-M tests for the lower-secondary study is analyzed in detail in order to show exemplarily what it means to measure mathematical knowledge for teaching. The objectives are, firstly, to increase the understanding of the nature of MCK and MPCK, which are still fuzzy domains (for an overview on the most recent discussion see Depaepe, Verschaffel, & Kelchtermans, 2013) by extending our work on the TEDS-M assessment of primary teachers (Döhrmann, Kaiser, & Blömeke, 2012). Secondly, we aim to provide a substantive background for interpretations of the TEDS-M test results by examining whether some educational traditions may be more accurately reflected in the test items than others. For this purpose, the TEDS-M items that have been released by the IEA are presented and analyzed.

OBJECTIVES AND DESIGN OF TEDS-M

The main research questions of TEDS-M were:

> What is the level and depth of the mathematics and related teaching knowledge attained by prospective primary and lower secondary teachers? How does this knowledge vary across countries? (Tatto et al., 2008, p. 13)

Similarly to the COACTIV-study, TEDS-M is based on the competency approach by Weinert (2001), who described the professional competencies of teachers as the specific ability to cope with the professional demands of teaching and is strongly related to action-oriented approaches:

> The theoretical construct of action competence comprehensively combines those intellectual abilities, content-specific knowledge, cognitive skills, domain-specific strategies, routines and subroutines, motivational tendencies, volitional control systems, personal value orientation, and social behaviors into a complex system. Together, this system specifies the prerequisites required to fulfill the demands of a particular professional position. (p. 51)

Departing from the theoretical approach by Shulman (1987), like most projects and theoretical approaches described above, TEDS-M describes MCK and MPCK as essential cognitive components underlying teacher performance in the classroom, complemented by general pedagogical knowledge, personality traits and beliefs.

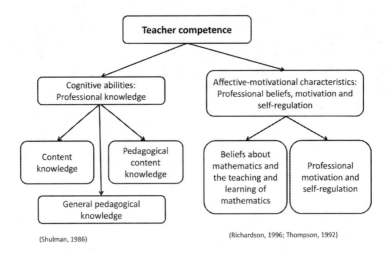

Figure 4. Conceptual model of teachers' professional competencies (based on Döhrmann, Kaiser, & Blömeke, 2012, p. 327)

MCK and MPCK were assessed with paper-and-pencil-tests (Tatto et al., 2008). The underlying conceptual framework was the result of a long and intense discussion between the participating countries, in which international acceptance was accomplished. In order to achieve this, national specifications of what was meant by MCK or MPCK had – by necessity – to be left out.

As TEDS-M is the first international large-scale study on teacher education, the theoretical conceptualization of MCK and MPCK as well as developing proficiency tests necessitated extensive work and an enormous amount of time previous to the implementation of the study. In 2002, representatives from the countries participating in TEDS-M met for the first time to discuss their nationally and culturally shaped conceptions on the professional knowledge of mathematics teachers. The result emerging from this process was a definition of MCK and MPCK that predominantly focused teachers' tasks rather than normative – often implicit – curricular requirements. Thus, a teacher's mathematical knowledge was expected to cover at least the mathematical content of the grades the teacher will teach from a higher and reflected level. In addition, a teacher was considered to be able to integrate the educational context as well as to connect the mathematics content to following, higher levels of education. Therefore, the conceptualization of MCK considers the content areas used by TIMSS 2007 where the international discussions lead to the distinction of the four subdomains of number, algebra, geometry and data as essential for school mathematics. The MCK and MPCK tests were oriented towards these subdomains. The overall reliability, validity and credibility of the items have already been demonstrated (Blömeke, Suhl, Kaiser, & Döhrmann, 2012; Blömeke, Suhl, & Kaiser, 2011; Senk et al., 2012). Now, we can look beyond what was accomplished in order to meet further research needs.

In order to provide insight into the nature of the TEDS-M tests, special aspects of the items released and their requirements are featured. This detailed item analyses is partly based on ACER documents, as these provide the percentage of correct answers as indicators of the countries' range of proficiency. In addition, we provide background information about the items and an analysis from a mathematics education point of view. The complete set of TEDS-M Lower-secondary items released by the IEA together with coding guides is available: http://www.acer.edu.au/research/projects/iea-teacher-educationdevelopment-study-teds-m/. As displayed in the diagram on the teachers' professional competencies, beliefs play an important role in TEDS-M. Using well-known scales, various kinds of epistemological beliefs were evaluated such as the beliefs on the genesis of mathematics knowledge and its nature. Due to the focus of this paper on the mathematical knowledge for teaching, this aspect is not described in this paper, but was covered in the evaluation.

ANALYZING TEDS-M ITEMS

We start by analyzing the items that are supposed to assess MCK, and refer to the subdomains of algebra, geometry, number and data. After that, an analysis follows that covers the items that are supposed to assess MPCK, and which refer to the knowledge displayed prior to a lesson in terms of planning but also enacted knowledge in terms of student-teacher interaction. Several item examples are presented and discussed in detail.

Assessing Mathematics Content Knowledge (MCK)

The MCK test consists of 76 items in a multiple-choice and constructed-response format. It covers topics dominating mathematics education all over the world and which mainly come from algebra, number and geometry. Data and probability items are scarcely represented in the test. This reflects their low importance in the mathematics curricula of schools and teacher education in the participating countries. As a consequence, the number of items for the subdomains algebra, number and geometry were nearly uniformly distributed in the test with four items per subdomain. In addition to these subdomains, three cognitive domains were defined: knowing, applying and reasoning (according to TIMSS). The cognitive as well as the content domains constituted a heuristic tool for the item development.

All items were categorized into levels of difficulty arising from the item's curricular level. In detail, the *novice* level of difficulty indicates mathematics content that is typically taught at the grades the future teacher will teach. The *intermediate* level of difficulty indicates content that is typically taught one or two grades beyond the highest grade the future teacher will teach and finally, the *advanced* level of difficulty indicates content that is typically taught three or more years beyond the highest grade the future teacher will teach (Tatto et al., 2008, p. 37).

Subdomain Algebra

The algebra items of the MCK test mostly belong to the field of functions. Proportional relations as well as linear, quadratic and exponential functions, and the absolute value function are typical mathematical topics in most secondary schools all over the world. These are also represented in the test. The required competencies include, for example, identifying the graph of a given function class, identifying the quantitative relation within a given realistic context, and determining a given function as appropriate to model the relation as well as judging the adequacy of given examples for the definitions of a continuous function.

In the test for the future secondary school teachers, patterns play a less role than in the primary test (Döhrmann, Kaiser, & Blömeke, 2012). Only one MCK item refers to patterns (and also 3 MPCK items). Here, the test persons have to compare and determine various patterns of growth. Another item refers to sequences and requires higher skills which are usually taught in university. For this item, the future teachers need to know the concept of convergence and the limit of a sequence.

Skills concerning functions are also partly needed for items in the subdomain geometry. Here, some items refer to equations and require, for example, to use equations to represent and solve a given contextualized problem as well as determine the set of solutions for a given equation especially on the set of complex numbers.

Overall the subdomain algebra emphasizes the concept of functions as well as the language of formulae and their application to contextualized problems and problems within mathematics. Structural algebraic concepts such as groups or rings are not covered by the items.

The following item example shows one algebra item concerning functions on an intermediate level.

Prove the following statement:
If the graphs of linear functions

$$f(x) = ax + b \text{ and } g(x) = cx + d$$

intersect at a point P on the x-axis, the graph of their sum function

$$(f+g)(x)$$

must also go through P.

IEA Teacher Education and Development Study. Source: ACER (2011)

For this item, participants have to prove that the graph of the sum of two linear functions f(x) and g(x) also goes through the point P on the x-axis if both functions f(x) and g(x) intersect the x-axis at that point. This proof could be realized with or without using the function expressions of f and g. A complete and accurate answer would be for example the following sentences: "Suppose f(x) and g(x) intersect at point (p, 0) on the x-axis. Then f(p) = 0, g(p) = 0. Then (f + g)(p) = f(p) + g(p) = 0 + 0 = 0. Therefore f+g also goes across point (p, 0)."

Knowledge about linear functions is essential for solving this task as well as knowledge about the intersection of two functions and the sum function which is constituted by summing both function values (actually, for any functions, this statement is true). Only 10 percent of the future secondary school teachers tested in TEDS-M were able to adequately and completely formulate this proof. Another 8 percent of those teachers received partial credit because their proof was valid to some extent but incomplete. Internationally, this task showed an enormous deviation. 99.7 percent of the future teachers in Chile did not achieve any solution while 69 percent of the future teachers in Taiwan solved the task completely. Prior to the test, this task was classified as intermediate level of difficulty but empirically it showed to be more complex. This may attribute to difficulties in formulating adequate proof.

Subdomain Number

In the subdomain, number, there is only one item that was classified as novice level. This item requires only a simple operation, but knowledge about the concept of the arithmetic mean is essential as well. The other items in this subdomain refer to mathematics topics that are usually not taught in secondary school. The subdomain number mostly focuses on the cognitive domains knowing and reasoning while algebra predominantly focuses on applying knowledge. The future mathematics teachers need, among other things, to judge if statements about irrational numbers are true, assign the solution of equations to the set of numbers it belongs to, and judge the adequacy of given examples as a proof of a statement about number theory.

The test requirements in this subdomain are quite high. While computations with numbers are not requested, most of the items deal with statements about numbers and its characteristics. Here, the test persons have to apply and compare the properties of numbers and number systems. The following item shows an item-example of the subdomain number that was classified as intermediate level and belongs to the cognitive domain reasoning.

You have to prove the following statement:

If the square of any natural number is divided by 3, then the remainder is only 0 or 1.

State whether each of the following approaches is a mathematically correct proof.

Check one box in each row.

		Yes	No
A.	Use the following table: Number: 1,2,3,4,5,6,7,8,9,10 Square: 1,4,9,16,25,36,49,64,81,100 Remainder when divided by 3: 1,1,0,1,1,0,1,1,0,1	☐	☐
B.	Demonstrate that $(3n)^2$ is divisible by 3 and for all other numbers, $(3n \pm 1)^2 = 9n^2 \pm 6n + 1$ which always has a remainder of 1 once it has been divided by 3.	☐	☐
C.	Choose a natural number n, find its square n^2, and then check whether the statement is true or not.	☐	☐
D.	Check the statement for the first several prime numbers and then draw a conclusion based on the Fundamental Theorem of Arithmetic.	☐	☐

IEA Teacher Education and Development Study. Source: ACER (2011)

In this task, different ideas are presented to the future teachers who need to judge whether the given arguments represent correct or incorrect proofs. Only the idea given in B illustrates a mathematically correct proof while the ideas presented in A and C are only based on examples and thus they are not mathematically correct proofs. In addition, the idea presented in D is unsuitable to prove the given statement. Besides basic knowledge about number theory, the future teachers predominantly need abilities to prove mathematical statements as well as knowledge about the criteria that contribute to a correct and complete proof in order to solve this task correctly.

Regarding the international average of all countries participating in TEDS-M, the items A and C describing possible arguments for the statement (see item above) were solved very differently. Item A which is based on ten examples was solved correctly by 45 percent of the future teachers while item C, which is only based on one example, was detected as an incorrect proof by 57 percent. Again, there was an enormous range that varied from 18 percent correct answers in Malaysia to 84 percent in Taiwan concerning item A. Regarding item C, the range varied from 18 percent in Chile to 92 percent correct answers in Taiwan. Identifying item B as a correct proof was easiest to the future secondary school teachers. On average, 62 percent of the test persons succeeded ranging from 25 percent in Chile to 91 percent in Taiwan. Item D was solved correctly by 54 percent in average. In every country, more than one fourth of the teachers correctly answered this task (Chile: 28%) but no country reached more than 80 percent correct answers (Taiwan: 79%). This may be due to the implicit reference to the proving concept of induction that shows through the item as well as its reference to the fundamental theorem of number theory which is a powerful theorem in the field of number theory but its inappropriateness for the given proof is not immediately obvious.

Subdomain Geometry

Only a few items in the subdomain geometry refer to geometric measurement of two- and three-dimensional objects. For one task, the future teachers have to determine the area of a pictured irregular shape. For another, they have to estimate the surface area and the volume of a represented three-dimensional object. For a third one, they have to compare the properties of three-dimensional objects. These tasks were assigned to a basic level of difficulty prior to the testing and may as well be content of a secondary school-mathematics textbook. The most difficult task of the entire test belongs to the subdomain geometry though and refers to the axiom of the uniqueness of a parallel line. In order to solve this task, the future teachers need to decide if statements are equivalent to the axiom of the uniqueness of a parallel line. This requires mathematical knowledge that is usually taught in university. The task includes four items, three of which were classified as the most difficult items of the entire MCK-test as measured by the international Rasch difficulty.

In addition, two more tasks that were assigned to a high level of difficulty refer to university mathematics. One of these tasks, which refers to analytical geometry, demands the future teachers to interpret the solutions of a linear function geometrically, while they have to interpret the properties of a geometrical function in another task. The high level of difficulty that was assigned to the tasks prior to the testing was also approved empirically, but its solution required a more general understanding of geometry than the task on axiom of the uniqueness of a parallel line which involved more specialized knowledge.

Two tasks refer to the topic of transformations. In one of these tasks, the future teachers have to determine the number of symmetry axes for different shapes, in the other task they had to identify the transformation that was applied to an object. The other items refer to relations between lines and angles in geometrical figures. Here, for example, the future teachers need to judge if a statement is true by using the theorem of intersecting lines.

The following task illustrates the requirement to compare the properties of three-dimensional objects.

Two gift boxes wrapped with ribbon are shown below. Box A is a cube of side-length 10 cm. Box B is a cylinder with height and diameter 10 cm each.

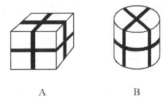

A   B

Which box needs the longer ribbon? _____

Explain how you arrived at your answer.

IEA Teacher Education and Development Study. Source: ACER (2011)

Here, it is required to identify whether the ribbon of the cube or the cylinder is longer and to explain the choice adequately. Answers are regarded as correct and complete if box A is identified and the choice is explained by correctly calculating both ribbon lengths as shown in the following example: "Box A requires 3*40cm = 120cm ribbon. Box B requires 2*40cm = 80cm and the length of the circumference 2*π*5cm ≈ 31cm. In total, Box B needs 111cm ribbon and thus less than Box A." But teachers could also focus their reasoning on comparing the circumferences of the circle and the square regarding that the remaining lengths are equal for both boxes. An example for this explanation would be the following: "Box A needs more ribbon because the circumference of a circle with a diameter of 10 cm is smaller than the circumference of a square with 10 cm side lengths and all remaining sizes are equal."

Incomplete answers as well as answers with minor errors are valued as partial solutions. Answers are considered wrong if either the lengths of both ribbons are calculated incorrectly or teachers expect the same length for both ribbons. In addition, answers are considered wrong if a correct answer did not include an explanation because this was the main focus for this task. Finally, answers are considered wrong if misconceptions become apparent (for example with recourse to area or volume calculation).

In advance to the test, this task was classified as novice level of difficulty but the international average proved that it was more difficult regarding completely correct answers. Only 33 percent of the future teachers reached a complete and correct solution ranging from three percent correct solutions in Chile and the Philippines to 75 percent in Taiwan. Almost all the teachers chose the approach to calculate both ribbon lengths and compare the results. Apart from basic additions of the cube's side lengths, this only requires knowledge about calculating the circle's circumference by multiplying the diameter and π. But this is probably the greatest difficulty in most country and resulted in less complete and correct answers than expected. Substantial parts of teachers who reason conceptually by comparing the square's and the circle's circumference and refer to equality of all other sides only appeared with about five to ten percent in Singapore, Taiwan, Russia, Germany and Switzerland.

On average of all TEDS-M countries, 21 percent of the future teachers succeeded in developing a partial solution ranging from five percent in Chile to 38 percent in Germany. In Chile, only eight percent developed at least an explanatory approach while future teachers in Germany easily managed to develop at least a partial solution. In Germany a correct or partial correct solution was achieved by 79% of the future secondary school teachers.

Subdomain Data

Only four MCK-items and two MPCK-items refer to the subdomain data. Two of the four MCK-items deal with probability, the two other items with statistics. Here, the future teachers have to calculate the probability of an event in a Laplace experiment or interpret and compare data with regard to their standard deviation.

For one item the future teachers have to calculate the conditional probability of an event. This item was classified as advanced level of difficulty in advance to the test and proved to be empirically difficult as well. Internationally, the subject area "data" is not part of every school curriculum and the item requires knowledge that goes beyond basic abilities in the field of probability.

One item of this subdomain that combines MCK and MPCK is presented in the following description of the MPCK test.

Assessing Mathematics Pedagogical Content Knowledge (MPCK)

The part of the test that is supposed to assess MPCK consists of only 27 items. The lower number of items results especially from the difficulties to obtain an internationally accepted consensus about MPCK that is universally required by future mathematics teachers. Compared to MCK this was an even greater challenge. In this regard, theories and developments are affected even stronger by traditions and culture. The conceptualization of MPCK therefore was oriented towards the teacher's core task of teaching. For TEDS-M, two subdomains of mathematics pedagogical content knowledge were differentiated: (a) *curricular knowledge and knowledge of planning for mathematics teaching and learning* and (b) *knowledge of enacting mathematics for teaching and learning*.

Subdomain Curricular Knowledge and Knowledge of Planning for Mathematics Teaching and Learning

Three tasks which were categorized to this subdomain refer to curricular knowledge. The future teachers had to identify consequences for the planning of teaching due to a thematic change of the curriculum and determine the required precognition for a mathematical content. The item example below shows a task of this category. Another task that required the future teachers to decide if some given story-problems are an adequate representation of mathematical content was categorized as knowledge of planning for mathematics teaching and learning.

A mathematics teacher wants to show some students how to prove the quadratic formula.

Determine whether each of the following types of knowledge is needed in order to understand a proof of this result.

Check one box in each row.

		Needed	Not needed
A.	How to solve linear equations.	☐	☐
B.	How to solve equations of the form $x^2 = k$, where $k > 0$.	☐	☐
C.	How to complete the square of a trinomial.	☐	☐
D.	How to add and subtract complex numbers.	☐	☐

IEA Teacher Education and Development Study. Source: ACER (2011)

For this task, the future teachers have to determine the required precognition for proving the quadratic formula. The knowledge described in A, B and C is needed, but the knowledge described in D is not. Mathematical knowledge of the quadratic formula and methods to prove this formula are essential in order to solve this task correctly. The future teachers easily solved items A, B and D. On international average, item A was answered correctly by 77 percent of the future teachers, item B by 76 percent and item D by 62 percent. Item C was merely solved correctly by 48 of the test persons. Probably, the term trinomial is not very familiar and is not directly associated with solving quadratic equations. Merely 35 percent of the future teachers in Singapore, who reached the third highest score in the MCK test in total, answered item C correctly while at least 85 percent of them answered the other three items correctly. Then again, 80 percent of the future teachers on the Philippines correctly solved item C but only achieved low scores in the MCK test in total and only 31 percent of them correctly solved item D. This may indicate cultural differences regarding the familiarity of the term trinomial as well as knowledge about proving the quadratic formula.

THE CONCEPTION OF MATHEMATICS KNOWLEDGE FOR TEACHING

Subdomain Knowledge of Enacting Mathematics for Teaching and Learning

15 items were categorized as knowledge of enacting mathematics for teaching and learning. For one item, the future teachers have to argue why the level of difficulty for two tasks with a similar context is different. The other items refer to students' solutions. Here, the future teachers have to evaluate given verbal or illustrated solutions. They need to decide if some given statements are appropriate responses to a student's solution or they have to analyze a student's solution with regard to typical students' misconceptions. One task that combines this subdomain with the MCK subdomain data is shown below.

The following graph gives information about the adult female literacy rates in Centra and South American countries.

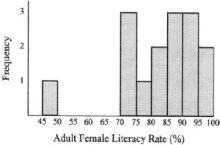

Suppose you ask your students to tell you how many countries are represented in the graph. One student says, "There are 7 countries represented."

Check one box.
Right   **Wrong**

a) Is the student right or wrong? ☐₁   ☐₂

b) In your opinion, what was the student thinking in order to arrive at that conclusion?

IEA Teacher Education and Development Study. Source: ACER (2011)

The first item requires the future teachers to judge whether the given interpretation of the bar chart is correct or not (MCK). In the following MPCK item, they need to analyze the reason for the student's interpretation. In order to give a correct answer to the first item, the concept of frequency needs to be known and the future teachers have to understand the bar chart as a frequency distribution. The correct answer is "wrong" because the bar chart represents 15 countries instead of only seven. Regarding the MPCK item, responses are accepted that indicate that the student thought each bar represents one country. This task is a positive example for successfully linking the MCK and the MPCK items. Both items refer to the same context while the MPCK item can be answered correctly even if the MCK item was answered incorrectly.

Both items were classified as novice level of difficulty in advance to the test. This task requires mathematical knowledge that is attained in secondary school in most of the countries and may already be part of primary school mathematics education. On international average, 72 percent of the future teachers correctly answered the MCK item while 70 percent gave a correct response to the MPCK item. Thus, these items proved to be empirically easy as well. The least correct answers for both items were given by future teachers in Georgia. Here, the MCK item reached

---

When you multiply 3 consecutive natural numbers, the product is a multiple of 6.

Below are three responses.

[Kate's] answer

A multiple of 6 must have factors of 3 and 2.
If you have three consecutive numbers, one will be a multiple of 3.

Also, at least one number will be even and all even numbers are multiples of 2.

If you multiply the three consecutive numbers together the answer must have at least one factor of 3 and one factor of 2.

[Leon's] answer

$1 \times 2 \times 3 = 6$

$2 \times 3 \times 4 = 24 = 6 \times 4$

$4 \times 5 \times 6 = 120 = 6 \times 20$

$6 \times 7 \times 8 = 336 = 6 \times 56$

[Maria's] answer

n is any whole number

$n \times (n+1) \times (n+2) = (n^2 + n) \times (n+2)$

$= n^3 + n^2 + 2n^2 + 2n$

Canceling the n's gives $1 + 1 + 2 + 2 = 6$

Determine whether each proof is valid.

Check one box in each row.

		Valid	Not valid
A.	[Kate's] proof	☐	☐
B.	[Leon's] proof	☐	☐
C.	[Maria's] proof	☐	☐

IEA Teacher Education and Development Study. Source: ACER (2011)

40 percent correct responses and the MPCK item 19 percent. The highest solution frequency regarding the MCK item was achieved by future teachers in Singapore (95% correct responses) while future teachers in Switzerland reached the highest solution frequency for the MPCK item (91% correct responses).

For another task the future teachers have to judge if three student responses are valid proofs of a statement of number theory (see below).

With the approval of Healy and Hoyles, the task was adapted and used in the TEDS-M study in 2008 (Healy & Hoyles, 1998). It refers to the domain number and requires interactive knowledge on a novice level. Three different student solutions are presented that involve different ideas of proof and need to be verified regarding validity. To begin with, the participating future teachers need to understand and analyze the student solutions.

Kate's solution involves a complete and generally valid proof that can be classified as correct. In contrast, Leon's solution only uses examples because he only analyzes four special cases. Thus, his proof is not generally valid. Maria initially uses an adequate approach to prove the statement but undertakes an invalid reduction and, thus, does not give valid proof either.

The three items presented proved to be of different empirical difficulty. As classified in advance, identifying Kate's proof as invalid was easy. On international average, 74% identified this student solution as invalid, ranging from 51% in Botswana to 97% in Taiwan.

The future secondary school teachers had more difficulties in classifying Maria's solution. On average, of all TEDS-M participating countries, 59% of the future teachers classified the proof as invalid ranging from 44% in Chile to 92% in Taiwan. Interestingly, classifying Leon's solution caused the greatest difficulties across the 15 participating countries of the TEDS-M study. Only 45% rejected this student solution. In Botswana, merely 3% of the future mathematics teachers identified that the proof is based on examples. Obviously, the formalized procedure was confused with general validity.

SUMMARY, DISCUSSION AND CONCLUSIONS

The mathematical requirements in the TEDS-M secondary level test are much higher than in the primary level test (Döhrmann, Kaiser, & Blömeke, 2012). There are tasks in all three subdomains, algebra, number and geometry, which require MCK that is usually acquired in university courses. Such tasks only sporadically appear in the primary level test while the secondary level test's subdomains, number and geometry, include several of these items. Again, the subdomain, algebra, contains additional tasks that can be solved by secondary school students as well.

Similar to the MCK primary level test, the subdomain, data, is merely represented by few items. Statistics and probability are unequally implemented into the mathematics curricula of schools and teacher education in the participating countries while algebra, number and geometry belong to the standard repertoire of mathematics education all over the world (cf. KMK, 2004; NCTM, 2000; NGA & CCSSO, 2010; Schmidt, McKnight, Valverde, Houang, & Wiley, 1997).

The secondary level test more strongly focusses on the conceptual understanding of mathematics and understanding of mathematical structures as compared to the primary level test. Computations are of lower relevance while argumentations and proofs are strongly fostered. Like in the primary level test, heuristic problem solving, modelling of non-routine problems and the use of technology are areas that were mostly left out of the test. This leads to the conclusion that more traditional notions of mathematics influenced the conceptualization of the cognitive domains of MCK in TEDS-M.

It is by definition probably impossible to design MPCK items without any mathematical content. However, it must be acknowledged that in order to answer some of the MPCK items in the TEDS-M test correctly, mathematical knowledge is required. The MPCK item presented above is a representative example for this effect. In addition, the items' different levels of difficulties can hardly be explained by mathematics pedagogical content knowledge and skills. Instead, they may be caused by different *mathematical* requirements. The MPCK test focuses on an analysis and evaluation of students' responses while other *didactical* requirements of the framework, such as identifying the key ideas in learning programs, establishing appropriate learning goals, choosing assessment formats and predicting typical students' responses, were less often considered.

As already stated for the primary level test, it also extends to the secondary level test of TEDS-M that the conceptualization of MPCK was oriented towards the teacher's core task of teaching. The test refers to various abilities and skills that are essential to concretely plan and realize mathematical lessons. These abilities and skills can be described as an internationally accepted common core of MPCK that is universally required by future mathematics teachers. This also includes analyzing and evaluating students' responses.

National characteristics of MPCK from individual participating countries had to be excluded, of course. The framework, for example, did not include didactical concepts, the promoting of process-related competencies based on mathematical contents, strategies for dealing with children's heterogeneity, theoretical knowledge about preschool age mathematical knowledge development or the knowledge about research in mathematics pedagogy. As already indicated by findings from the primary school study, the conceptualization of MPCK in TEDS-M is thus guided by curriculum theory and educational psychology which dominates in English-speaking countries. In contrast, continental European traditions rather focus on subject-related reflections, called *Didaktik* in German or *didactique* in French. Subject-related didactics describe the pedagogical transformation of disciplinary content to teaching content, taking into account the whole teaching-and-learning-process (Pepin, 1999). These differences in basic orientations of the countries participating in TEDS-M need to be explored in further studies although it may be difficult to test corresponding knowledge and skills on a large scale.

NOTE

[1] TEDS-M was funded by the IEA, the National Science Foundation (REC 0514431), and the participating countries. In Germany, the German Research Foundation funded TEDS-M (DFG, BL 548/3–1). The instruments are copyrighted by the TEDS-M International Study Center at MSU (ISC). The views expressed in this chapter are those of the authors and do not necessarily reflect the views of the IEA, the ISC, the participating countries or the funding agencies.

REFERENCES

An, S., Kulm, G., & Wu, Z. (2004). The pedagogical content knowledge of middle school mathematics teachers in China and the U.S. *Journal of Mathematics Teacher Education, 7*, 145–172.

Australian Council for Educational Research for the TEDS-M International Study Center. (2011). *Released items – future teacher Mathematics Content Knowledge (MCK) and Mathematics Pedagogical Content Knowledge (MPCK) – primary.* Paris: IEA.

Ball, D. L., Thames, M. H., & Phelps, G. (2008). Content knowledge for teaching: What makes it special? *Journal of Teacher Education, 59*, 389–407.

Baumert, J., & Kunter, M. (2013). The COACTIV model of teachers' professional competence. In M. Kunter, J. Baumert, W. Blum, U. Klusmann, S. Krauss, & M. Neubrand (Eds.), *Cognitive activation in the mathematics classrooms and professional competence of teachers* (pp. 25–48). New York, NY: Springer.

Blömeke, S. (2012). Content, professional preparation and teaching methods: How diverse is teacher education across countries? *Comparative Education Review, 56*, 684–714.

Blömeke, S., & Kaiser, G. (2012). Homogeneity or heterogeneity? Profiles of opportunities to learn in primary teacher education and their relationship to cultural context and outcomes. *ZDM Mathematics Education, 44*, 249–264.

Blömeke, S., Suhl, U., & Döhrmann, M. (2013). Assessing strengths and weaknesses of teacher knowledge in Asia, Eastern Europe and Western countries: Differential item functioning in TEDS-M. *International Journal of Science and Mathematics Education, 11*, 795–817.

Blömeke, S., Suhl, U., & Kaiser, G. (2011). Teacher education effectiveness: Quality and equity of future primary teachers' mathematics and mathematics pedagogical content knowledge. *Journal of Teacher Education, 62*, 154–171.

Blömeke, S., Suhl, U., Kaiser, G., & Döhrmann, M. (2012). Family background, entry selectivity and opportunities to learn: What matters in primary teacher education? An international comparison of fifteen countries. *Teaching and Teacher Education, 28*, 44–55.

Cole, Y. (2012). Assessing elemental validity: The transfer and use of mathematical knowledge for teaching measures in Ghana. *ZDM Mathematics Education, 44*, 415–426.

Depaepe, F., Verschaffel, L., & Kelchtermans, G. (2013). Pedagogical content knowledge: A systematic review of the way in which the concept has pervaded mathematics educational research. *Teaching and Teacher Education, 34*, 12–25.

Döhrmann, M., Kaiser, G., & Blömeke, S. (2012). The conceptualisation of mathematics competencies in the international teacher education study TEDS-M. *ZDM Mathematics Education, 44*, 325–340.

Fennema, E., & Franke, L. M. (1992). Teachers' knowledge and its impact. In D. A. Grouws (Ed.), *Handbook of research on mathematics teaching and learning* (pp. 147–164). Reston, VA: National Council of Teachers of Mathematics.

Hill, H. C., Ball, D. L., & Schilling, S. G. (2008). Unpacking pedagogical content knowledge: Conceptualizing and measuring teachers' topic-specific knowledge of students. *Journal for Research in Mathematics Education, 39*, 372–400.

Hill, H. C., Rowan, B., & Ball, D. (2005). Effects of teachers' mathematical knowledge for teaching on student achievement. *American Educational Research Journal, 42*, 371–406.

Hill, H. C., Sleep, L., Lewis, J. M., & Ball, D. L. (2007). Assessing teachers' mathematical knowledge. In F. K. Lester (Ed.), *Handbook for research on mathematics education* (2nd ed., pp. 111–155). Charlotte, NC: Information Age.

Kaiser, G. (1995). Results from a comparative empirical study in England and Germany on the learning of mathematics in context. In C. Sloyer, W. Blum, & I. Huntley (Eds.), *Advances and perspectives on the teaching of mathematical modelling and applications* (pp. 83–95). Yorklyn, DE: Water Street Mathematics.

Kaiser, G. (1999). *Unterrichtswirklichkeit in England und Deutschland: Vergleichende untersuchungen am beispiel des mathematikunterrichts*. Weinheim: Deutscher Studien Verlag.

Kaiser, G. (2002). Educational philosophies and their influences on mathematics education: An ethnographic study in English and German classrooms. *Zentralblatt für Didaktik der Mathematik, 34*, 241–256.

KMK. (Ed.). (2004). *Bildungsstandards im fach mathematik für den mittleren schulabschluss (Jahrgangsstufe 10)*. München: Wolters Kluwer.

Kunter, M., Baumert, J., Blum, W., Klusmann, U., Krauss, S., & Neubrand, M. (Eds.). (2013). *Cognitive activation in the mathematics classrooms and professional competence of teachers*. New York, NY: Springer.

Meredith, A. (1995). Terry's learning: Some limitations of Shulmans' pedagogical content knowledge. *Cambridge Journal of Education, 25*(2), 175–187.

National Council of Teachers of Mathematics (NCTM). (2000). *Principles and standards for school mathematics*. Reston, DE: National Council of Teachers of Mathematics.

National Governors Association Center for Best Practices & Council of Chief State School Officers (NGA & CCSSO). (2010). *Common core state standards for mathematics*. Retrieved from http://www.corestandards.org/math

Ng, D. (2012). Using the MKT measures to reveal Indonesian teachers' mathematical knowledge: Challenges and potentials. *ZDM Mathematics Education, 44*, 401–413.

Paine, L. (1997). Chinese teachers as mirrors of reform possibilities. In W. K. Cummings & P. G. Altbach (Eds.), *The challenge of eastern Asian education* (pp. 65–83). Albany, NY: SUNY Press.

Paine, L., & Ma, L. (1993). Teachers working together: A dialogue on organizational and cultural perspectives of Chinese teachers. *International Journal of Educational Research, 19*, 675–697.

Pepin, B. (1999). Existing models of knowledge in teaching: Developing an understanding of the Anglo/American, the French and the German scene. In B. Hudson, F. Buchberger, P. Kansanen, & H. Seel (Eds.), *Didaktik/Fachdidaktik as science(s) of the teaching profession* (Vol. 2, pp. 49–66). Lisbon: TNTEE Publications.

Rowland, T., Huckstep, P., & Thwaites, A. (2005). Elementary teachers' mathematics subject knowledge: The knowledge quartet and the case of Naomi. *Journal of Mathematics Teacher Education, 8*, 255–281.

Rowland, T., & Ruthven, K. (Eds.). (2010). *Mathematical knowledge in teaching*. Dordrecht: Springer.

Schoenfeld, A. (2011). *How we think: A theory of goal-oriented decision making and its educational applications*. New York, NY: Routledge.

Senk, S. L., Tatto, M. T., Reckase, M., Rowley, G., Peck, R., & Bankov, K. (2012). Knowledge of future primary teachers for teaching mathematics: An international comparative study. *ZDM Mathematics Education, 44*, 307–321.

Shulman, L. S. (1986). Those who understand: Knowledge growth in teaching. *Educational Researcher, 15*(2), 1–22.

Shulman, L. S. (1987). Knowledge and teaching: Foundations of the new reform. *Harvard Educational Research, 57*, 1–22.

Tatto, M. T., Schwille, J., Senk, S., Ingvarson, L., Peck, R., & Rowley, G. (2008). *Teacher Education and Development Study in Mathematics (TEDS-M): Policy, practice, and readiness to teach primary and secondary mathematics conceptual framework*. East Lansing, MI: Teacher Education and Development International Study Center, College of Education, Michigan State University.

Turner, F., & Rowland, T. (2011). The knowledge quartet as an organising framework for developing and deepening teachers' mathematical knowledge. In T. Rowland & K. Ruthven (Eds.), *Mathematical knowledge in teaching* (pp. 195–212). Dordrecht: Springer.

Weinert, F. E. (2001). Concept of competence: A conceptual clarification. In D. S. Rychen & L. H. Saganik (Eds.), *Defining and selecting key competencies* (pp. 45–65). Seattle, WA: Hogrefe & Huber.

Westbury, I. (2000). Teaching as a reflective practice: What might Didaktik teach curriculum? In I. Westbury, S. Hopman, & K. Riquarts (Eds.), *Teaching as a reflective practice: The German Didaktik tradition* (pp. 15–40). Mahwah, NJ: Lawrence *Erlbaum* Associates.

Zeichner, K. (1980). Myth and realities: Field based experiences in pre-service teacher education. *Journal of Teacher Education, 31*(6), 45–55.

Martina Döhrmann
Institute of Mathematics
University of Vechta, Germany

Gabriele Kaiser
Faculty of Education
University of Hamburg, Germany

Sigrid Blömeke
Centre for Educational Measurement (CEMO)
University of Oslo, Norway

PART II

ACQUIRING AND IMPROVING MATHEMATICS KNOWLEDGE FOR TEACHING THROUGH TEACHER PREPARATION

SHU XIE, YUN-PENG MA AND WEI CHEN

5. ELEMENTARY MATHEMATICS TEACHER PREPARATION IN CHINA

INTRODUCTION

Large scale curriculum reform has swept across the world since the early 1990s (Sahlberg, 2011). The success of these reforms depends heavily on the quality of teachers (Hargreaves & Shirley, 2009). Therefore, it is important to enhance teacher quality. Teachers' knowledge, literacy and ability structure have been found to have significant impact on the quality of their work (Moreno, 2005). Pre-service teacher education is the first step of the professional development path to equip young teachers with the essential knowledge, values and beliefs for effective teaching.

This chapter aims to reveal and evaluate how pre-service elementary mathematics teachers in Mainland China are trained. Researchers collected data from the curriculum designers, teacher educators, and pre-service teachers themselves through questionnaire surveys and interviews. Curriculum documents were also collected and analyzed. It was found that the teacher training program was structured and subject matter knowledge was emphasized. Teaching practices have helped pre-service teachers learned to integrate and apply their mathematics knowledge for teaching. The questionnaire survey showed that the pre-service mathematics teachers' subject matter knowledge was not sufficient and their pedagogical content knowledge was the weakest component among all the elements of their professional knowledge. The findings indicate that the preparation of student teachers for teaching of mathematics needs to be improved. The learning of mathematics subject matter knowledge structure and pedagogical content knowledge should be strengthened. Teaching practicum is an effective means for student teachers to develop their teaching competencies.

BACKGROUND OF ELEMENTARY PRE-SERVICE MATHEMATHICS TEACHER EDUCATION IN CHINA

Up to the late 1990s, elementary mathematics teachers were trained through sub-degree programs. In 1999, the Ministry of Education of P.R. China authorized five universities to run degree programs to prepare elementary mathematics teachers. Since then, the pre-service mathematics teacher training programs has

Y. Li & R. Huang (Eds.), How Chinese Acquire and Improve Mathematics Knowledge for Teaching, 85–108.
© 2018 Koninklijke Brill NV. All rights reserved.

proliferated. By the end of 2013, there were 166 universities or colleges offering pre-service elementary teacher training programs.

In Mainland China, all 4 year long for elementary teacher education, the degree program, follow three different models (Ma, Xie, Zhao, & Li, 2008):

> Discipline-based model. This model aims to prepare subject specialists in one particular subject. Pre-service teachers are required to select one subject such as Mathematics, Chinese, English, Science, Music, Arts and ICT (information communication technology), during the training period according to their interests, and they would be qualified to teach one particular subject in the future. Therefore, the teaching training program emphasizes the teaching of subject matter knowledge.

> Comprehensive model. This model prepares student teachers to teach several subjects (mainly focus on Mathematics, Chinese, Science, Social Studies, etc.). This model aims at training teachers' comprehensive teaching skills and research abilities. Elementary teachers are also supposed to pay considerable attention to the basic knowledge of various subjects, educational theory and skills.

> Stream model. This model is a combination of the two models mentioned above. In this model, elementary teachers are required to possess comprehensive abilities as well as choose one of the strands, liberal arts or sciences, as their teaching area in the future. Liberal arts include subjects as Chinese (key part), History and Social Studies, and the sciences cover Mathematics (key part), Science and ICT.

Although the training models vary, they all aim to prepare elementary teachers with the necessary knowledge, skills and beliefs.

METHODOLOGY

Focus of the Study Framework

To be effective, teachers must have teachers' need to master the thorough understanding of this subject (Kennedy, 1998). They need to master knowledge about what to teach, why to teach and how to teach. Shulman (1986, 1987) proposed categories of teachers' professional knowledge which has had great impact on research in teacher knowledge. Since then, a number of scholars have proposed other ways of classifying teachers' professional knowledge. An analysis of these classifications show that they all have differences, but they have five core elements of professional teacher knowledge: general pedagogical knowledge, curriculum knowledge, subject matter knowledge, pedagogical content knowledge, and knowledge of learners.

In the area of mathematics education research, many scholars have focused on mathematics subject matter knowledge (see for example, Kilpatrick, Swafford, &

Findell, 2001; Ma, 1999; Simon, 1997) and mathematics pedagogical content knowledge (An, Kulm, & Wu, 2004; Fennema & Franke, 1992; Ma, 1999; Marks, 1990; Wong, 2009). Ball and her colleagues (2008), based on Shulman's initial teachers' knowledge categories, proposed the constructs of teachers' mathematics knowledge for teaching (MKT). They hypothesize that Shulman's categories of content knowledge can be subdivided into common content knowledge (CCK), specialized content knowledge (SCK) and horizon content knowledge (HCK); while the pedagogical content knowledge (PCK) can be subdivided into knowledge of content and students (KCS), knowledge of content and teaching (KCT), and knowledge of content and curriculum (KCC) (Ball, Thames, & Phelps, 2008; Hill, Ball, & Schilling, 2008).

This study employed Ball's framework of MKT to examine how elementary math student teachers acquire professional knowledge in their pre-service teacher training program. Specifically, the following three research questions are investigated:

1. What is the current state of pre-service elementary mathematics teachers' MKT?
2. What are the sources for pre-service mathematics teachers to acquire MKT?
3. What are the possible ways to improve pre-service elementary teachers' MKT?

Methods of Data Collection

To answer these research questions, a number of data collection methods are adopted. They include a questionnaire to assess student teachers' level of knowledge,

1. *Questionnaire:*
   There are two parts in the questionnaire:
   i. *Teachers' Professional Knowledge Questionnaire.* The study adapted a survey instrument developed by Ma (2010) to gauge student teachers' MKT. The instrument contains three parts: background information, teachers' professional knowledge (mathematics curriculum knowledge, mathematics subject matter knowledge, mathematics pedagogical content knowledge), and source of teachers' professional knowledge. In this research, mathematics curriculum knowledge, mathematics subject matter knowledge, and mathematics pedagogical content knowledge are major components of MKT. Analysis of the data focused on the above aspects.
   The following are sample questions of the three types of MKT:
   a. Mathematics curriculum knowledge:
   Sample: ( ) What is the sequence of understanding shape according to content design in the learning phase I.
   A. be able to learn 2-dimensional (2D) shapes first, and then 3-dimensional (3D) shapes;
   B. be able to learn 3-dimensional (3D) shapes first, and then 2-dimensional (2D) shapes;

C. be able to learn 2-dimensional (2D) shapes and 3-dimensional (3D) shapes at the same time;
D. None.

b. Subject matter knowledge:

According to the *Compulsory Education Mathematics Curriculum Standards* (Ministry of Education of Peoples of Republic of China, 2001), the content areas of Number and Algebra, Space and Shape, Statistics and Probability are the key parts in Mathematics learning. The test items also take into consideration of the level of understanding. Based on Bloom et al.'s (1956) Taxonomy of Educational Objectives, Anderson's (2000) taxonomy for learning, teaching and assessing, and teaching objectives of cognition in Chinese teaching standards, the instrument focuses on assessing three cognitive levels including memorizing, understanding, and applying. The third dimension considered when developing items is based on He's (2001) classification of subject matter knowledge. According to the connection between teaching and subject matter knowledge, subject matter knowledge could be evaluated through three levels: external knowledge, supporting knowledge, and core knowledge (Figure 1).

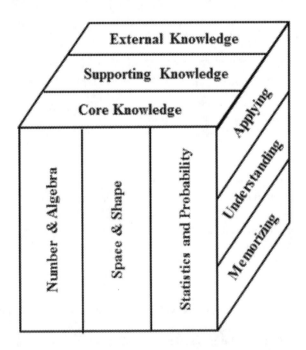

Figure 1. Three-dimension framework of elementary mathematics teachers' subject matter knowledge

One of the items for assessing mathematics subject matter knowledge (about equation thinking) is shown below:
(　) What kinds of balls might be put into the black covered container?

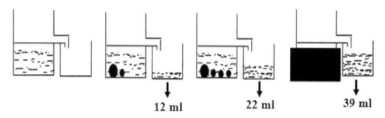

A. 1 big ball, 7 small balls;   B. 2 big balls, 3 small balls;
C. 2 big balls, 4 small balls;   D. 2 big balls, 5 small balls;

c. Pedagogical content knowledge: The pedagogical content knowledge part of the questionnaire contained four questions. Two of them were adapted from Li-ping MA (1999), who mainly evaluated teachers' understanding and representation toward students' errors and relationship among mathematics concepts. The other two questions focused on evaluating teachers' responses to algorithm, and teaching strategies dealing with common errors in students' work.

Sample I:
A student computed 26×53, the vertical form on the right was used. How do you explain his/her error? Write down what you might say to this student.

$$\begin{array}{r} 26 \\ \times\ 53 \\ \hline 78 \\ 130 \\ \hline 208 \end{array}$$

Sample II:
When you teach 38 + 49, students might use various computing methods. Could you state three methods that your students might use? If the method used by a student is not what is suggested in the textbook, nor what you consider as the best way, how would you deal with that? Why?

ii. *Sources of professional knowledge questionnaire*

Survey on sources of teachers' professional knowledge was adapted from Fan's (2003) and Han's (2011) research in light of the characteristics of the current pre-service elementary teacher programs. Student teacher respondents were asked to rate the importance of sources where they had learnt the mathematics curriculum knowledge, subject matter knowledge and pedagogical content knowledge. There were six sources: educational theory courses (such as psychology), mathematics teaching methodology course, school experience, teaching practicum, micro-teaching, extra-curricular clubs, and private tutoring. In the subject matter knowledge part, pre-university mathematics courses and university mathematics courses were included in addition to the above six types.

The questionnaire was piloted and modifications were made according to the results. After the questionnaire was finalized, it was administrated to students in the elementary BA program at A1 and C3[1], which are all 211 Project universities. A1 was administered by the metropolitan government, and its elementary teacher program follows the discipline-based model. University C3 is under the administration of the National Ministry of Education and its program comprehensive model. A total of 252 questionnaires were distributed, and 241 completed questionnaires were collected (56 in Univ. Year 1, 51 in Year 2, 100 in Year 3, and 34 in Year 4). The response rate was 96%.

Interview. In order to get insiders' view of the design of the pre-service teacher preparation program, and also the implementation patterns and problems 14 deans, 9 faculty members closely involved in mathematics teaching and research in universities, 5 pre-service teacher mentors in elementary schools, and 10 student teachers from elementary education programs were interviewed. This set of data helps us to understand how the pre-service teacher training program functioned, its contribution to the learning of MKT, and how it prepares student teachers to teach mathematics.

Document analysis. Nationally recommended programs were used as the criteria to evaluate the quality of teachers' training. Thus, all 16 nationally distinguished program plans in primary education were collected. The courses were classified into three groups: university common courses, professional educational courses (such as philosophy of education, learning psychology, pedagogy, teaching practicum), and subject related courses (including courses on mathematic and teaching of mathematics), to reveal the structure and coverage of the program.

Methods of Data Analysis

Analysis of questionnaire. The questionnaire contains multiple choice items and short answer questions. The items on assessing student teachers' understanding of curriculum knowledge and subject matter knowledge were multiple choice questions. All of these were judged as correct or not and were given a score of 1 or 0. Context-based short answers were used to evaluate pedagogical content knowledge. A team of experts were invited to rate the quality of the answers. Pass rate[2] was used when reported pre-service teachers' current status.

The sources of knowledge were investigated by multiple choices through 3 levels: most important, more important, and unimportant.

Analysis of interview. Constant comparison was used to code data. Categories, key themes and their relationships emerged and were constructed during several rounds of analysis. Triangulation methods were performed to ensure the trustworthiness of the data.

Document analysis. The mathematics programs were classified and analyzed to identify the weighting and sequence of different types of the courses.

HOW DO PRE-SERVICE ELEMENTARY MATHEMATICS TEACHERS ACQUIRE MATHEMATICS KNOWLEDGE FOR TEACHING?

The Characteristics of the Pre-service Elementary Mathematics Teachers' Curriculum Structure

The aims of all 16 programs analyzed are basically the same, namely to train pre-service teachers to be highly qualified elementary education staff (teachers, administers, and researchers) who are committed to elementary education and possess good moral qualities, systematic and well-organized professional knowledge, strong educational and teaching ability, as well as research capability. Ten programs explicitly point out that "pre-service teachers will be trained on both elementary teaching and researching aspects" while six programs emphasized the training of pre-service teachers as elementary teachers and school managers. The other four programs referred to training pre-service teachers to be "elementary teachers, researchers and school managers." Similar to other teacher training programs, pre-service teachers were also expected to have good political quality.

All the curricula of the elementary mathematics teacher training programs were structured in a logical manner. The courses of the four year programs were divided

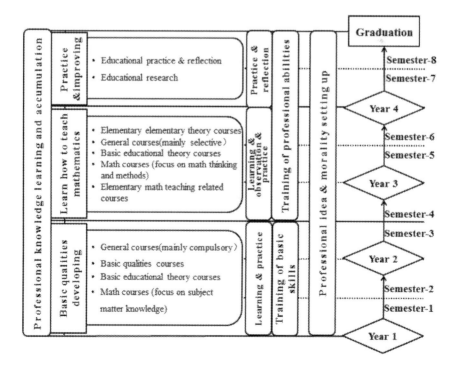

Figure 2. Curriculum structure of pre-service elementary mathematics teachers program

into three phases based on the various content focuses, which are semesters 1 to 3, semesters 4 to 6, and semesters 7 to 8 (Figure 2). Some student teachers are enrolled as general educational major students. When they get into the third semester, some of them will transfer from their general education major to elementary education or mathematics/science major. For example, student teachers in Univ. C6 are allowed to join elementary education program after they finish 10 credits of professional courses (educational theory, general psychology, modern educational technology, children's literature, mandarin). All programs, regardless of the training model adopted, demand student teachers to go through comprehensive learning and subject learning in order to be well prepared for future teaching in certain subjects.

Table 1 sums up the weighting of each type of courses. It is found that the total credit points required in pre-service training ranged from 197 to 143. Most of the courses are compulsory and only small portions are elective courses.

Table 1. Statistic of courses in pre-service elementary teacher education

Training model		Subject model	Medium model	Comprehensive model
Average credits		176.4	161.8	162.2
Types of courses		Percentage (%)		
University common courses		25.7	30.9	29.6
Professional Educational courses	Basic educational theory	6.1	3.2	17.3
	Comprehensive quality	9.6	5.1	7.7*
	Elementary education theory	3.7	8.3	5.2**
	Educational methods/skills	10.5	9.9	9.4
	Educational practice	12.3	12.0	12.9
Subject related courses	Mathematics courses	15.6	13.3	3.0
	Primary mathematics teaching related courses	7.0	6.2	4.8
	Subject courses (none-math)	7.6	10.1	11.7
Others***		1.9	1	1.2

*C2 doesn't have this type of courses, 7% is average of the other 4 programs.
**C3 doesn't have this type of courses, 5.2% is average of the other 4 programs.
***Others contain some non-educational related practical courses and activities.

At present, there are no standard guidelines from the Ministry of Education on the curriculum structure. The curriculum structure of the programs is influenced by the program designers' educational ideals and curriculum orientation, which to some extent, is one of the reasons leading to fairly wide variation of courses between universities.

Due to the differences among training models, the weighting of the types of courses, especially the mathematics knowledge courses varied. The amount of mathematics knowledge courses in the Discipline-based and Stream Models is nearly 10% more than in the Comprehensive Model. The Comprehensive Model only offers two or three mathematics courses such as Advanced Mathematics, Foundations of mathematics and Elementary Number Theory. In contrast, the other two models supply more compulsory courses and elective courses about professional mathematics, which intend to help student teachers build a solid base of mathematics subject matter knowledge in areas such as Mathematical Analysis, Spatial Analytic Geometry, Projective Geometry, Non-Euclidean Geometry, Theory of Probability, The Structure of Algebra, a Brief History of Mathematics, Mathematical Modelling, Advanced Algebra, Mathematical Thinking Methods etc.

Mathematics related courses are more elementary-based. For example, the contents which are highly connected with elementary mathematics teaching in Elementary Number Theory, Probability and Statistics Theory, Foundation of Modern Algebra etc., are offered as the main mathematics related courses while the parts associated with elementary mathematics teaching also play an important role in content teaching. Furthermore, courses like Calculus, Linear Algebra, 3D Analytical geometry, which are helpful in the formation of core principles of scientific mathematics, are added to the curriculum system with appropriate reduction of their level of difficulties.

Different levels are shown in mathematics related courses, especially in the programs adopting the discipline-based and Stream models because "It is supposed to be available for each student to get access to fundamental mathematics knowledge. Different students would achieve different development" (Interview, A1-1). These mathematics courses are classified as follows:

- *Basic courses.* Mainly about basic theory of advanced mathematics and elementary mathematics teaching. These courses could help students master the general theory within certain area of mathematics (such as, calculus, algebra, geometry, equation, probability and statistics). The content of these courses are general and broad, and helps to build up student teachers' general mathematical abilities.
- *Professional courses.* These would supply students with systematic professional knowledge for future mathematics teaching (e.g. Mathematical Analysis, Advanced Algebra, Analytic Geometry, Probability and Statistics, Elementary Number Theory). The content of these courses are profound and in depth, which aims at developing students' mathematics thinking and research ability in their future field of teaching.
- *Advanced courses.* These courses intend to help the student teachers develop their potentialities, energy, and interests in mathematics, and enhance their professional knowledge and abilities. These courses are usually elective modules, such as Combinatorics, Mathematical Modelling, Mathematics History, Projective Geometry, Non-Euclidean Geometry, Mathematical Thinking and Methodology, etc.

The proportion of mathematics related courses in the Comprehensive Model is a little lower than the other two models. Courses mainly focus on basic theory (e.g. Theory of Elementary Mathematics Curriculum and Instruction, Study of Elementary Mathematics Learning Psychology), case studies (e.g. Analysis about Elementary Mathematics Curriculum Standards and Textbooks, Analysis about Elementary Mathematics Concept, Study on High-Qualified Classroom Teaching, Study on Mathematics Problem-Solving, etc.), and skills and application (e.g. Elementary Mathematics Teaching Design and Practice, Mathematics Micro-teaching, etc.). All these courses aim to help pre-service teachers acquire the knowledge about how to teach. Subject knowledge, knowledge about student, curriculum knowledge and pedagogical knowledge based on certain content are integrated in the curriculum. All of the courses were taught by integrating practice-based case studies and materials.

In the interviews, university instructors admitted that they are aware of the limitations of the traditional way of teaching. Student teachers' learning is quite passive in some classes. Many courses used paper-and-pencil tests to evaluate students' achievement, which is deemed as imitation memory. A student from Univ. F2 said, "When the exam is over, the knowledge is given back to my teachers. I would like to say, I did not learn much in the courses." Therefore, changing learning mode and forms of evaluation is essential. The courses should motivate students to develop their potential abilities, foster creative ability and critical thinking. Continuous authentic assessment could prompt students to think and improve. Assignments could be differentiated to include presentations, reports on research, and case studies, as to provide openings to inquiry info topics such as mathematics problem solving.

The entire pre-service teacher training program includes practical courses for student teachers to experience the work of teachers in real context, and to develop their comprehensive practical skills. This professionalization process would help students to transform the theoretical knowledge they have learned into their own practical knowledge. In addition, student teachers have the opportunity to gradually integrate subject matter knowledge, students' knowledge, curriculum knowledge, and pedagogical knowledge together. Thus, their practical abilities and wisdom about education would be developed.

Teaching practicum consists of three parts: school experience, classroom teaching, and final thesis. The credit points for teaching practicum vary between programs. The minimum credits in this part are 14 and the maximum is 33. During school experience, student teachers are required to observe teachers, mostly good ones, as they work. This is usually arranged from semester 3 to 6. The duration and timing of school experience differs between programs. The average duration is 4.6 weeks. In most universities, student teachers have to attend discussion sessions after school experience activities to reflect on what they observe in the school and classrooms.

In most programs, classroom teaching practicum is arranged in Semester 7. However, some institutions arrange classroom practice sessions in Semesters 6, 7, and 8. Generally, the average duration is 13.5 weeks, and student teachers are required to teach two to four times.

It is suggested that school experience is to be carried out in excellent schools so that student teachers can observe good teachers and thus can learn more there. Meanwhile classroom teaching practice is arranged in less endowed schools so that the students would receive more opportunities to take on more roles in the classroom. Univ. D2 employed various types of schools for student teaching, such as rural schools, urban schools, rural migrant children schools, etc. Such schools offer students diverse experiences, which help them to understand education and elementary mathematics teaching through different viewpoints.

The third component of the teaching practice is that student teachers are required to complete a research paper. The research questions on their final paper is often generated during the practicums. It is expected that student teachers would develop the competencies to carry out educational research through conducting a small scale research study.

How Do Pre-service Elementary Mathematics Teachers Acquire Mathematics Knowledge for Teaching?

Surveys investigating the sources of MKT were carried out in two universities adopting the comprehensive and discipline-based models. Results showed: significant difference between the two models. Therefore, separate analysis was performed. The sources of MKT were classified into 3 levels, namely most important, more important and unimportant. Results from the comprehensive model are rendered is below (Table 2).

Table 2. Attitude of pre-service teachers (taking the comprehensive model program) toward the importance of sources of MKT

Sources	Mathematics subject matter knowledge	Pedagogical content knowledge	Mathematics curriculum knowledge
The most important	• Math pedagogical courses • Educational observation & practicum • Pre-university math courses	• Educational observation & practicum • Math pedagogical courses	• Educational observation & practicum • Math pedagogical courses
More important	• Private tutoring • Clubs • Micro-teaching	• Educational courses • Private tutoring • Micro-teaching	• Private tutoring • Educational courses • Micro-teaching
Unimportant	• University math courses	• Clubs	• Clubs

From Table 2, pre-service elementary teachers deemed educational observation, student teaching practice, and mathematics pedagogical courses as the most important sources of MKT, and micro-teaching and Private tutoring as the more important. In addition, educational courses were more important to both curriculum knowledge and PCK. Significantly, for pre-service teachers within the comprehensive model, experience of mathematics learning before university was the most important to their subject matter knowledge; in contrast, university mathematics courses played the least important role in pre-service elementary teachers' subject matter knowledge learning.

Table 3 shows the attitude of pre-service teachers from the discipline-based model toward the importance of sources of MKT.

Table 3. Attitude of pre-service teachers (taking the discipline-based model program) toward the importance of sources of MKT

Sources	Mathematics subject matter knowledge	Pedagogical content knowledge	Mathematics curriculum knowledge
The most important	• Educational observation & practicum • Tutoring • Math pedagogical courses	• Educational observation & practicum • Math pedagogical courses • Tutoring	• Educational observation & practicum • Tutoring • Math pedagogical courses
More important	• Pre-university math courses • Micro-teaching • University math courses	• Educational courses • Micro-teaching	• Educational courses • Micro-teaching • Clubs
Unimportant	• Clubs	• Clubs	-----------

Results were similar to the comprehensive model in some aspects, for example, pre-service teachers in the subject model considered educational observation, practice and mathematics pedagogical courses most impactful in professional knowledge development. In contrast, they thought private tutoring was also most important to this knowledge. Micro-teaching was more important, and clubs were least. Moreover, mathematics courses, both before and in university, were viewed as more important to mathematics subject matter knowledge.

Data from interviews supports the conclusion from questionnaire about contrary attitudes towards university mathematics courses between these two models. Mathematics courses in Univ. C3 were taught by mathematicians (but none had experience in elementary mathematics teaching or research). The courses were normally deep and difficult. Most of pre-service student teachers used to be arts majors in senior secondary schools. Therefore, they felt that it was very difficult, and sometimes boring. Mathematics courses in Univ. A1 were taught by mathematics educators with experience in elementary mathematics teaching or research. These teachers were able to select contents which would be related to future mathematics teaching in elementary schools. Therefore, pre-service teachers could try to build connection between subject matter knowledge they learned in university and subject matter knowledge in elementary mathematics teaching. They clearly understood the value and orientation of content they learned. Obviously, pre-service student teachers valued highly practice-oriented courses, which they viewed as the most important sources of MKT.

Pre-Service Elementary Mathematics Teachers' Mathematics Knowledge for Teaching

The Teacher Professional Knowledge Questionnaires were marked and rated by experts. The scores of the pre-service teachers who participated in the survey were compared with the full mark. A ratio of 0.6 was rated as a "pass" which denotes a satisfactory level of pre-service teachers' knowledge. Based on the data from the survey and One-Way ANOVA, significant differences were detected among student teachers in different years on mathematics subject matter knowledge ($F = 16.35$, $p < 0.01$), mathematics curriculum knowledge ($F = 24.99$, $p < 0.01$), and pedagogical content knowledge ($F = 25.23$, $p < 0.01$). The level of mathematics subject matter knowledge and mathematics curriculum knowledge were increasing as they advanced to senior years.

Mathematics subject matter knowledge was not strong enough. The survey showed that student teachers' mathematics subject matter knowledge was less desirable than expected. Their average score was only 0.59, which means that they only got 59% of the full score. There was almost no difference among the three areas of subject matter knowledge, namely Theory of Numbers and Algebra, Statistics and Probability, and Space and Shape, for pre-service teachers. The score for Theory of Numbers and Algebra, and Statistics and Probability were 0.60. The score for Space and Shape was 0.57.

Pre-service teachers' subject matter knowledge could also be analysed according to three levels, from memorizing, cognition and understanding, to applying. Analysis shows that the level of cognition and understanding showed the best results with a score of 0.78 while the score of memorizing was 0.57, and applying showed the lowest pass rate of 0.53.

According to the results of pre-service teachers from Year1 to Year 4, there was a significant difference among different years in areas of Number and Algebra, and Statistics and Probability ($F = 13.92$, $p < 0.01$; $F = 19.32$, $p < 0.01$). There was no significant difference among different years in memorizing and understanding ($F = 1.81$, $p < 0.87$), but significant difference in the applying level ($F = 28.80$, $p < 0.01$). Pre-service teachers from senior years exhibited better ability in application in comparison to junior years. With respect to applying level, the pass rate of pre-service teachers in Year 1 was 0.38, while 0.41 for Year 2. These data suggest that, pre-service courses have a positive impact on improving subject matter knowledge.

Mathematics pedagogical content knowledge level was lowest. The pass rate of mathematics curriculum knowledge was 0.49. While the pass rate of mathematics pedagogical content knowledge was 0.41, which was the lowest compared with other types of knowledge. Analysis of the PCK task showed presence of errors on judgement and explanation. Many student teachers emphasized operational rules, or lacked the ability to give clear explanation or clearly point out established rules (Table 4).

Table 4. Evaluation of pre-service teachers' pedagogical content knowledge (N = 241)

	Types of responses in PCK	%
Item I	Type1: only emphasize on operational rules;	37.1
	Type2: able to point out problems but not focus on the key part and no clear explanation;	28.3
	Type3: able to explain operational theory, but not use proper mathematics language;	6.6
	Type4: able to clearly explain but not clearly point out established rules or give an example;	28.3
	Type5: perfect answers.	2.8
Item II	Type1: errors on judgement and explanation;	20.9
	Type2: improper explanation;	19.2
	Type3: able to give explanation without counterexamples;	28.9
	Type4: perfect answers.	1.2

Although pre-service courses have positive influence on MKT, student teachers' subject matter knowledge and pedagogical content knowledge were weak as reflected in the survey data. This needs to be improved. Teacher educators should try their best to help pre-service teachers to achieve deeper understanding of MKT.

Teacher Educators' Considerations When Planning the Teacher Training Program

In this study, deans, teacher educators, pre-service teachers and mentors have been interviewed to better understand how they planned the curriculum and what factors they took into consideration during development. The major findings are summarized below.

Mathematics knowledge of teaching courses focus on developing pre-service teachers' understanding of mathematics subject thinking and pedagogical content knowledge of elementary mathematics. In the interviews, many teacher educators shared their aim of training pre-service teachers' mathematics subject thinking. Ball (1991) categorized mathematics subject knowledge into knowledge of mathematics and knowledge about mathematics. The former is about conceptual knowledge and procedural knowledge and the latter is about understanding the nature of mathematics. Pre-service training courses were considered as having to pay attention not only to subject content knowledge, but also to the understanding of the nature of the subject. Some of deans interviewed have noted that advanced mathematics knowledge is useful to form a theoretical structure to understand the nature of subject matter knowledge. This could help student teachers improve their competency in teaching the application of mathematics in real world context (Interview, A1-2, C4-1, and C6-1). For example, a mathematical modelling course can help pre-service teachers observe daily life through mathematical perspective and use mathematical methods to solve daily life problems. If the content selection and teaching were carried out properly, this goal could be achieved. For instance, teachers who taught the topic about "classification" in Modern Algebra Foundation introduced some classification concepts in elementary mathematics textbooks, compared the differences between mathematic subject and elementary mathematics on this topic, and discussed what should be noticed when teaching this topic in elementary (Interview, D1-2). C3-2 said, when she taught "limit" in calculus, she discussed with students about how to derivate the formula for the area of a circle, and how to illustrate $0.999...=1$. All of these would help students understand the nature of certain knowledge and its application in elementary teaching and gradually build up students' mathematics beliefs.

Some deans said they were in the process of reforming the teacher training curriculum and pedagogy to tackle the criticism that advanced mathematical knowledge is useless to elementary school teachers. Mathematics teacher educators who are teaching mathematics courses are encouraged to review their own teaching which would form the basis for improving the programs.

Secondly, teacher educators need to build connection between the mathematics subject courses and elementary mathematics teaching. They hoped to help student teachers to view elementary mathematics teaching from the subject knowledge perspective. Despite this effort, many students commented negatively on the courses. "*Some of courses are so boring, also difficult*" (A1-S2, C3-S1, C3-S3), "*exams are tough, some are difficult to do*" (A1-S1, A1-S3, C3-S1, C3-S4), "*high pressure*" (A1-S2, C3-S1, C3-S3, C3-S5) "*I don't know why we need to learn these. Useless*" (C3-S1, C3-S3). Some deans agreed that these negative views were the result of failing to teach difficult concepts in proper ways (interview, A1-2, C4-1, and C6-1). Certainly some aspects needed to be improved. For example, some teachers mentioned, "*Some of courses are never offered because no one can teach them.*" (A1-1, C2-1, C3-2, F2-1) and courses offering has not considered pre-service teachers' preferences. Rather, it was determined by the numbers of enrolled students and the availability of suitable teaching staff.

Thirdly, the academics interviewed reported that considering the nature of elementary education, curriculum must contribute to elementary teachers' professional development, meet the needs of future teaching, as well as integrate the practical and theoretical themes in the courses. Situated in particular contexts, case-based learning and reflection were seen as able to promote pre-service teachers' professional development and develop their knowledge and skills in applying theoretical knowledge in teaching. The courses, such as Case Studies on Elementary Mathematics Teaching, Study on Elementary Problem Solving, etc., combined academic and systematic knowledge module with cases, context, and exercise module together, which strengthens application value in educational practice. Students are trained to learn how to use proper teaching representation based on the understanding of elementary mathematics knowledge, subject curriculum knowledge, and children's thinking.

Teacher educators shared in the interview that teacher educators should employ multiple teaching methods and develop teaching resources that scaffold learning to improve student teachers' teaching skills. A teacher said,

> In my opinion, the most important thing is how to teach this high level and difficult mathematics knowledge, how to teach the tacit knowledge like pedagogical content knowledge. They are all referred to as teaching strategy. In my classroom, I changed the 'teaching-oriented' into 'learning-oriented' approach. Students take more responsibilities in teamwork, inquiry learning, and task-based activities. This is much more useful for students to understand the nature of the subject and teaching. (Interview, A1-3)

Some teacher educators use their research findings to teach. For example,

> We are doing research on the following areas: the core knowledge of elementary mathematics, students' errors in certain content they were learning, and strategies which would help to deal with these errors. These findings are quite beneficial resource for student teachers to learn and discuss. (Interview, C3-1)

This type of teaching helps pre-service teachers to understand principles of teaching and learning, within real contexts, accumulate SMK, understand systematic errors that students might have, and appreciate how to use various teaching strategies.

Multiple task-based educational practices and reflections help pre-service student teachers integrate and apply MKT. Classroom observation and teaching practicum are not only important parts of the entire pre-service education, but also the central parts of the teaching practice component. It helps pre-service teachers to form educational belief and develop their sense of professional responsibility, as well as achieve and improve educational knowledge and skills. Each student teacher is normally supervised by a teacher from the university/college, and a mentor from the elementary school where the student teacher is placed. Pre-service teachers are encouraged to identify some questions and apply knowledge they have learnt into activities they participate in, such as classroom observation, correcting pupils' work, group study, designing lesson plan, lesson review, seminar, reflection, etc. They are also given certain tasks to do in their practicum field. In this "task-based educational practice" student teachers will focus on one or two themes each week and mainly try to solve those issues, such as what are the main points should the teacher pay attention to when they prepare lessons, how to supervise group work, how to teach multiplication, how to design mathematics task for students, and so on. The spiralling process of "practice-reflect-practice-reflect-" is a way of accumulating educational experience and assimilating practical and theoretical knowledge. In the process, student teachers have opportunities to develop critical thinking, strengthen their internal motivation of self-improving, overcome frustration and build confidence.

DISCUSSION AND CONCLUSION

In light of the findings reported above, it could be concluded that the present teacher training program could be improved in a number of ways. Revamping and improving the present program is timely as the Ministry of Education of China has enacted the *Teacher Education Curriculum Standards (Experimental Version)*, the first national standards for teacher education in China in Oct., 2011. These standards, including the basic requirements for teacher education curriculum, form the framework and standards of professional teacher training. In February 2012, the Ministry of Education issued *Elementary Teacher Professional Standards (Experimental Version)* which clearly stated the standards of qualified elementary teacher, such as professional ethics and morals, knowledge, skills. The development of these documents are milestones in the Chinese teachers' professional development. In response to these two official documents, Chinese universities and colleges have been reforming and revamping their elementary teacher's training programs and curricula.

Pre-Service Education Should Help Pre-Service Teachers Enrich Their MKT Structure and System

A meta-analysis of research studies on the impact of the characteristics of teachers on students' achievement (Fennema & Franke, 1992) showed that when teachers studied advanced mathematics, there was only a 10% positive main effect on students' achievement. Monk (1994) also found that the mathematics pedagogical courses teachers attended had a greater impact on their students' achievement compared with mathematics subject knowledge courses. Advanced mathematics only had 0.04% effect on students' achievement. These findings illustrate that mathematics teachers with rich advanced mathematics subject matter knowledge are not necessarily effective mathematics teachers. However, this does not mean that subject matter knowledge is not important. Indeed, subject matter knowledge had significant correlation with pedagogical content knowledge. Quite a number of scholars commented that the subject matter knowledge in pre-service programs was not enough to equip student teachers to be effective teachers. Other studies (Gabel, Samuel, & Hunn, 1986; Haidar, 1997; Kikas, 2004; Smith, 1999) showed, pre-service teachers have misconceptions in subject matter knowledge. Unless this anomaly was rectified in the pre-service program, it would likely have a negative influence on their teaching. Hence, it is important for universities and colleges to review and improve the mathematics subject matter courses to ensure that student teachers grasp a good understanding of the subject matter knowledge essential to effective teaching in the classroom.

Mathematics Pedagogical Content Knowledge Should be Emphasised More and Taught in a Proper Way

PCK is important to teachers' professional development and also deemed as a key part of teacher education (Shulman, 1986). PCK can help teachers improve their teaching quality. In our research study, pre-service teachers from Univ. A1 and C3 did not achieve the researchers' expected level of understanding in terms of PCK although a number of courses, for the learning of PCK have been included in the programs. The reason for failing to properly prepare student teachers' PCK is due to the courses having not been taught properly. In view of this, teacher educators need to realize the key role of PCK and find strategies to combine the content and related pedagogy together. A previous study on pre-service teachers showed they lacked basic teaching skills, ICT, and research ability, organization of classroom teaching, etc. (Zhang, 2011). In addition, they lacked integrating knowledge (Ma, 2008) and were underachieving in areas such as subject matter knowledge structure, subject thinking methods and subject framework (Shao & Yuan, 2011). Moreover, they lacked understanding of students and their learning which resulted in strategies or representations that were not appropriate (Gess-newsome, 1999; Li, 2006; Veal & Kubasko, 2003). These inadequacies are all related to PCK.

PCK is a key factor of effective teaching. It affects the way and effectiveness of the communication between teachers and students. Teachers who have good PCK can adopt appropriate strategies and representations to help students understand and develop subject knowledge (Driver & Scott, 1996; Osborne & Wittrock, 1983; Posner et al., 1982; Wang, Duan, & Zhang, 1998). Furthermore, these teachers can improve their classroom learning quality (Magnusson, Krajcik, & Borko, 1992). All in all, PCK should be treated as one of the most important parts in pre-service teacher education and teachers' professional development. Teacher education curriculum should pay more attention to PCK and its related content. Moreover, teacher educators also need to explore how to teach PCK effectively

Teaching MKT through Practice-Oriented Strategies

Teachers' professional competencies are rooted in the real world teaching context (Xu, 2008). Contextualized learning is effective in teacher development (Darling-Hammond, 2005). It is no surprise the Ministry of Education promulgated "practice-oriented" strategies in teacher education program (Ministry of Education of Peoples of Republic of China, 2011). When universities adopt and implement practice-oriented strategies teacher educators should have practice-related backgrounds. Furthermore, mentors in elementary schools should possess rich practical resources.

It is important for pre-service student teachers to synthesize theoretical and practical knowledge, and develop insights into ways of applying knowledge in classroom teaching. To achieve this, teacher education programs should provide learning opportunities for pre-service teachers to integrate elementary mathematics knowledge, theories of primary student learning and knowledge of teaching strategy, through in-depth exploration of primary mathematics key teaching topics. Moreover, student teachers need to have opportunities to extend and integrate the different areas of subject matter knowledge to build up a web of subject matter knowledge. They should also accumulate a range of teaching strategies and representations of the key primary mathematics curriculum content. Teaching experience which could improve PCK should be the focus of attention.

The planning and delivering of the teacher education programs should be informed by the latest news and research to make sure the knowledge is up to date. For example, *Mathematics Curriculum Standards of Compulsory Education* was updated and released in 2011. The teacher training programs should be revamped to achieve the targets set.

Improving MKT Quality Through Cooperation from Various Aspects

The teaching quality of pre-service training is influenced by teacher educators' understanding, preparation, and implementation of curriculum. Most of the interviewees (teacher educators, mentors, teachers in elementary schools, pre-service teachers) mentioned pre-service teachers, as a whole, have not acquired enough knowledge and skills when they graduate from universities/colleges. They further noted that the pre-service teachers do not have enough theoretical and practical knowledge about elementary mathematics teaching. To address this, several suggestions are put forward:

Employing expert elementary mathematics teachers to strengthen the teacher education team. Some universities have employed expert elementary mathematics teachers to deliver some of the courses (e.g. A1, C2, C3, D2 etc.). Analysis of Elementary Mathematics Curriculum Standards and Textbooks, Pedagogy of Elementary Mathematics in C3 invited several expert teachers to join. These expert teachers had both strong theoretical training as they are all doctors or doctoral candidates with rich front line teaching experience and over 10 years teaching experience. They are also very committed to education and teaching as well as active in mathematics. The feedback from the student teachers to their teaching was highly positive as they could enable pre-service teachers to master SMK and construct a knowledge network. If all the institutions adopt this practice of involving expert teachers to deliver the courses, the quality of student teacher learning will be improved.

Sharing teaching resources among universities. Nowadays, some institutes work together to develop MOOC (massive open online courses) courses. Some courses allowed students to learn and earn credits. Its purpose is to narrow the gap between the quality of the institutions. For instance, a MOOC course on "Analysis of Elementary Mathematics Curriculum Standards and Textbooks" has been done by a team comprising professors, elementary mathematics teachers, and researchers.

Cooperation between institutes and elementary schools. The supervisors from universities/colleges and mentors from elementary schools offer various viewpoints which can help student teachers learn to teach. In the survey, practice was deemed as the most important source to mathematics teaching related knowledge. Thus, quality of the learning experience in teaching practice schools should be given more attention. The *Triangle Model* of cooperative supervision in pre-service practicums can be put in practice where mentors in universities and elementary schools get together and supervise pre-service teachers with different perspectives. The joint efforts could help pre-service teachers learn how to teach properly. Through professional practice and reflections on certain teaching strategies, they could manage to relate SMK to the knowledge and experience In order to give more practical opportunities for student teachers, schools could employ them as part-time teachers. This way, in-service teachers could gain time to participate in professional development programs, while student teachers would get more opportunities to practice.

Pre-Service Education Curriculum Should Pay Attention to Pupils Development: Thoroughly Understand Their Mathematics Learning

The core business of teaching is to help children develop. To achieve this goal, teachers must have good understanding of what learning means, and how children develop (Zhong, 2008). Elementary teachers also need to study students' physical and mental development characteristics from both a practical and theoretical perspective. Primary school teachers are supposed to study how students learn mathematics and how students construct or develop mathematics knowledge. Study of learning about learning could help pre-service teacher transform adult mathematics into children's and tailor the curriculum to fit students.

Ausubel (1968) opined that the only key factor which impacts learning is what the learner had already known. Students' misunderstandings and learning difficulties could influence teaching plan, decision, and implementation (Koirala, Davis, & Johnson, 2008; Lu, 2008; Park & Oliver, 2008). Thus, primary teacher's training programs should place the understanding of primary students' learning as the core. Teachers should accumulate knowledge about elementary students' mathematics learning (e.g. previous knowledge, misunderstanding, learning difficulties, typical errors and learning needs within certain content).

Equipping Pre-Service Teachers with Lifelong Learning Capacities

Lifelong learning and reflection abilities have become one of the important goals of pre-service teacher education in many parts of the world such as the USA, Japan and the United Kingdom. In the *National Teacher Professional Standards* (*Experimental Version*) (Ministry of Education of Peoples of Republic of China, 2012), it is also pointed that teachers need to practice-reflect-practice-reflect-… so as to constantly and continuously transform knowledge, theory and understanding of students into their own personal cognition. Constantly improving one's professional abilities would help teachers enter into children's world and understand them better. Nowadays, teachers are not seen as technicians with teaching skills, but experts with reflective ability. Pre-service training should be the foundation of lifelong sustainable development. "Consciousness and ability of independent development, sense and ability of reflection, research ability, attitude and the ability to cooperate" have been identified as key factors which might impact the level of teachers' quality by the teacher educators interviewed in this study. These qualities are also the foundations for teachers' sustainable development ability. Reflection is an effective way for sustainable development and is quite important to continuous professional development (Driel, Jong, & Verloop, 2002; Loughran, Mulhall, & Berry, 2004; Tuan, Chang, Wang, & Treagust, 2000). Posner's famous saying, "experience + reflection = growth," is the rule for teacher development. Many scholars have opined that PCK could be achieved and improved by reflection and educational practice (Shulman, 1987; Veal, 1999; Driel, 2012). In the process of curriculum design and implementation, we can enhance student teachers' abilities of realizing, understanding, reflecting, constructing and accumulating in practice. Furthermore, we can also improve the formation of professional thinking attitudes and ways of thinking. Moreover, students' willingness and ability to perform independent study should be developed. More practice opportunities for independent study should be given to student teachers and we should aim to train "reflective practitioners" who possess theoretical based, research competencies.

ACKNOWLEDGEMENTS

We would like to express our gratefulness to Professors LAM Chi Chung, Rongjin Huang, and Yeping Li for their comments, and also want to thank Dr Bo-wei Zhang, Dr Rui Ding and Cynthia Zhu for their assistance in the preparation of this chapter. We are also grateful to all the interviewees for sharing their experiences, time, ideas and efforts. We appreciate The National Social Science Foundation of China (No. CHA130167) and Northeast Normal University Social Science Foundation (No. 13QN008) for supporting this research.

NOTES

[1] Sample codes: Capital letter represents for the batch they were approved; the first number represents for random order. The second number when showed means the code of interviewee in that institution. E.g. C3 means the No.3 university/college which was in the third batch approved to be National special and model program. C3-1 means No.1 teacher educator interviewee in C3. C3-S1 means No.1 pre-service teacher interviewee in C3. C3-M1 means No.1 teacher mentor interviewee of C3.

[2] Pass rate is average score divided by full score.

REFERENCES

An, S., Kulm, G., & Wu, Z. (2004). The pedagogical content knowledge of middle school mathematics teachers in China and the US. *Journal of Mathematics Teacher Education, 7*, 145–172.

Anderson, L. W., & Krathwohl, D. (2000). *A taxonomy for learning, teaching, and assessing: A revision of Bloom's taxonomy of educational objectives.* Boston, MA: Allyn and Bacon.

Ausubel, D. P., Novak, J. D., & Hanesian, H. (1968). *Educational psychology: A cognitive view.* New York, NY: Holt, Rinehart and Winston.

Ball, D. L. (1991). Research on teaching mathematics: Making subject-matter knowledge part of the equation. In J. Brophy (Ed.), *Advances in research on teaching* (pp. 1–8). Greenwich, CT: Jai Press.

Ball, D. L., Thames, M. H., & Phelps, G. (2008). Content knowledge for teaching: What makes it special? *Journal of Teacher Education, 59*, 389–407.

Bloom, B. S., Engelhart, M. D., Furst, E. J., Hill, W. H., & Krathwohl, D. R. (1956). *Taxonomy of educational objectives: The classification of educational goals. Handbook I: Cognitive domain.* New York, NY: David McKay Company.

Borko, H., & Livingston, C. (1989). Cognition and improvisation: Differences in mathematics instruction by expert and novice teachers. *American Educational Research Journal, 26*, 473–498.

Darling-Hammond, L., & Baratz-Snowden, J. C. (2005). *A good teacher in every classroom: Preparing the highly qualified teachers our children deserve.* San Francisco, CA: John Wiley & Sons.

Ding, R., Ma, Y. P., & Wang, Y. (2012). Analysis on mathematics knowledge current station and source of pre-service teacher of primary education. *Journal of Northeast Normal University, 7*, 194–199.

Driver, R., & Scott, P. H. (1996). Curriculum development as research: A constructivist approach to science curriculum development and teaching. In D. F. Treagust, R. Duit, & B. J. Fraser (Eds.), *Improving teaching and learning in science and mathematics* (pp. 94–108). New York, NY & London: Teachers College Press.

Fan, L. H. (2003). *Study on teachers' pedagogical knowledge development.* Shanghai: East China Normal University Press.

Fennema, E., & Franke, M. L. (1992). Teacher knowledge and its impact. In D. Grouws (Ed.), *Handbook of research on mathematics teaching and learning* (pp. 147–164). New York, NY: Macmillan.

Gabel, D., Samuel, K., & Hunn, D. (1986). Understanding the particulate nature of matter. *Journal of Chemical Education, 64*(8), 695–697.

Gess-Newsome, J., & Lederman, N. G. (1999). *Examining pedagogical content knowledge.* Boston, MA: Kluwer Academic Publishers.

Grossman, P. L. (1990). *The making of a teacher: Teacher knowledge & teacher education.* New York, NY: Teachers College Press.

Haidar, A. H. (1997). Prospective chemistry teachers' conceptions of the conservation of matter and related concepts. *Journal of Research in Science Teaching, 34*, 181–197.

Han, J. W., Ma, Y. P., Zhao, D. C., & Huang, Y. Y. (2011). Survey on sources of teachers' knowledge of middle school mathematics teachers. *Teacher Education Research, 3*, 66–70.

Hargreaves, A. P., & Shirley, D. L. (Eds.). (2009). *The fourth way: The inspiring future for educational change.* Thousand Oaks, CA: Corwin Press.

He, C. X. (2011). Evaluation and reflection on chemical teachers' subject matter knowledge structure. *Chemical Education, 5*, 22–25.

Hill, H. C., Ball, D. L., & Schilling, S. G. (2008). Unpacking "pedagogical content knowledge": Conceptualizing and measuring teachers' topic-specific knowledge of students. *Journal for Research in Mathematics Education, 39*, 372–400.

Kennedy, M. (1998). Ed schools and the problem of knowledge. In J. D. Raths & A. C. McAninch (Eds.), *Advances in teacher education: What counts as knowledge in teacher education?* (pp. 29–45). Stamford, CT: Ablex.

Kikas, E. (2004). Teachers' conceptions and misconceptions concerning three natural phenomena. *Journal of Research in Science Teaching, 41*, 432–448.

Kilpatrick, J., Swafford, J., & Findell, B. (2001). *Adding it up: Helping children learn mathematics.* Washington, DC: The National Academies Press.

Koirala, H. P., Davis, M., & Johnson, P. (2008). Development of a performance assessment task and rubric to measure prospective secondary school mathematics teachers' pedagogical content knowledge and skills. *Journal Math Teacher Education, 11*, 127–138.

Li, Q., Ni, Y. Q., & Xiao, N. B. (2006). Elementary school mathematical teachers' pedagogical content knowledge: Its features and relationship with subject matter knowledge. *Journal of Educational Studies, 4*, 58–65.

Lin, C. D. (1996). Structure and training of teacher quality. *Journal of the Chinese Society of Education, 6*, 16–22.

Loughran, J., Mulhall, P., & Berry, A. (2004). In search of pedagogical content knowledge in science: Developing ways of articulate and documenting professional practice. *Journal of Research in Science Teaching, 41*(4), 370–391.

Lu, J. L. (2008). *Comparative study on missing professional knowledge of elementary mathematics teachers in Shanghai and Hong Kong* (Doctoral dissertation). East China Normal University, Shanghai.

Ma, L. P. (1999). *Knowing and teaching elementary mathematics.* Mahwah, NJ: Lawrence Erlbaum Associates.

Ma, Y. P., Xie, S., Zhao, D. C., & Li, Y. P. (2008). Research on primary teacher training mode in four-year undergraduate programs. *Journal of Higher Education, 4*, 73–78.

Ma, Y. P., Zhao, D. C., & Han, J. W. (2010). Survey and analysis on teachers' professional knowledge. *Educational Research, 12*, 70–76.

Magnusson, S., Krajcik, J., & Borko, H. (1999). Nature, sources, and development of pedagogical content knowledge for science teaching. In J. Gess-Newsome & N. G. Lederman (Eds.), *Examining pedagogical content knowledge* (pp. 95–132). Dordrecht: Kluwer Academic Publishers.

Marks, R. (1990). Pedagogical content knowledge: From a mathematical case to a modified conception. *Journal of Teacher Education, 41*, 3–11.

McDiarmid, G. W., Ball, D. L., & Anderson, C. W. (1989). Why staying one chapter ahead doesn't really work: Subject-specific pedagogy. In M. Reynolds (Ed.), *The knowledge base for the beginning teacher* (pp. 193–205). Elmsford, NY: Pergamum.

Ministry of Education of Peoples of Republic of China. (2001). *Compulsory education mathematics curriculum standards.* Beijing: Beijing Normal University Press.

Ministry of Education of Peoples of Republic of China. (2011). *Teacher educational curriculum standards.* No. [2011]6.

Ministry of Education of Peoples of Republic of China. (2012). *National teacher professional standards (Experimental Version).* No. [2012]1.

Monk, D. H. (1994). Subject area preparation of secondary mathematics and science teachers and students achievement. *Economics of Education Review, 13*(2), 125–145.

Moreno, J. M. (2005). *Learning to teach in a knowledge society (Final Report).* Washington, DC: The World Bank.

Osborne, R., & Wittrock, M. (1983). Learning science: A generative process. *Science Education, 67*, 489–508.

Park, S., & Oliver, J. S. (2008). Revisiting the conceptualisation of Pedagogical Content Knowledge (PCK): PCK as a conceptual tool to understand teachers as professionals. *Research in Science Education, 38*(3), 261–284.

Posner, G., Srike, K., Hewson, P., & Gertzog, W. (1982). Accommodation of a scientific conception: Toward a theory of conceptual change. *Science Education, 66*, 211–227.

Sahlberg, P. (2011). The fourth way of Finland. *Journal of Educational Change, 12*(2), 173–185.

Shao, Z. H., & Yuan, X. T. (2011). Research on subject thinking training in geography teaching. *Journal of Northeast Normal University, 3*, 262–264.

Shulman, L. S. (1986). Those who understand knowledge growth in teaching. *Educational Researcher, 15*(2), 4–14.

Shulman, L. S. (1987). Knowledge and teaching: Foundations of the new reform. *Harvard Education Review, 1*, 1–22.

Shulman, J. H. (1992). *Case methods in teacher education.* New York, NY: Teachers College Press.

Simon, A. M. (1997). Developing new models of mathematics teaching: An imperative for research on mathematics teacher development. In E. Fennema & B. Scott-Nelson (Eds.), *Mathematics teachers in transition* (pp. 55–86). Mahwah, NJ: Lawrence Erlbaum Associates.

Smith, D. C. (1999). Changing our teaching: The role of pedagogical content knowledge in elementary science. In J. Gess-Newsome & N. G. Lederman (Eds.), *Examining pedagogical content knowledge: The construct and its implications for science education* (pp. 163–97). Dordrecht: Kluwer Academic Publishers.

Tuan, H., Chang, H., Wang, K., & Treagust, D. F. (2000). The development of an instrument for assessing students' perceptions of teachers' knowledge. *International Journal of Science Education, 22*(4), 385–398.

Van Driel, J. H., & Berry, A. (2012). Teacher professional development focusing on pedagogical content knowledge. *Educational Researcher, 41*(1), 26–28.

Van Driel, J. H., Jong, O. D., & Verloop, N. (2002). The development of preservice chemistry teachers' pedagogical content knowledge. *Science Education, 86*(4), 572–590.

Van Driel, J. H., Verloop, N., & de Vos, W. (1998). Developing science teachers' pedagogical content knowledge. *Journal of research in Science Teaching, 35*(6), 673–695.

Veal, W. R., & Kubasko, D. (2003). Biology and geology teachers' domain-specific pedagogical content knowledge of evolution. *Journal of Curriculum and Supervision, 18*(4), 334–352.

Veal, W. R., Tippins, D. J., & Bell, J. (1999). The evolution of pedagogical content knowledge in prospective secondary physics teachers. *ERIC*. ED443719.41. Retrieved from https://eric.ed.gov/?id=ED443719

Wang, G. H., Duan, X. L., & Zhang, H. B. (1998). Middle school students' perceptions of science teachers' subject teaching. *Journal of Science Education, 12*, 363–381.

Xie, S. (2013). *Structure and characteristics analysis on primary mathematics teachers' pedagogical content knowledge* (Doctoral dissertation). Northeast Normal University, Changchun.

Xu, B. Y. (2008). *Multiple methods of teachers' professional development*. Shanghai: Shanghai Education Press.

Zhang, D. Z., & Li, X. H. (2007). Mathematics: The academic for mand the educational form. *Mathematics Teaching, 8*, 27.

Zhang, J. (2011). Study on current situation, problem and strategies of pre-service teachers' professional abilities. *Morden Educational Science, 5*, 15–18.

Zhong, Q. Q. (2008). Issues on innovation of Chinese teacher educational system. *Peking University Education Review, 3*, 46–59.

Shu Xie
Faculty of Education
Northeast Normal University
Changchun, China

Yun-peng Ma
Faculty of Education
Northeast Normal University
Changchun, China

Wei Chen
School of Education
Harbin University
Harbin, China

YINGKANG WU AND RONGJIN HUANG

6. SECONDARY MATHEMATICS TEACHER PREPARATION IN CHINA

INTRODUCTION

The significance of the role that teachers have played in students' growth and development has been acknowledged widely. For instance, in the best-selling book, *The World is Flat* (Friedman, 2005), the author argues that nowadays the greatest survival skill is the ability to learn how to learn, the best way to learn how to learn is to love to learn, and the best way to love to learn is to have great teachers who inspire. Teachers in China are highly respected and regarded as engineers of human souls. Pre-service teacher education has been traditionally mentioned as "Shi Fan" (师范) education. Shi means teacher and Fan means model. Great learning makes a teacher, and moral integrity makes a model (学高为师，身正为范). This statement defines the goal of teacher education and characteristics of being a teacher. In fact, it has been used by many normal universities and colleges in China as a motto to encourage and inspire prospective teachers.

Teacher preparation system in China has been established and developed over decades (Huang, Peng, Wang, & Li, 2010). To prepare teachers to meet the requirement of the new curriculum reform starting from the beginning of the 21st century, China central government has issued a series of official documents including *Curriculum Standard for Teacher Education Program (Experimental)* (Ministry of Education of China, 2011a), *Professional Standards for Teachers* at different levels (Ministry of Education of China, 2012), *Interim Rules for Teacher Qualification Examination* (Ministry of Education of China, 2013a), and *Interim Rules for Regular Registration of Teacher Qualification Certificate* (Ministry of Education of China, 2013b). These documents not only highlight the importance of teacher education but also provide mechanisms and strategies for improving the quality of teacher education. Adopting the curriculum model of TIMSS (Mullis, Martin, Ruddock, O'Sullivan, & Preuschoff, 2009), this chapter aims to examine the characteristics of secondary mathematics teacher preparation program in China from the perspectives of intended curriculum and obtained curriculum. Specifically, this chapter intends to address the following three research questions:

1. What are the characteristics of preparation program for secondary mathematics teachers, as reflected in the *Curriculum Standard for Teacher Education Program (Experimental)* and a sample teaching program?

2. What are the requirements of being a secondary mathematics teacher, as viewed from the syllabi of national secondary mathematics teacher qualification examination (Ministry of Education of China, 2011b)?
3. What are the characteristics of prospective mathematics teachers' knowledge for teaching, as revealed in a survey on knowledge of algebra for teaching?

Answers to questions 1 and 2 will depict the expectations of the reformed teacher education program in alignment with the *Curriculum Standard for Teacher Education Program (Experimental)* and the syllabi of national secondary mathematics teacher qualification examination. Answers to question 3 will provide information about content knowledge of prospective secondary mathematics teachers obtained from the current teacher preparation program.

BACKGROUND AND THEORETICAL CONSIDERATION

This section includes two aspects. First, we briefly introduce research findings on traditional teacher preparation programs in mathematics in China. Then, we summarize theoretical perspectives on mathematics teacher knowledge for teaching.

The Characteristics of Traditional Mathematics Teacher Preparation Program

Pre-service secondary school teacher preparation programs in China are specialized and discipline-based (Ding et al., 2014). In a national survey on pre-service teacher education programs in normal universities and colleges in China (Ding et al., 2014), it was found that proportion of credits on educational courses including subject-related educational courses in different normal universities and colleges is from 10% to 30%, which is obviously much less than that of content courses. This result supports the finding that secondary mathematics teacher preparation in China highly values the importance to mathematics content knowledge so as to build strong content knowledge base and problem solving skills for pre-service teachers, with less attention to pedagogical knowledge and limited experience in student teaching (Li, Huang, & Shin, 2008; Li & Huang, 2009; Yang, Li, Gao, & Xu, 2012).

This traditional teacher preparation approach has been criticized. In the above mentioned national survey, the participant pre-service teachers spent less time and effort in learning educational courses compared to content courses. The participant administers from employing schools pointed out that the most pressing aspect pre-service teachers should improve is on teaching skills, although they were generally satisfied with the graduates' current teacher education program (Ding et al., 2014). This general research finding is in line with the finding specific to the preparation of mathematics teachers. It was concluded from a questionnaire survey on MKT of 392 pre-service mathematics teachers from five different normal universities or colleges in China (Pang, 2011), that these pre-service teachers' performance on subject matter knowledge was significantly better than their pedagogical content knowledge, and their specialized content knowledge as well as knowledge of content and students, and knowledge of content and teaching were very limited. Based on questionnaire and interview survey results, Fan (2013) reported that the pre-service secondary mathematics teachers from a province-level normal university were lacking of knowledge on school mathematics curriculum and did not have appropriate knowledge of school students. Yan (2014) analyzed 64 pre-service secondary mathematics teachers' ability to identify, explain and correct students' errors in dealing with mathematics problems via a survey. He found that although a majority of the pre-service teachers could successfully identify the errors, only about 30% of them could explain the errors precisely and deeply, and about 20% of them could help students to correct these errors using appropriate teaching techniques. The above research findings indicate that pre-service secondary mathematics teachers may not have acquired enough pedagogical content knowledge for teaching mathematics. This might reflect the current practice of "emphasizing on content and deemphasizing on pedagogy" in secondary mathematics teacher preparation program in China. Therefore, it is rather crucial to rethink how to teach secondary mathematics teachers so that they acquire sufficient knowledge for teaching mathematics.

Theoretical Perspectives on Mathematics Knowledge for Teaching

In the past three decades, discussions on the kind of knowledge teachers need to possess has drawn much attention. In elaborating the components that comprise the knowledge base of teaching, Shulman (1986) introduced the concept of pedagogical content knowledge (PCK), subject matter knowledge (SMK) and curriculum knowledge. Following Shulman's idea, many studies (e.g., Ball, Thames, & Phelps, 2008; Baumert et al., 2010) have been carried out, within mathematics education, to define, theorize, and measure SMK and PCK, as well as to investigate relationship between mathematics teachers' knowledge and students' learning. It was found that although pedagogical content knowledge was inconceivable without sufficient content knowledge, content knowledge couldn't substitute for pedagogical content knowledge (Adler & Venkat, 2014). Therefore, content knowledge and pedagogical content knowledge are both indispensable in teaching mathematics. This leads to the widely accepted conclusion that both content knowledge and pedagogical content knowledge should be included as a part of teacher education.

Besides providing theoretical knowledge such as SMK and PCK, preparation for skillful teaching practice is also proposed as crucial in pre-service teacher education program (Ball, 2013; Cohen, 2014). For example, Ball (2013) suggested certain key practices of teaching such as eliciting and interpreting individual students' thinking, explaining and modeling core content, and formulating and proposing questions in mathematics classroom teaching. This viewpoint has its root in the nature of teacher knowledge, which is regarded as contextualized, tacit, individualized and practical

(Shao, 2011). Teachers learn how to teach from their own teaching experiences rather than simply by applying their theoretical knowledge into the actual teaching at their school practice. In a contextualized teaching situation, they experience the process of searching for strategies to tackle the problems, which occurred in this specific teaching moment by integrating their theoretical knowledge with the contextualized situation. In this way, teachers construct, develop and justify their practical knowledge for teaching through their actual teaching processes (Zou & Chen, 2005). Zhang (2005) proposed an idea of three forms of mathematics: primitive, academic and educational. In its academic form featuring with rigor deduction and logical reasoning, mathematics is precise but cold, with its hot and primitive thinking process hidden behind formalized symbols and signs. In its educational form, mathematics is inspiring, interesting, beautiful and easy to access. It is the teachers' duty to transform mathematics from its academic form to its educational form. Theoretical knowledge for teaching could provide information on the academic and educational form of mathematics, but it could not inform adequately how to make the transformation. This piece of practical knowledge could be achieved through reflecting on one's own actual teaching practice.

From the above discussion, it is clear that theoretical knowledge, which includes content knowledge and pedagogical content knowledge, and teaching practice training are both important in teacher preparation program. The theoretical knowledge for teaching is the prerequisite for skillful teaching practice, while the actual teaching practice provides a platform to develop individualized and contextualized practical knowledge for teaching.

In addition, to get deeper understanding of what pre-service teacher achieved after completion of their first three or four years of teacher preparation programs, we reported the findings of teachers' knowledge of algebra for teaching based on a second analysis of data taken from a larger study (Huang, 2014). Since algebra is an important body of school mathematics (National Council of Teachers of Mathematics [NCTM], 2000; NMAP, 2008), researchers have proposed different models defining teachers' knowledge for teaching algebra (e.g., Artigue, Assude, Grugeon, & Lenfant, 2001; Even, 1993; Ferrini-Mundy, McCrory, & Senk, 2006). Specifically, Ferrini-Mundy and her colleagues (2006, 2012) have developed a framework that describes mathematics knowledge of algebra for teaching. According to this model, the knowledge for teaching algebra includes three types of knowledge: school algebra knowledge, advanced algebra knowledge, and teaching algebra knowledge. *School algebra knowledge* refers to the algebra covered in the K-12 curriculum. The content consists of two major themes (1) expressions, equations/inequalities, and (2) functions and their properties. *Advanced algebra knowledge* includes calculus and abstract algebra that are related to the school algebra; and *teaching algebra knowledge* refers to typical errors, canonical uses of school algebra knowledge, and topic trajectories in curriculum. This framework informed the design of the survey on knowledge of algebra for teaching that was used by Huang (2014).

PREPARATION PROGRAM FOR SECONDARY MATHEMATICS TEACHERS: ANALYSIS FROM THE CURRICULUM STANDARD FOR TEACHER EDUCATION PROGRAM

The *Curriculum Standard for Teacher Education Program (Experimental)* (known as the Curriculum Standards) was released in October 2011 by the Ministry of Education of China. It is the first curriculum standard for teacher education in China. This document regulates the educational curriculum for pre-service teachers from kindergarten to secondary school, without including the other two components of teacher education curriculum, that are, subject content curriculum and knowledge of common basic curriculum. It clarifies the purpose, educational curriculum structure, and credit requirements without specifying concrete teacher education courses. It aims to set the basic requirements of educational curriculum for teacher preparation. According to the document, teacher education institutions could establish their own teacher education program based on the Curriculum Standard. The following sections describe its basic ideas, purposes, as well as curriculum structure and credit requirements for secondary teacher preparation.

Basic Ideas and Purposes

The core value of teacher education curriculum is to promote teacher professional development (Ministry of Education of China, 2013c). In order to actualize this core value, teacher education curriculum follows three basic principles, *whole-person development based*, *practice oriented*, and *life-long learning*. The three principles are established to deal with issues in the current teacher education curriculum. *Whole-person development based* refers to the idea of respecting students, understanding students, and caring for students, indicating the focus of teacher education curriculum is shifted from basic educational knowledge and theories to school students and their learning and development. *Practice oriented* is concerned with the deemphasizing practical training feature of the current teacher education curriculum. It emphasizes the importance of developing practical knowledge for teaching, identifies the connection between practical knowledge and theoretical knowledge for teaching, and stresses that teacher education curriculum needs to address problems from real teaching practice. *Life-long learning* refers to the idea that the teacher, as a professional, needs to continue learning and development, and teacher education curriculum is required to help teachers to become independent, active and life-long learners.

The content of the curriculum has a two dimensional setup, with the vertical represented by the levels from kindergarten to secondary school, and the horizontal referring to, *educational beliefs and responsibility*, *educational knowledge and ability*, and *educational practice and experiences*. Each aspect includes specific purposes and requirements. In general, pre-service teachers are expected to establish appropriate views toward students, teachers, and education, to acquire knowledge and skills in understanding and educating students and in self-developing. They also need to obtain experiences in observing teaching, implementing lessons and studying educational practice (Ministry of Education of China, 2011a). The statements of the purpose are essentially the same across levels, with slight difference in the descriptions of the detailed requirements. It is emphasized that secondary teachers have comprehensive, systematic and solid content knowledge related to the subject they are going to teach.

Curriculum Structure and Credit Requirements for Secondary Teacher Preparation

The teacher education curriculum structure in the Curriculum Standards is featured as module-based, selectivity-based and practice-oriented. The curriculum contains six learning fields, which are the same across different school levels. Among the six learning fields, only two, subject pedagogy and guidance on educational activity and teaching practice, are subject relevant, with the other four fields, adolescent development and learning, educational base, psychological healthy and moral education, as wells career ethic and professional development, are on general education and pedagogy. This indicates that pedagogical content knowledge has not received much attention in the Curriculum Standards.

The duration of teaching practice, which includes school fieldwork and student teaching, is expected to be 18 weeks, which is much longer than the currently common practice. This change, on one hand, indicates a full understanding of the significance of practical knowledge played in teacher professional development, on the other hand, it shows that it is necessary and urgent for teacher education institutions to elucidate requirement for school field work and student teaching and prepare correspondent guidelines, to guide pre-service teachers on how to observe and learn from experienced teachers' classroom teaching, and how to reflect on and in their own teaching practices.

The credit requirements for educational courses are suggested to be 12 and 14 points for a three-year and a four-year program respectively, without taking the credit points for teaching practice into account. One credit point requires 18 teaching periods on a certain course, and one period is normally taken as 45 minutes. Based on a survey result (Ministry of Education of China, 2013c), the average credit points in total required for graduation from a three-year and for a four-year teacher preparation program are 127 and 164 respectively. Although, it is stated that the Curriculum Standards is about teacher educational curriculum without regulating subject content curriculum and common basic curriculum, we could have a flavor of the balance between teacher educational curriculum and the other two curricula from the suggested credit points for teacher educational courses. It implies that teacher educational curriculum still occupies a small portion of the total credit points required for teachers' graduation.

An Example of Teaching Program for Preparing Secondary Mathematics Teachers

In order to give a more complete picture of the course requirement for preparing secondary school mathematics teachers and to compare the suggested teacher education curriculum in the Curriculum Standards with the current teacher education curriculum implemented in teacher education institutions, an example of a four-year teaching program from a prestigious normal university in China is provided as shown in Table 1. This normal university is chosen due to the fact that its mathematics teacher education program is regarded as superior in China, which makes the selected teaching program representative of a high quality.

The secondary mathematics teacher preparation program shown in Table 1 has been implemented since 2012. A total of 156 credit points is required for graduation. Among the required total credit points, 60 points (38.4%) are composed of common basic curriculum, 72 points (46.2%) of mathematics curriculum, and 24 points (15.4%) reflect the teacher education curriculum. The credit point of mathematics curriculum is three times that of teacher education curriculum, showing that mathematics content knowledge is highly emphasized in preparing secondary mathematics teachers. This is consistent with the characteristics of preparing secondary mathematics teachers in China reported elsewhere (e.g., Li, Huang, & Shin, 2008; Yang, Li, Gao, & Xu, 2009).

Table 1. A program for preparing secondary mathematics teachers

Category	Name of the courses	Credit points
Common basic curriculum	English I, II, III, IV	12
	Physical Education	4
	Fundamentals of Computer	5
	Political and ideological courses, such as Mao Zedong Thought and Theory of Socialism, Outline of History of China	14
	Military Theory and Training	2
	Career Guidance	1
	Chinese Language and Literature/Classical Sinology Reading	2
	Free elective courses: One series from five branches: language, arts and physical education, humanity, social science, teacher literacy)	6
	Science courses for student teachers: Physics, Chemistry, Geoscience, Life Science	14
	Subtotal	*60*

(Continued)

Table 1. (Continued)

Category	Name of the courses	Credit points
Mathematics curriculum	Analytic Geometry	3
	Higher Algebra I, II	9
	Mathematical Analysis I, II, III	15
	Ordinary Differential Equations	3
	Classical Geometry	2
	Complex Analysis	3
	Probability & Statistics	4
	Abstract Algebra I, II	6
	Differential Geometry	3
	Number Theory	3
	Real Analysis	3
	Combinatorics and Graph Theory	4
	Mathematical Experiment	3
	Mathematical Modeling	3
	Thesis	8
	Subtotal	72
Teacher educational curriculum	Pedagogy	2
	Psychology	2
	Language for Teaching	1
	Information-assisted Teaching practice & Design	1
	Microteaching	1
	School field work (2 weeks)	2
	Student teaching (12 weeks)	6
	Electives for all student teachers of the university, such as Arts of school classroom teaching, Integration of ICT and curriculum, Theory and practice of learning science, Study on student behavior etc.	4
	Mathematics Pedagogy	2
	Electives for pre-service secondary mathematics teachers	3
	Subtotal	24
Required total points for graduation		*156*

One can observe from the table that under common basic curriculum category one series of the five branches of free elective courses (6 credit points) is teacher literacy, which includes courses like research method in education, classroom management, development of teacher knowledge and skills, evaluation of teachers, and so on. This shows that the total credit point for teacher educational curriculum could reach 30 if pre-service teachers choose the series of teacher literacy as their elective courses. Hence, the maximum credit point of 22 for teacher educational courses, excluding eight points for teaching practice, is higher than the suggested points of 14 in the Curriculum Standards. This proves that the program from this normal university emphasizes more the importance of teacher educational courses in preparing secondary mathematics teachers than the Curriculum Standards.

A careful examination on the teacher educational courses in the program leads to two findings. First, the duration of student teaching and school field work is 14 weeks in total. Although it is four-week shorter than the suggested duration in the Curriculum Standards, it is eight-week longer than the previous requirement of six weeks. This shows that the importance of teaching practice has been noticed and an initial step to improve it has been made. However, whether or not to extend the duration to the suggested weeks needs a deeper consideration. It involves adjustment of course arrangement and cooperation from corresponding secondary schools. Secondly, the various teacher educational courses, no matter compulsory or elective, are relatively aligned with the six learning fields mentioned in the Curriculum Standards. Moreover, various elective courses on mathematics education are provided, including courses to deepen pre-service teachers' content knowledge and problem solving skills on secondary school mathematics such as *Modern mathematics and secondary school mathematics*, *Mathematics methodology*, and *Problem solving and mathematics competition*, and courses related to pedagogical issues in mathematics classroom teaching and assessment such as *Instructional design of secondary school mathematics*, *Assessment in mathematics teaching* and *Technology in mathematics teaching*, as well as courses involving the development and evolvement of mathematics such as *Culture and history of mathematics*. In addition, some experienced secondary school master teachers are invited to deliver courses like *Graphic calculator and secondary mathematics teaching* and *Teaching excellent high school mathematics students* to pre-service mathematics teachers. In spite of the variety on the mathematics educational courses offered, only five credit points are required in total and *Mathematics pedagogy* is the only compulsory course. Thus, pre-service teachers are not inspired or encouraged adequately to prepare themselves more intensely in pedagogical content knowledge although various learning opportunities are provided.

In short, although this program demonstrates the traditional features of preparing secondary mathematics teachers, that is, emphasizing content knowledge and deemphasizing pedagogical content knowledge, it is worthy to notice the shift toward emphasizing teaching practice. The comparison of the teacher educational curriculum with the Curriculum Standards shows good alignment in general, but reflects the same inclination of not taking sufficient consideration into subject-relevant teacher educational courses.

ENTRY REQUIREMENT FOR SECONDARY MATHEMATICS TEACHERS: ANALYSIS FROM NATIONAL TEACHER QUALIFICATION EXAMINATION

In order to ensure teaching quality, the Ministry of Education of China has piloted a national teacher qualification examination in six provinces since 2011, including Zhejiang, Hubei, Shanghai, Hebei, Guangxi and Hainan (Wu & Ge, 2015). Based on experiences from pilot examinations, the Ministry of Education released *Interim Rules for Teacher Qualification Examination* in 2013. This official document clearly states that the national teacher qualification examination is to evaluate whether or not applicants have acquired a basic literacy and an ability to be qualified to be school teachers. It regulates application prerequisites, content, form and implementation issues of the examination, as well as organization and management issues. The national teacher certificate examination is mandatory for anyone who wants to become a teacher.

To become secondary mathematics teachers, applicants need to take three written examinations, i.e., synthesis quality (secondary level), educational knowledge and skills (secondary level), and mathematics knowledge and instructional skills (junior secondary or senior secondary), and pass one oral examination. The sections below discuss the written and oral examinations, with a focus on the examination of mathematics knowledge and instructional skills.

Synthesis Quality

The purpose of the written examination of synthesis quality is to assess whether or not applicants have advanced educational knowledge, knowledge of laws and regulations on education and professional ethics, literacy on history, arts, sciences, literature etc., and basic abilities in communication, reading comprehension, logical reasoning and information management. The examination consists of multiple choice questions (47%), and material analysis and essay writing (53%), which is unified for all applicants of secondary level in different subjects.

Educational Knowledge and Skills

This examination is to assess applicants' basic knowledge in education and pedagogy, student guidance and classroom management. It includes eight content modules; basic educational knowledge and theories, secondary curriculum, pedagogy on secondary graders, secondary students' learning psychology, secondary students' developmental psychology, secondary students' psychological consultation, secondary school moral education, and secondary school classroom management and teacher psychology. The examination includes multiple choice questions (about 30%), differentiation and analysis questions (which requires judging true or false first and then giving reasons for the judgment), short answer questions, and material analysis questions (about 70%). For example, "Negative reinforcement equals punishment" is a sample of differentiation and analysis type question. This is a unified exam for all applicants of secondary level.

Mathematics Knowledge and Instructional Skills

This examination is to assess whether or not applicants have mastered essential knowledge of college mathematics and school mathematics and be able to apply the knowledge in their mathematics teaching, have grasped of and able to use school curriculum knowledge, and have acquired basic knowledge and skills for mathematics teaching. It includes five types of question, namely, multiple choice (about 27%), short answer, problem solving, scenario discussion, and lesson planning as well (about 73%). The examination for junior secondary and senior secondary teacher qualification applicants is slightly different in terms of mathematics knowledge and curriculum knowledge.

Table 2 presents the detailed requirements of this examination. It covers four aspects of content, namely, mathematics content knowledge, curriculum knowledge, and knowledge and skills needed for teaching mathematics. For mathematics content knowledge, applicants are required to possess knowledge of the content of college mathematics which includes mathematical analysis, advanced algebra, analytical geometry, probability and statistics and so on, as well as the content of school mathematics they are expected to teach. Figure 1 illustrates two sample questions related to mathematics knowledge taken from the examination syllabi. Question 1 asks applicants to prove an inequality, which could be solved by applying the concept of concavity. Concavity and convexity of functions are fundamental concepts related to derivative in calculus. Question 2 is an open-ended question. It requires graph understanding, making connections between the graph and relevant context, and mathematical representation and communication skills.

Figure 2 gives two sample questions on mathematical pedagogical content knowledge, which covers knowledge of curriculum and knowledge and skills needed for mathematics teaching. Question 1 is to assess understanding of junior secondary mathematics curriculum. Question 2 assesses applicants' knowledge and skills for mathematics teaching by giving two different instructional designs on the same topic. Applicants are required to use their pedagogical content knowledge related to this specific mathematics topic to compare and analyze the two teaching cases.

Table 2. Detailed requirements of the examination of mathematics knowledge and instructional skills

Aspect	Senior secondary	Junior secondary
Mathematics knowledge (41%)	College mathematics Senior secondary school mathematics	College mathematics Junior secondary school mathematics
Curriculum knowledge (18%)	Familiarity with senior secondary mathematics curriculum	Familiarity with junior secondary mathematics curriculum
Mathematics teaching knowledge (8%)	• Grasp teaching methods including lecturing, classroom discussion, guided discovery etc. • Grasp basic mathematics teaching knowledge such as how to teach mathematics concepts, theories and rules. • Know the teaching process, which includes lesson planning, classroom implementation, homework and quiz marking, mathematics extra-curricular activity, and evaluation of mathematics teaching. • Grasp basic knowledge and approaches to evaluate mathematics teaching.	
Mathematics teaching skills (33%)	• Planning a lesson. Be able to connect teaching content with what students have already learned; be able to determine teaching goals, key points and difficult points according to the curriculum and student cognitive characteristics; master the mathematics content and understand the developmental process and nature of mathematical concepts, rules and conclusions, and permeate mathematical ideas and methods as well as application and innovation consciousness in classroom teaching; be able to choose appropriate teaching methods and techniques to arrange teaching content so as to deliver the lesson plan in time. • Conducting a lesson. Be able to create mathematics teaching context, to motivate students' mathematics learning, and to advise students' own mathematics exploration, guessing and cooperation; be able to apply appropriate teaching methods and approaches to conduct a mathematics lesson effectively; be able to deal with various contextualized teaching issues. • Making assessment. Be able to assess students' mathematics knowledge and skills, processes and methods, and affects, attitudes and values as well by various methods and approaches; be able to evaluate teachers' mathematics classroom teaching; be able to promote teaching and students' learning via assessment.	

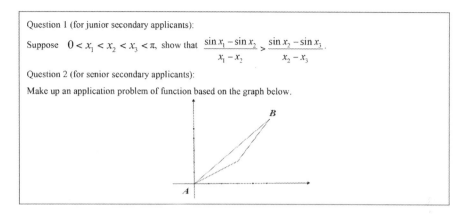

Figure 1. Example questions on mathematics knowledge

Oral Examination

Applicants who have passed written examination are eligible for taking the oral examination. The purpose of the oral is to examine professional ethics, dispositions, manners, communication skills and teaching skills. The oral lasts 40 minutes and consists of two parts. In the first part, applicants randomly select their oral questions that include one structured interview question and one mathematics topic for mini teaching, and have twenty minutes to prepare. In the second part, applicants are given twenty minutes to orally present to a panel of interviewers. Specifically, applicants are assigned five minutes to answer the structured interview question, ten minutes to conduct a mini teaching on the given topic, and five minutes to answer questions from the panel.

The structured interview questions are very diverse, involving classroom management, communication skills with school students, views toward teaching as a professional and so on. For example, suppose there was one student in your class whose academic scores were low, and he surprisingly performed very well on a recent test. His classmates thought he was cheating on the test. How would you deal with this situation? The topic for mini teaching is chosen from a secondary school textbook and some specific requirements for the teaching are given. Figure 3 shows one example. Rubrics for evaluating applicants' performance during the oral is constructed from eight dimensions, namely, profession awareness, disposition, manners, oral communication, trait of thinking, design of teaching, implementation of teaching, and evaluation of teaching.

In summary, the national teacher qualification examination, featuring with diversity of examination content, practice emphasized, and professional oriented, sets the entry requirement for being a secondary mathematics teacher. According to the data from 2011 to 2013 (Yu & Zhao, 2015), the average passing rate in these

Question 1 (for junior secondary applicants):

In junior secondary mathematics curriculum, the content of function is sequenced after the content of algebraic expressions and equations. What is your view about such arrangement?

Question 2 (for senior secondary applicants):

Read the two teaching cases about the inequality $ab \leq \dfrac{a^2+b^2}{2}$, and answer the questions that follow.

Teaching case 1:

Activity 1: Let students choose specific numbers for a and b, and check whether or not the given inequality is valid.

Activity 2: Discuss the geometrical interpretation of $ab, \dfrac{1}{2}a^2, \dfrac{1}{2}b^2$ referring to the graph given below.

Discussion 1: What is the relationship among the three figures which represent the three expressions in the graph below?

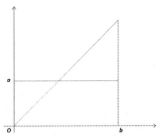

Discussion 2: When is the equal sign valid, and when is the unequal sign valid?

Activity 3: Strictly prove the above inequality.

Discussion 3: If there are three numbers: $a>0$, $b>0$, $c>0$, what might be the corresponding inequality?

Teaching case 2:

Activity: Group work on how to prove the given inequality.

Discussion: Students show their works and whole class discussion.

Please answer the questions below:

1. Analyze the intention of teaching case 1.
2. Describe how to guide students to experience the process from plausible to deductive reasoning based on the above of teaching cases.
3. Compare the instructional rationale behind the two teaching cases.

Figure 2. Example questions on mathematics pedagogical content knowledge

three years is 35% for the written examinations and 70.9% for the oral. Applicants are allowed to take the oral only when they have passed all the written examinations. They have chances to repeat both written and oral examinations if they fail. Applicants obtain the teacher qualification certificate only when they pass both the written examinations and the oral. Therefore, it is challenging, in general, to acquire

---

Design a lesson plan on linear functions and implement it (Textbook: Mathematics by People Education Press, Grade 8, Chapter 11.2 Linear function)

You are specially required to

1. Provide appropriate board-writing.
2. Devise questions to ask in your teaching.
3. Make formative assessment in your teaching.
4. Answer the following question; When a student cannot answer your question or provides a wrong answer to the question, what shall you do?

Figure 3. An example interview question

a teacher qualification certificate. This implies the professional character of teachers and promises recruitment of qualified school teachers.

OBTAINED MATHEMATICS KNOWLEDGE FOR TEACHING: A SURVEY OF PRE-SERVICE TEACHERS' KNOWLEDGE OF ALGEBRA FOR TEACHING

This part reports findings of prospective secondary mathematics teachers' knowledge based on second analysis of the data taken from a larger study on prospective mathematics teachers' knowledge for teaching (Huang, 2014).

Methodology

Instrument. Based on a questionnaire by Floden and McCrory (2007), the adapted instrument includes 17 multiple choice items and eight open-ended items to measure three types of knowledge: school algebra knowledge (SM, 7 items), advanced algebra knowledge (AM, 8 items), and teaching algebra knowledge (TM, 10 items). Figure 4 shows an example item to measure TM.

On a test, a student marked both of the following as non-functions

(i) $f: R \rightarrow R$, $f(x) = 4$, where R is the set of all the real numbers.

(ii) $g(x) = x$ if x is a rational number, and $g(x) = 0$ if x is an irrational number.

(a) For each of (i) and (ii) above, decide whether the relation is a function;

(b) If you think the student was wrong to mark (i) or (ii) as a non-function, decide what he or she might have been thinking that could cause the mistake(s). Write your answer in the Answer Booklet.

Figure 4. An example item to measure TM

Data collection. 376 participants from five purposely-selected universities in China completed the survey in the Spring of 2009. The survey was administrated within a normal class period (around 45 minutes) for junior and senior students in mathematics education preparation programs. About 56% of participants were junior students, while 44% were senior students.

Data analysis. The data analysis included three phases: (1) quantifying the data: developing a five-point rubric for quantifying the open-ended items; (2) analyzing the items and structure of the KTA; and (3) analyzing open-ended items qualitatively: problem solving methods or mistakes, and flexibility in taking appropriate perspectives of function. For each multiple-choice item (items 1 to 17), the correct choice was scored as 1, while the wrong choice was scored as 0. For each open-ended item, we developed a five-point rubric for scoring the answers: 0 point refers to blank or providing useless statements; 1 point means providing several useful statements without a chain of reasons for the correct answers; 2 points refers to giving a correct answer but the explanations or procedures with major conceptual mistakes; 3 points means giving a correct answer and appropriate explanations or procedures, with some minor mistakes; and 4 points presents giving a correct answer and appropriate explanations and procedures.

Results

Characteristics of knowledge of algebra for teaching (KAT). There are 7 items (1, 3, 6, 14, 17, 19, & 23) in school mathematics, 8 items (4, 8, 9, 12, 13, 16, 20, & 24) in advanced mathematics, and 10 items (2, 4, 7, 10, 11, 15, 18, 21, 22, & 25) in teaching mathematics. The means and standard deviation (SD) of the multiple-choice items with a brief description of the contents are listed in Table 3.

The table shows that the participants performed very well on items 1, 2 3, 5, 8, 16, and 17 (higher than 90%), while they performed poorly on items 6, 9, 10, 11, 12, 13, and 15 (less than 70%). Examining the content of the items, we found that the high performing items are related to using algebraic expressions to present quantitative relationships (item 1, 90%), find the value of expression and function (item 3, 96%; item 17, 96%), solve equations (Item 2, 96%; item 5, 95%), make sense of a graph (regarding velocity, time, and distance; Item 8, 90%), find derivative and slope of function (item 16, 92%). Take item 8 for example, as shown in Figure 5. 90% of the participants got the correct answer (B), which require them to make judgment based on speed and time relationship and the graph.

The participants' scores were low in the following areas: using geometrical representation to presenting fraction and algebraic formula (item 6, 38%), the number of roots of $\tan x = x^2$ (item 9, 47%), judge perpendicular relationship of two lines by using their slopes (item, 10, 68%), multiple ways to introduce the concept of slope of a line (item 11, 63%), operation rules in different number systems (item 12, 47%), mathematical induction (item 13, 64%), multiple ways to help expand an algebraic expression formula $(x + y + z)^2$ (item 15, 66%).

Table 3. Means and SD of multiple-choice items with theme description

Item	Theme	Mean	SD
1	Express quantitative relationship in word problems using algebraic expressions	.90	.30
2	Solve quadratic equations $2x^2=6x$ (losing roots)	.96	.20
3	Given a quadratic function $f(x)$, find $f(x+a)$	.96	.20
4	Transform $f(x) = \log_2 x^2$	.78	.41
5	Solve equation: $9^x - 3^x = 0$ using substitution method (adding roots)	.95	.23
6	Represent fraction, percent, and algebraic expressions such as 3/5, 60%, and $a(b + c) = ab + ac$ using the area of rectangle.	.38	.49
7	Given two points, find the functions whose graphs passing these two points.	.83	.38
8	Given a graph representing speed vs. time for two cars, judge the position of the two cars	.90	.30
9	Judge the number of root of equation: $\tan x = x^2$	.47	.50
10	Judge perpendicular relationship of two lines by using their slopes	.68	.47
11	Multiple ways to introduce the concept of slope of line	.63	.49
12	Judge the proposition "For all a and b in S, if $ab = 0$, then either $a = 0$ or $b = 0$" in different number systems	.47	.50
13	Meaning of mathematical induction	.64	.48
14	Roots of irrational equations $\sqrt{x-2} = \sqrt{1-x}$ (adding roots)	.87	.33
15	Expand algebra expressions by area relationship	.66	.47
16	Find derivative and slope of a function	.92	.27
17	Find value of a composition function	.96	.20

Consider item 6 for an example. Only 35 % of the participants answered it correctly (E). Interestingly, 45% of them selected D. It seemed that less than half of the participant did not realize the equation $\frac{3}{5} = 60\%$ could be represented by the area of a rectangle. That means the participants were not skilled in linking algebraic (arithmetic) and geometrical representations.

The means of open-ended problems items 18 to 25 (4 points of each), and subscales of SM (13 points), AM (14 points) and TM (22 points) and KAT are displayed in Table 4.

The table shows that the participants performed well (higher than 70%), except on items 22 and 25 (less than 70%). It is impressive that more than 90% of the participants can correctly provided two methods of solving quadratic inequalities

125

The given graph represents speed vs. time for two cars. (Assume the cars start from the same position and are traveling in the same direction.) Use this information and the graph below to answer
What is the relationship between the *position* of car A and car B at $t = 1$ hour?

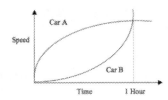

A. The cars are at the same position B. Car A is ahead of car B
C. Car B is passing car A D. Car A and car B are colliding
E. The cars are at the same position and car B is passing car A.

Figure 5. Item 8

Item 6, which of the following can be represented by areas of rectangles?

i. The equivalence of fractions and percentage, e.g. $\dfrac{3}{5} = 60\%$

ii. The distributive property of multiplication over addition:
For all real numbers a, b, and c, we have $a(b + c) = ab + ac$

iii. The expansion of the square of a binomial: $(a+b)^2 = a^2 + 2ab + b^2$

A. ii only B. i and ii only C. i and iii only
D. ii and iii only E. i, ii, and iii

Figure 6. Item 6

(item 19). More than 80% of the participants could use counterexamples to disprove if $A * B = O$ (A and B are matrixes, O presents zero matrix), then either $A = O$ or $B = O$ (item 20, 87%). They can also use algebraic operations to find a quadratic function and maximum value of the function (item 23, 82%) or prove an algebraic proposition (item 24). However, the participants were relatively weak in examining the impacts of parameters of a, b, c on graph of the function (item 22, 66%) and creating real life situations which would correspond to the characteristics of a given graph (item 25, 56%).

In addition, the participants performed better on School Mathematics (85%) and Advanced Mathematics (78%) than on Teaching Mathematics (69%).

Table 4. Means of open-ended items and subscales

Item	Theme	Mean	SD	Correct rate
18	Definition of functions and students' misconceptions	2.92	1.11	0.73
19	Solve quadratic inequality using two methods (algebra and graphical methods)	3.66	.84	0.92
20	Judge and explain if $A\Delta B = O$ (A and B are matrixes, O zero matrix), then either A = O or B = O?	3.47	1.16	0.87
21	Judge the number of roots of a quadratic function with certain constraints by graphical methods.	2.97	1.44	0.74
22	Judge the impact of changing parameters of quadratic functions on its graph translations	2.64	1.47	0.66
23	Given three specific points, find the maximum of a quadratic function whose graphs passing these points	3.28	1.17	0.82
24	Given f(x) and g(x) intersect at a point P on the x-axis, prove the graph of their sum function (f + g) (x) must also go through P.	3.24	1.33	0.81
25	Given a figure, find a daily situation, which corresponds to the figure.	2.23	1.38	0.56
SM	School mathematics	11.02	2.01	0.85
AM	Advanced mathematics	10.88	2.59	0.78
TM	Teaching mathematics	15.30	4.17	0.69
KAT	Knowledge of algebra for teaching	37.19	7.16	0.78

In sum, with respect to specific content areas, the participants performed better in traditional algebra contents such as using algebraic expressions to present quantitative relationship, evaluating expressions and functions, solving various equations and inequalities; they are also stronger in making sense of a graph, reasoning using algebra operations and properties. However, the participants seemed to be relatively weak in the connections between different concepts, and between mathematics and daily life situation such as making connections between geometrical, numerical and algebraic representation, multiple ways to introducing a concept (e.g., slope), discussing families of functions regarding multiple parameters, and connections between graphs and daily life situations. In the following sections, we will illustrate strategies that student used in responding open-ended items.

Fluency of teacher knowledge for teaching the function concept. Items 18, 24 and 25 were designed for measuring knowledge of understanding and applying function concept from different perspectives (*process* and *object*). The score distribution of the three items is shown in Table 5.

Table 5. Score distribution of items related to the adaption of function perspectives

Item	0(%)	1(%)	2(%)	3(%)	4(%)
Item 18	4.8	5.9	19.4	32.2	37.8
Item 24	8.8	6.6	6.9	7.2	70.5
Item 25	18.9	7.4	28.2	23.1	22.3

About 70% of the participants got roughly correct answers to Item 18 as shown in Figure 4 with appropriate explanations as follows:

Let two real number sets A, B, if for any a x belongs to set A, there is only one b in the set B corresponding to a, then this corresponding relationship f from A to B is a function. According to this definition, (i) and (ii) are functions.

To prove the following proposition (item 24), it is necessary to have an appropriate perspective of function concept. 77 % of the participants got correct answer (3 or 4 points). Among those, 71% of them adopted the object perspective as follows: Let $f(x)$ and $g(x)$ intersect at x-axis (p, 0), then, $f(p) = 0$, $g(p) = 0$. So, $(f+g)(p) = f(p) + g(p) = 0 + 0 = 0$. Thus, $(f+g)(p) = 0$.

When introducing the functions and the graphs in a class of middle school (14-15 years-old), tasks that consist of drawing graphs based on a set of pairs of numbers contextualized in a situation or from an equation were used. One day, when starting the class, the following graph was drawn on the blackboard and the pupils were asked to find a situation to which it might possibly correspond.

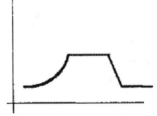

One student answered: 'it may be the path of an excursion during which we had to climb up a hillside, the walk along a flat stretch and then climb down a slope and finally go across another flat stretch before finishing.'
How could you elaborate on this student's suggestion? What do you think may be the cause of this comment? Can you give any other explanations of this graph?

Figure 7. Item 25

In the following item, participants were required to identify real situations that correspond to the graph. It is necessary for participants to have a full understanding of two perspectives of functions. 45% of the participant got roughly correct answers (3 or 4 points). Multiple ways were utilized to explain or interpret the graph.

In summary, the participants demonstrated fluency in solving these problems and flexibility in selecting appropriate perspectives of function concept, namely, *process and object*. But only a half of the participant can provide appropriate situation to illustrate the graph.

Flexibility in using different representations. The items 19, 21, 22, and 23 are deliberately designed and used for measuring knowledge for applying quadratic functions/equations/inequalities through flexible use of multiple representations. Score distributions of these items are shown in Table 6.

Table 6. Score distribution of items related to use of multiple representations

Item	0 (%)	1(%)	2(%)	3(%)	4(%)
Item 19	2.1	1.1	8.0	6.1	82.7
Item 21	11.4	9.8	7.4	14.0	58.2
Item 22	11.7	14.9	18.1	8.8	46.5
Item 23	4.3	5.3	16.8	5.8	67.8

The figure showed that 83% participants gave two essentially different solutions to Item 19 that asked to solve the inequality $(x - 3)(x + 4) > 0$ in two essentially different ways; algebraically and by using a graphing method. The participants provided multiple methods to solve the inequality.

About three-fourths (70%) of the participants provided roughly appropriate explanations of students' mistakes and a correct answer to Item 21 as shown in Figure 8.

If you substitute 1 for x in the expression $ax^2 + bx + c$ (a, b, and c are real numbers), you get a positive number, while substituting 6 for x gives a negative number. How many real solutions does the equation $ax^2 + bx + c = 0$ have?

One student gives the following answer:

According to the given conditions, we can obtain the following inequalities:

$a + b + c > 0$, and $36a + 6b + c < 0$.

Since it is impossible to find fixed values of a, b and c based on previous two inequalities, the original question is not solvable.

What do you think about possible reason for the student's answer? What are your suggestions to the student?

Figure 8. Item 21

More than half (58%) of the 376 participants gave fully correct answers that indicated their strong knowledge and skills in shifting between symbolic and graphic representations when solving this problem, compared with previous findings (Even, 1998).

More than half (55%) of the participants gave the roughly correct choice and appropriate explanations (3 or 4 points) to Item 22 as shown in Figure 9. These participants either explained by analyzing the effects of changes of *a*, *b*, and *c* on the changes of the graphs or analyzed the symmetrical features.

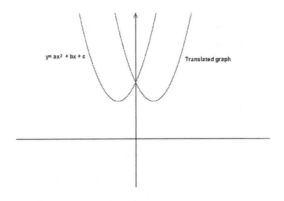

Mr. Seng's algebra class is studying the graph of $y = ax^2 + bx + c$ and how changing the parameters of a, b, and c will cause different translations of the original graph.

Which of the following is a correct translation of the original graph $y = ax^2 + bx + c$ to map the translated graph?

A. Only the value **a** changed.   B. Only the value **c** changed.
C. Only the value **b** changed.   D. At least two of the parameters hanged.
E. You cannot generate the translated graph by changing any of the parameters.

Figure 9. Item 22

Sixty-eight of the participants correctly solved Item 23. Typically, they used standard form of a quadratic (i.e., $y = ax^2 + bx + c$) to find the function expressions, and then transformed it into a vertex form (i.e., $y = a(x - h)^2 + k$) to find the maximum. However, multiple strategies were used to find the solutions, which included using factored form of a quadratic: $y = a(x - x_1)(x - x_2)$, using the standard form $y = ax^2 + bx + c$, and using Vièta's theorem: $x_1 \cdot x_2 = \frac{c}{a}$, $x_1 + x_2 = -\frac{b}{a}$. Overall, the participants not only demonstrated a sound knowledge needed for teaching the concept, but also showed the flexibility in using representations appropriately.

> Given quadratic function $y = ax^2 + bx + c$ intersects x-axis at (−1, 0) and (3, 0), and its y-intercept is 6. Find the maximum value of the quadratic function.

Figure 10. Item 23

Summary

Based on this survey, we formulated several conclusions. Firstly, it was evident that the participants demonstrated an appropriate understanding of knowledge and skills in algebra for teaching. They were strong in traditional algebra areas such as algebraic expressions, equations and inequalities, and quadratic functions. However, the participants revealed several weakness such as unable to use the concept of the slope from multiple perspectives; unable to express numerical and algebraic formula using geometrical representations; unable to build connections between function, graph and daily life situations. Secondly, the participants demonstrated sound knowledge and skills needed for teaching the concept of function. The participants were able to provide diverse interpretations of their solutions, and also showed their fluency in dealing with problems of quadratic equation/function, and flexibility in using multiple representations (graphic and algebraic).

CONCLUSIONS AND DISCUSSION

The analysis further supports the claim (e.g., Li et al., 2008) that secondary mathematics teacher preparation program in China greatly emphasizes mathematics content knowledge with less attention to pedagogical knowledge and student teaching. Since the implementation of *Curriculum Standard for Teacher Education Program (Experimental)* (Ministry of Education, China, 2011), as indicated in the examined program, the duration for field experiences has extended as suggested in the Curriculum Standard. A tendency to make a balance between content knowledge and practice-based knowledge in mathematics teacher preparation seems to emerge. Before moving too far away from the tradition of Chinese mathematics teacher of emphasizing content knowledge development (both advanced and school mathematics), we need to take a second consideration. First, cross-culturally, Zhou and colleagues (2006) found that Chinese elementary mathematics teachers outperformed U.S. counterparts in content knowledge but no difference in pedagogical content knowledge was noted, and Huang (2014) revealed that Chinese secondary mathematics prospective teachers outperformed their U.S. counterparts in both content knowledge (advanced and school algebra) and teaching knowledge.

The strength in content knowledge and pedagogical content knowledge of Chinese prospective teachers in mathematics echoes the superb performance in content and pedagogical content knowledge of prospective Chinese Taiwan middle school teachers in Teacher Education and Development Study in Mathematics (TEDS-M) (Tatto et al., 2012). It is consensus that it is impossible to develop pedagogical content knowledge without appropriate content knowledge. Within the tradition of emphasizing content knowledge in China, how to develop prospective teachers' strong pedagogical content knowledge is an issue needed a further exploration. Secondly, as indicated by researchers (Huang et al., 2010) there are multiple-tired and systematic teacher professional systems in China supporting teachers advancing from novice, to qualified and to excellent teachers. The strong subject knowledge gained from teacher preparation may be an asset for teachers' continued growth within such a supportive and self-motived professional development system. Thus, we need to study the legitimacy, strengths and weaknesses of mathematics teacher preparation practices before we take extreme changes.

Recognition of the weakness of PCK of prospective mathematics teachers as indicted by researchers (e.g., Fan, 2013; Yan, 2014) and the low passing rate (around one third) of teacher entry examination (Yu & Zhao, 2015), it is necessary to take imperative strategies to improve. At a system level, although the Curriculum Standards are crucial for regulating educational courses and student field experience, it is urgent to have a Mathematics Curriculum Standards for Teacher Education. For example, American Mathematical Society in cooperation with Mathematical Association of America newly released a document about The Mathematical Education of Teacher II (Conference Board of the Mathematical Sciences, 2012). The document defines "essentials in the mathematical preparation", "important addition mathematics content that can be learned in undergraduate electives" and "essentials mathematical experiences" for elementary, middle and high school mathematics teachers (CBMS, 2012, p. 54). Moreover, National Council of Teachers of Mathematics developed Standards for Mathematics Teacher Preparation (NCTM CAEP, 2012) for reviewing mathematics teacher preparation programs in the United States. The Standards include seven standards: content knowledge, mathematical practices, content pedagogy, mathematical learning environment, impact on student learning, professional knowledge and skills, and secondary mathematics field experience and clinic practices.

At course design levels, how to develop high quality courses by integration of mathematics content and practice-based mathematic pedagogical knowledge is a long-standing challenge. Recently, researchers have emphasized the importance of using high-leverage practices in helping prospective and novice teachers learn to teach (Ball & Forzani, 2011; Ball, Sleep, Boerst, & Bass, 2009). According to Ball and colleagues (2009), high-leverage practices "are those that, when done well, give teachers a lot of capacity in their work. They include activities of teaching that are essential to the work and that are used frequently, ones that have significant power for teachers' effectiveness with pupils" (pp. 460–461). In addition, positive roles of video clips in teachers' learning have been widely documented (Borko, Jacobs, Eiteljorg, & Pittman, 2008; Sherin & van Es, 2009). Through Chinese lesson study approach (Huang, Su, & Xu, 2014), practicing teachers could develop exemplary lessons focusing on high-leverage practice. Thus, the videos of exemplary lessons that include "micro" high-leverage practices could be valuable materials for prospective teachers to develop their PCK.

In conclusion, mathematics teacher preparation in China has undergone a significant transition from solely focusing on content knowledge to balancing content and practice-based knowledge in order to prepare teachers for teaching in the 21st Century. The tradition of emphasizing content knowledge still exhibits, but an attempt to emphasize practice-based pedagogical knowledge has been made by extending the duration of field experience. How to keep the tradition of strong subject knowledge while emphasizing the development of practice-based content pedagogical knowledge is a new challenge still facing mathematics teacher preparation programs in China.

REFERENCES

Adler, J., & Venkat, H. (2014). Mathematical knowledge for teaching. In S. Lerman (Ed.), *Encyclopedia of mathematics education* (pp. 385–388). Dordrecht: Springer.

Artigue, M., Assude, T., Grugeon, B., & Lenfant, A. (2001). Teaching and learning algebra: Approaching complexity through complementary perspectives. In H. Chick, K. Stacey, & J. Vincent (Eds.), *The future of the teaching and learning of algebra* (Proceedings of the 12th ICMI Study Conference) (pp. 21–32). Melbourne: The University of Melbourne.

Ball, D. L. (2013, October 2). *It is a moral imperative: Skillful teaching cannot be left to chance*. Presentation at the University of Delaware College of Education and Human Development, Newark, DE. Retrieved from http://www-personal.umich.edu/~dball/presentations/100213_UDEL.pdf

Ball, D. L., & Forzani, F. M. (2011). Teaching skillful. *Educational Leadership, 68*(4), 40–45.

Ball, D. L., Sleep, L., Boerst, T., & Bass, H. (2009). Combining the development of practice and the practice of development in teacher education. *Elementary School Journal, 109*, 458–474.

Ball, D. L., Thames, M. H., & Phelps, G. (2008). Content knowledge for teaching what makes it special? *Journal of Teacher Education, 59*, 389–407. doi:10.1177/0022487108324554

Baumert, J., Kunter, M., Blum, W., Brunner, M., Voss, T., Jordan, A., & Tsai, Y. M. (2010). Teachers' mathematical knowledge, cognitive activation in the classroom, and student progress. *American Educational Research Journal, 47*, 133–180. doi:10.3102/0002831209345157

Black, D. J. W. (2007). *The relationship of teachers' content knowledge and pedagogical content knowledge in algebra and changes in both types of knowledge as a result of professional development* (Unpublished doctoral dissertation). Auburn University, Auburn, AL.

Borko, H., Jacobs, J., Eiteljorg, E., & Pittman, M. E. (2008). Video as a tool for fostering productive discussions in mathematics professional development. *Teaching and Teacher Education, 24*, 417–436.

Cohen, D. (2014, March 17). *Why weaknesses of teacher evaluation policies are rooted in the weaknesses of teaching and teacher education*. Presentation at the University of Delaware College of Education and Human Development, Newark, DE.

Conference Board of the Mathematical Sciences. (2012). *The mathematical education teachers II*. Washington, DC: American Mathematical Society and Mathematical Association of American. Retrieved February 22, 2015, from http://www.cbmsweb.org/MET2/met2.pdf

Ding, G., Li, M., Sun, M., Li, Y., Chen, L., Yang, F. et al. (2014). *National survey and policy analysis on the pre-service teacher education of student teachers in normal universities and colleges*. Shanghai: East China Normal University Press. [in Chinese]

Even, R. (1993). Subject-matter knowledge and pedagogical content knowledge: Prospective secondary teachers and the function concept. *Journal for Research in Mathematics, 24*, 94–116.

Even, R. (1998). Factors involved in linking representations of functions. *Journal of Mathematical Behavior, 17*(1), 105–121.

Fan, Q. (2013). *Survey on pedagogical content knowledge among pre-service mathematics teachers from normal universities and colleges* (Unpublished master thesis). Shaanxi Normal University, Xian. [in Chinese]

Ferrini-Mundy, J., McCrory, R., & Senk, S. (2006, April). *Knowledge of algebra teaching: Framework, item development, and pilot results.* Research symposium at the research presession of annual meeting of National Council of Teachers of Mathematics, St. Louis, MO.

Floden, R. E., & McCrory, R. (2007, January). *Mathematical knowledge for teaching algebra: Validating an assessment of teacher knowledge.* Paper presented at 11th AMTE Annual Conference, Irvine, CA.

Friedman, T. L. (2005). *The world is flat.* New York, NY: Farrar, Straus and Giroux.

Huang, R. (2014). *Prospective mathematics teachers' knowledge of algebra: A comparative study in China and the United States of America.* Wiesbaden: Springer Spektrum.

Huang, R., Peng, S., Wang, L., & Li, Y. (2010). Secondary mathematics teacher professional development in China. In F. K. S. Leung & Y. Li (Eds.), *Reforms and issues in school mathematics in East Asia* (pp. 129–152). Rotterdam, The Netherlands: Sense Publishers.

Huang, R., Su, H., & Xu, S. (2014). Developing teachers' and teaching researchers' professional competence in mathematics through Chinese lesson study. *ZDM: The International Journal on Mathematics Education, 46*, 239–251.

Li, S., Huang, R., & Shin, Y. (2008). Discipline knowledge preparation for prospective secondary mathematics teachers: An East Asian perspective. In P. Sullivan & T. Wood (Eds.), *Knowledge and beliefs in mathematics teaching and teaching development* (pp. 63–86). Rotterdam, The Netherlands: Sense Publishers.

Li, Y., & Huang, R. (2009). Examining and understanding prospective mathematics teacher preparation in China from an international perspective. *Journal of Zhejiang Education Institutes, 1*, 37–44. [in Chinese]

McCrory, R., Floden, R., Ferrini-Mundy, J., Reckase, M. D., & Senk, S. L. (2012). Knowledge of algebra for teaching: A framework of knowledge and practices. *Journal for Research in Mathematics Education, 43*(5), 548–615.

Ministry of Education of China. (2011a). *Curriculum standard for teacher education program (experimental).* Retrieved from http://www.moe.edu.cn/publicfiles/business/htmlfiles/moe/s6136/201110/125722.html [in Chinese]

Ministry of Education of China. (2011b). *Syllabi of national teacher qualification examination.* Retrieved from http://www.ntce.cn/a/kaoshitongzhi/bishibiaozhun/ [in Chinese]

Ministry of Education of China. (2012). *Professional standards for teachers at different levels.* Retrieved from http://www.moe.gov.cn/publicfiles/business/htmlfiles/moe/s7232/201212/xxgk_145603.html [in Chinese]

Ministry of Education of China. (2013a). *Interim rules for teacher qualification examination.* Retrieved from http://www.moe.gov.cn/publicfiles/business/htmlfiles/moe/s7711/201309/xxgk_156643.html [in Chinese]

Ministry of Education of China. (2013b). *Interim rules for regular registration of teacher qualification certificate.* Retrieved from http://www.moe.gov.cn/publicfiles/business/htmlfiles/moe/s7711/201309/xxgk_156643.html [in Chinese]

Ministry of Education of China. (2013c). *Interpretation on curriculum standard for teacher education program (experimental).* Beijing: Beijing Normal University Press. [in Chinese]

Mullis, I. V. S., Martin, M. O., Ruddock, G. J., O'Sullivan, C. Y., & Preuschoff, C. (2009). *TIMSS 2011 assessment frameworks.* Chestnut Hill, MA: TIMSS & PIRLS International Study Center, Boston College. Retrieved from http://www.timssandpirls.bc.edu/timss2011/downloads/TIMSS2011_Frameworks.pdf

National Council of Teachers of Mathematics. (2000). *Principles and standards for school mathematics.* Reston, VA: Author.

National Mathematics Advisory Panel. (2008). *Foundations for success: The final report of the national mathematics advisory panel.* Washington, DC: U. S. Department of Education.

NCTM CAEP. (2012). *Standards for mathematics teacher preparation.* Retrieved February 22, 2015, from https://www.nctm.org/Standards-and-Positions/CAEP-Standards/

Pang, Y. (2011). *A study of prospective teachers' mathematical knowledge for teaching and how to develop it* (Unpublished master thesis). East China Normal University, Shanghai. [in Chinese]

Shao, G. (2011). *Research on teachers' professional knowledge.* Hangzhou: Zhejiang University Press. [in Chinese]

Sherin, M. G., & van Es, E. A. (2009). Effects of video club participation on teachers' professional vision. *Journal of Teacher Education, 60,* 20–37.

Shulman, L. (1986). Those who understand: knowledge growth in teaching. *Educational Researcher, 15*(2), 4–14.

Tatto, M. T., Schwille, J., Senk, S. L., Ingvarson, L., Rowley, G., Peck, R., & Rowley, G. (2012). *Policy, practice, and readiness to teach primary and secondary mathematics in 17 countries: Findings from the IEA Teacher Education and Development Study in Mathematics (TEDS-M).* Amsterdam: IEA.

Wu, L., & Ge, J. (2015). Examination standard for school teacher qualification: Its background, purpose and content. *China Examination, 1,* 25–31. [in Chinese]

Yan, L. (2014). *Research on pre-service mathematics teachers' ability to analyze and deal with students' errors* (Unpublished master thesis). Shanxi Normal University, Xian. [in Chinese]

Yang, Y., Li, J., Gao, H., & Xu, Q. (2012). Teacher education and the professional development of mathematics teachers. In J. P. Wang (Ed.), *Mathematics education in China: Tradition and reality* (pp. 205–238). Singapore: Cengage Learning Asia Pte Ltd.

Yu, R., & Zhao, X. (2015). Statistical analysis on results of national teacher qualification examination. *China Examination, 1,* 32–39. [in Chinese]

Zhang, D. (2005). Educational mathematics: The educational state of mathematics. *Journal of Mathematics Education, 14*(3), 1–4. [in Chinese]

Zhou, Z., Peverly, S. T., & Xin, T. (2006). Knowing and teaching fractions: A cross-cultural study of American and Chinese mathematics teachers. *Contemporary Educational Psychology, 41,* 438–457.

Zou, B., & Chen, X. (2005). Tracing to the source of the concept of knowledge for teachers. *Curriculum, Teaching Materials and Method, 24*(6), 85–89. [in Chinese]

Yingkang Wu
Department of Mathematics
East China Normal University
Shanghai, China

Rongjin Huang
Department of Mathematical Sciences
Middle Tennessee State University
Murfreesboro, Tennessee, USA

ZHIQIANG YUAN AND RONGJIN HUANG

7. PEDAGOGICAL TRAINING FOR PROSPECTIVE MATHEMATICS TEACHERS IN CHINA

INTRODUCTION

The "learning gap" in mathematics identified by many international comparative studies (e.g., Hiebert et al., 2003; Stevenson & Stigler, 1992) between East Asia and the West, has been explored throughout the "preparation gap" found in TEDS-M (Schmidt et al., 2007). Mathematics teachers in Asian countries have had a very different configuration of learning experiences as part of their teacher preparation (Schmidt et al., 2007). Multiple East Asia scholars have tried to unveil the mysteries of Asian countries' teacher preparation from the perspective of mathematics content knowledge (such as, Li, Huang, & Shin, 2008; Li, Ma, & Pang, 2008; Li, Zhao, Huang, & Ma, 2008) and mathematics pedagogical content knowledge (such as, Hsieh, Lin, & Wang, 2012; Lin, 2005). Several studies revealed that mathematics teacher preparation programs in Mainland China emphasized the development of teachers' content knowledge, with very limited experience in student teaching (e.g., Li et al., 2008). However, little is known about process and features of pedagogical training for prospective teachers. Although pedagogical courses and mathematics education courses are important for developing prospective teachers' knowledge for teaching (as discussed in Chapter 6), this chapter focuses on three types of practice-based pedagogical training for prospective mathematics teachers which have not be explored empirically.

PEDAGOGICAL TRAINING FOR MATHEMATICS TEACHERS IN CHINA

There are many types of pedagogical training for prospective and practicing teachers in China including field observation (教育见习) and student teaching (教育实习). According to various policy documents, Chinese teachers must experience pedagogical training both in their teacher preparation program and their professional career. For example, the Ministry of Education (MOE) of China issued the *Curriculum Standards for Teacher Education (trial version)* (CSTE) in 2011, which required that prospective teachers take educational practice courses. Each prospective teacher must participate in an 18-week-long student teaching (MOE, 2011). According to *Regulations on Continuing Education for Primary and Secondary School Teachers* (MOE, 1999), practicing teachers must complete at least 240 hours of professional learning every five years. In addition to attending to course-work and summer institutes, a majority of the continuing education activities have been practiced in the form of *Teaching Research Group* activities (Yang, 2009).

Several international studies on practicing teachers' professional development have been reported including Huang and his colleagues who introduced *exemplary lesson development* as one of pedagogical training for practicing mathematics teachers (Huang & Li, 2009; Huang, Li, Zhang, & Li, 2011). This pedagogical training improved teachers' expertise as well as Li, Tang, and Gong (2011), who introduced how a *master teacher work station* improved experienced mathematics teachers' expertise. Li and Qi (2011) introduced how an *online study collaboration* program improved experienced mathematics teachers' expertise. Han (2012) described an *apprenticeship practice* existing widely in Chinese schools, where less experienced teachers were mentored by expert teachers to develop their pedagogical skills. Li and Li (2009) argued that a *teaching contest* is a useful platform for pursing excellence in classroom instruction. Peng (2007) introduced *lesson explaining* as an important type of pedagogical training for practicing teachers. This showed how individual teachers developed their own content knowledge of probability and how the professional community promoted teachers to develop their pedagogical content knowledge during the professional activities based on the task of lesson explaining.

However, very few empirical studies focus on pedagogical training for prospective mathematics teachers in China (Li & Li, 2008; Tong, 2013). The process and features of pedagogical training for prospective mathematics teachers are largely unknown for the international audience. To this end, we conducted systematic investigations on the pedagogical training of prospective Chinese mathematics teachers through three major tools including field observations, microteaching, and student teaching.

THEORETICAL FRAMEWORK

To examine how prospective teachers learn to teach, Hammerness, Darling-Hammond, Bransford, Berliner, Cochran-Smith, Mcdonald, and Zeichner (2005) proposed a model. This model suggests that new teachers learn to teach in a learning community that enables them to develop a vision for their practice; a set of understandings about deep knowledge of content, pedagogy, students, and social context; dispositions about habits of thinking and action regarding teaching and children; practices that allow them to act on their intentions and beliefs; and tools (conceptual tools include learning theories and ideas about teaching and learning while practical tools include particular instructional approaches and strategies, and resources) that support their efforts.

Specifically, the Teacher Education Development Study in Mathematics (TEDS-M) (Tatto, Schwille, Senk, Ingvarson, Peck, & Rowley, 2008) developed a framework of teachers' professional competence, which includes teachers' professional knowledge and teachers' professional beliefs. The former includes mathematics content knowledge (i.e., basic factual knowledge of mathematics, conceptual knowledge of structuring and organizing principles of mathematics as a discipline), mathematics pedagogical content knowledge (i.e., knowledge of mathematical curricular, knowledge of planning for mathematics teaching and learning, and knowledge of enacting mathematics for teaching and learning), and general pedagogical knowledge. The latter includes beliefs about the nature of mathematics, learning mathematics, mathematics achievement, preparedness for teaching mathematics and the program effectiveness.

Thus, we consider prospective teachers' learning outcomes in terms of their content knowledge, pedagogical content knowledge, general pedagogical knowledge, instructional practice or repertories, and beliefs concerning mathematics and mathematics teaching and learning. As far as pedagogical training is concerned, multiple ways have been widely used to develop prospective teachers' clinical experience around the world, including student teaching and microteaching, performance assessments and portfolios, analysis of teaching and learning, case methods, autobiography, and practitioner inquiry (Hammerness et al., 2005). To help prospective teachers in China get clinical experience, three major types of pedagogical training are used: field observations, microteaching, and student teaching. All of them focus on classrooms instruction to a certain extent: from observing practicing teachers' teaching, to microteaching, to teaching of actual lessons.

Lesson study is a format of practice-based professional development that has been widely adapted for promoting both prospective and practicing teachers' learning (Hart, Alston, & Murata, 2011). Through lesson study, participants can develop their knowledge and beliefs, and develop a professional learning community (Lewis, Perry, & Hurd, 2009). In particular, if teachers in a lesson study are regarded as key stakeholders, then the lesson study could develop both their practical and scientific knowledge (Kieran, Krainer, & Shaughnessy, 2013). Recently, a Chinese lesson study approach called *Same Content Different Designs activities* (SCDD, or *Tong Ke Yi Gou* 同课异构 in Chinese, or parallel lesson study), has been adopted in microteaching and public lessons (Huang, Su, & Xu, 2014; Yuan & Li, 2015). During a typical SCDD activity, two or more teachers teach the same topic to different groups of students with distinct lesson designs, while their fellow teacher participants observe each of these lessons. After all lessons are completed, the teachers involved gather to discuss the lesson designs, classroom teaching practices, and make comments and suggestions for future revisions and improvements (Yuan & Li, 2015).

Although a number of courses also contribute to prospective teachers' pedagogical training, such as analysis of mathematics curriculum standards and textbooks, mathematics teaching design, mathematics teaching methods and so on, in this chapter, we only focus on how three major practice-based pedagogical trainings (field observations, microteaching, and student teaching) develop teachers' pedagogical knowledge, beginning teaching repertoires and relevant vision and beliefs. Specifically, the following two questions will be investigated: (1) *How are these pedagogical training conducted?* and (2) *what are the impacts of these pedagogical training on prospective mathematics teachers' professional development (knowledge, instruction practice, and beliefs)?*

FIELD OBSERVATIONS

Background

According to *CSTE* (MOE, 2011), field observations consists of: (1) observing classroom teaching to know about its norms and process; (2) observing classroom management to know about its contents and requirements and getting direct experience of communicating with students; and (3) visiting school to know about its organizational structure and operational mechanism. However, there is no unified arrangement for this pedagogical training in China. Some teacher preparation programs may provide more opportunities for prospective teachers' field observations than others or may have arranged fixed time and school locations for this pedagogical training.

Procedures and Methods

We first introduce the tasks and requirements for field observations of a secondary school mathematics teacher preparation program in a provincial teacher education institute located in southeast coast of China (we gave this university a pseudonym *Qishan University*). This pedagogical training was arranged for a week in the fall semester of the third year of a four-year bachelor's degree program. All prospective mathematics teachers were equally divided into several groups based on the number of mathematics teacher educators and the range of students. Each mathematics teacher educator was in charge of one or two of the groups. Since field observations, microteaching, and senior undergraduates' student teaching were arranged in the same semester, they were administrated based on an overall consideration. Usually, a mathematics educator is responsible for supervising microteaching and field observations of prospective mathematics teachers simultaneously. These prospective teachers visited the same school as their supervisor, who was in charge of senior undergraduates' student teaching. Table 1 shows the tasks and requirements for field observations for each prospective mathematics teacher in a week.

This framework has been used to guide field observations over the past four years. We collected 65 (13 males, 52 females) prospective mathematics teachers' summary reports for field observations over the four years. After a qualitative analysis on the impacts of prospective teachers' field observations based on Nvivo software, we found that prospective teachers had benefited from this pedagogical training in many aspects.

Table 1. The tasks and requirements for field observations

Task	Requirements (in a week)
observing classroom teaching	no less than eight 45-minitue lessons
grading students' homework	at least twenty copies
writing lesson plans	at least one copy
participating in Teaching Research Group activities	at least one time
managing students as an assistant class supervisor	every day
writing a summary for field observations	no less than 2000 words

Findings

Next, the results will be reported from three aspects: teachers' beliefs, teachers' knowledge and teachers' practices.

Perceptions of teachers. The most impressive reaction from the prospective teachers during their field observations was "It's not easy to be a good teacher." Sixty percent of the prospective teachers (39 of 65) mentioned this point. For example, Miss Qiu wrote:

> I used to think that it's very easy and relax as a teacher because he or she only needs to have a few lessons and repeat teaching the same content year after year. After I finished my field observations, I understand that it's VERY HARD to be a teacher. A teacher not only needs to master the content knowledge by himself/herself, but also is able to explain the knowledge to students in the ways that students are easy to access. It takes a lot of time for teachers to prepare lessons.

Approximately 19% of prospective teachers (12 of 65) showed great appreciation to those teachers who could teach the content both humorously and rigorously. About 17% of the teachers (11 of 65) realized that teachers should impart knowledge and educate people, as well as express their willingness of "striving to become a teacher."

Teachers' knowledge. Prospective teachers paid much more attention to pedagogical content knowledge and general pedagogical knowledge than content knowledge.

Specifically, prospective teachers noticed the following classroom practices, which are related to practicing teachers' pedagogical content knowledge: (a) analysing work-out examples and exercise problems carefully (19 teachers, 29.2%); (b) appropriate use of information technology (16 teachers, 24.6%); (c) noticing of students' understanding (16 teachers, 24.6%); (d) infiltrating mathematical thoughts and methods (9 teachers, 13.8%); and (e) being good at generalization and summarization (8 teachers, 12.3%). For example, when Miss Xie commented on the teachers' strategies for explaining exercise problems, she wrote:

> Some teachers can present the exercise problems very inspiring. For example, some of them compared with the previous similar exercise problems. Some of them gave a summary after presenting a type of exercise problems. Some of them gave the similar exercise problems to the students after finishing presenting a problem. Some of them gave different solutions for different types of problems. For example, fill-in-the-blank problems and multiple- choice problems can have both general and specific solutions.

Prospective teachers also noticed the following classroom practices, which are related to practicing teachers' general pedagogical knowledge: (a) designing good blackboard writing and drawing (21 teachers, 32.3%); (b) creating an active classroom atmosphere (17 teachers, 26.2%); (c) applying effective classroom management strategies (9 teachers, 13.8%); and (d) optimizing languages use for teaching (7 teachers, 10.8%). For example, when Miss Cai reflected on the teachers' blackboard writing and drawing skills, she wrote:

> A teacher should be a model and set a good example to his students. After observing almost all mathematics teachers' classrooms teaching during my field observations week, I found a common characteristic of these teachers. Namely, all teachers' board-writings are especially beautiful. However, none of my classmates can write mathematics board-writing beautifully. So it is very urgent to practice to write a good board-writing.

Although most of prospective teachers focused their attention on pedagogical content knowledge and general pedagogical knowledge, 9.2% of prospective teachers (6 of 65) still noticed the importance of content knowledge. For example, Miss Wu said:

> By observing the classrooms, I realized the importance of content knowledge. Only having a solid content knowledge, a teacher can manage the classroom freely. …I should do some remedial work for my content knowledge.

Student learning. Apart from observing the teachers' classroom practices, grading students' homework made a deep impression on prospective teachers, with approximately 50% of the prospective teachers (32 of 65) mentioning it. For example, Miss Ma wrote on her experience:

> In the following days, I observed several lessons and graded a lot of homework. I found the students have very different understandings' level. Even in a middle level class, it's very hard to find some excellent homework. Most of students did very badly. OMG (oh, my god), this is a class A level school, but the students did so badly! I found many of my previous conceptions are wrong. For example, I took for granted that the students are very clever when I had

my simulated teaching. So I omitted many important details that the students may be puzzled. I think I should put analysing students' understanding in the first place now.

Overall, prospective teachers appreciated the following experiences during their field observations: observing class teaching (17 teachers, 26.2%), observing "same content different designs" activities (16 teachers, 24.6%), observing how lessons were prepared collectively (14 teachers, 21.5%) and observing post-lesson discussions (5 teachers, 7.7%). In addition, they realized that teachers must prepare for the lessons carefully in order to teach the lesson effectively (10 teachers, 15.4%). Some of the prospective teachers could link their field observations with their microteaching (7 teachers, 10.8%). Additionally, prospective teachers found some dissatisfactory classroom teaching practices during their field observations (12 teachers, 18.5%).

Summary

Through field observations, most of prospective teachers realized the difficulty of being a good teacher, but they were willing to face the challenge. They noticed the importance of having sound content knowledge, rich pedagogical content knowledge and general pedagogical knowledge. They also realized the value of grading homework as a venue to understand students learning. These field experiences and observations lay the foundation for miro-teaching and further student teaching.

MICROTEACHING

Background

Microteaching is a scaled-down, simulated teaching encounter designed for the training of both prospective and practicing teachers. It has been used widely since its invention at Stanford University in 1960s by Dwright W. Allen and his colleagues (Allen & Wang, 1996). In 1994, the Ministry of Education (MOE) of China issued a document: *An outline for training prospective teachers' professional skills in normal universities (trial version)*, which required prospective teachers to get several types of trainings of professional skills, including classroom teaching skills training. Most universities followed these requirements and provided pedagogical training (microteaching) to their prospective teachers in the following classroom teaching skills: skills of introduction, board-writing and drawing, demonstration, presentation, questioning, response and reinforcement, closure, organization teaching, and variation (MOE, 1994).

However, the classroom is a complex system (Ricks, 2007). Classroom teaching consists of more than an assembly of teaching skills. Many normal universities deal with these classroom teaching skills comprehensively. For example, they may allow prospective teachers to select one topic to mock teaching the whole lesson. Depending on the size of the class, each prospective teacher may have one or more opportunities to teach in a simulated classroom, where the peers act as students. Other universities may require a *lesson explaining* (Peng, 2007) activity during microteaching.

Procedures and Methods

We will introduce the procedure and method of mathematics microteaching in *Qishan University* based on two empirical studies (Yuan & Li, 2012; Yuan & Li, 2015). Typically, microteaching is arranged in the fifth semester with each instructor in charge of one or two classes of prospective teachers (less than 15 persons for each class). Two 45-minitue lessons are arranged each week for a class. In the first author's microteaching class, the following training procedure and method were typically used: each prospective teacher had a teaching simulation twice during microteaching. The first teaching simulation was held in the form of a *Same Content Different Designs* (SCDD) activity (Yuan & Li, 2015). The same lesson topic, for example, *normal distribution* (Yuan & Li, 2012) or *logarithmic functions and their properties* (Yuan & Li, 2015), was taught in a simulation classroom by all prospective teachers (less than 15 persons). The instructor chose these topics based on their importance and representativeness.

In the process of this SCDD activity, each prospective teacher went through five steps: (1) preparing for and writing lesson plans individually; (2) teaching simulation and peer observations; (3) explaining and evaluating lessons; (4) observing self-videotapes and writing reflective journals; and (5) revising lesson plans. Sometimes, not all prospective teachers were selected to simulate teaching due to time constraints. After the first teaching simulation, prospective teachers were expected to master the methods of preparing for and teaching the lessons.

In the second teaching simulation, different lesson topics were taught by different prospective teachers. Prospective teachers could select any topic at any levels based on their interests. Many times these topics would come from the same textbooks for the purpose of preparing for student teaching. After the second teaching simulation, prospective teachers were expected to be familiar with multiple mathematics topics at the different school levels, which would be helpful for their student teaching and teachers' recruitment.

Findings

In our study on developing prospective teachers' technological pedagogical content knowledge (TPACK, Mishra & Koehler, 2006) through learning TPACK course and participating in SCDD activities (Yuan & Li, 2012), thirteen (2 males, 11 females) prospective mathematics teachers took a 10-lecture mathematics education course, which integrated technology (Fathom dynamic data software), pedagogy, and content (statistical distribution) knowledge. They also experienced three rounds of teaching simulation and one round of authentic teaching based on SCDD activities. Three major prospective teachers (T1, T2 and T3), were all getting positive results. The first result included how the three prospective teachers' overarching conceptions about the purposes of integrating information technology into mathematics teaching had *changed significantly*. T1 changed from focusing on *interest* to *understanding*. T2 changed from focusing on *teachers* to both *teachers* and *students*. T3 changed from *vaguely* paying attention to *interest* and *understanding* to *clearly* focusing on *interest* and *understanding*. The second result was how the three prospective teachers' knowledge of curriculum resources and organization for integrating information technology into mathematics teaching had *changed significantly*. T1, T2 and T3 all changed from mainly using general information technology (for example, *PowerPoint*) to subject-specific information technology (for example, *Fathom dynamic data software*) and T1 and T2 could organize the curriculum content effectively by using real-world situations. The third result included how the three prospective teachers' knowledge of instructional strategies and representations for integrating information technology into mathematics teaching had *changed significantly*. T1, T2 and T3 could all adopt some instructional strategies, which are much more in alignment with students' recognition rules. They all could use pictures and real-world situations more effectively as representations. The fourth result was how the three prospective teachers' knowledge of students' understandings and misconceptions for integrating information technology into mathematics teaching had *changed a little*. T1, T2 and T3 could recognize the importance of understanding students' thinking, yet they have little knowledge about students' typical understandings and misconceptions (Yuan & Li, 2012). Although microteaching based on SCDD activities is only a part of interventions of this experiment, it is vital for its success.

Summary

We found that prospective teachers can learn how to teach by learning and teaching the same topic in the first teaching simulation because they spend a long time on the same topic. This model for teacher learning is both collaborative and reflective. The prospective teachers can easily compare different instructional strategies and instructional representations with each other under the hint of the instructor. Secondly, the teaching simulation provided them the opportunity to practice what they had learned from the first SCDD activity. These experiences helped prospective teachers change their conceptions of good mathematics teaching and helped to develop their knowledge of curriculum, pedagogy and student learning.

STUDENT TEACHING

Background

According to *CSTE* (MOE, 2011), student teaching includes: (1) designing and implementing classroom teaching under the supervision of mentors; (2) participating in tutoring students, managing classes, organizing activities, and getting experience with communicating with families and communities under the supervision of mentors; and (3) participating in a variety of teaching research group activities and getting the opportunity to directly communicate with other teachers. The cumulative time for educational practice (including field observations and student teaching) should be not less than 18 weeks.

Procedures and Methods

As an example, we will introduce the specific schedule and requirements for student teaching of *Qishan University*. All prospective teachers are arranged for student teaching in the seventh semester of their four-year program. Typically, they get assigned to a school in the second week of September and leave from the school in the last week of November, lasting about 12 weeks. In the first week of the semester for student teaching, they will experience another round of microteaching. After student teaching, they will have 4 weeks to summarize and communicate their experience with others. The cumulative time for educational practice (including one week of field observations happened in the fifth semester) is exactly 18 weeks.

According to the guideline for student teaching, prospective teachers in *Qishan University* should fulfil the requirements mentioned by *CSTE* (MOE, 2011). Additionally, they must finish some specific tasks and requirements as shown in Table 2.

Table 2. The tasks and requirements for student teaching

Task	Requirements
taking notes when observing classroom teaching	at least six copies of notes
filling in table on teachers' classroom behaviour	three copies
filling in table on students' classroom behaviour	three copies
writing lesson plans and teaching these lessons	at least 8 copies of lesson plans
grading students' homework	every day
participating in Teaching Research Group activities	every week
teaching in "open lesson" activities	at least one "open lesson" for the entire team
managing students as an assistant class supervisor	every day
organizing topic class meeting/activities	three times
participating in educational administration and writing a report	at least 2000 words
participating in educational survey and writing a report	at least 3000 words
writing a summary for student teaching	at least 1500 words

Two supervisors are in charge of prospective teachers' professional development in mathematics. One is from the university and the other is from the local school. During student teaching, prospective teachers will mainly observe their school supervisors' classroom teaching and teach in their particular classes. Since Teaching Research Group activities are very popular in Chinese schools, prospective teachers also have the opportunity to participate in activities including preparing collectively for lessons and observing and discussing open lessons. Typically, one or two excellent prospective teachers on behalf of the prospective teachers' team in that school, delivers an open lesson to be observed by prospective and practicing teachers.

To ensure the quality of public lessons and to promote prospective teachers' learning from developing the lessons, we developed a six-step model of SCDD activity: (1) preparing for and writing lesson plans individually; (2) conducting open lessons; (3) explaining and evaluating lessons; (4) interviewing students; (5) watching self-videotaped lessons and writing reflective journals; and (6) revising lesson plans (Yuan & Li, 2015).

Findings

In our empirical study on how SCDD can promote prospective mathematics teachers' professional development (Yuan & Li, 2015), two prospective mathematics teachers and one practicing teacher participated in a SCDD activity during student teaching. We found that a prospective teacher, Nadine, improved both her knowledge and beliefs.

From the perspective of content knowledge, Nadine developed a more focused and clearly structured network of the core mathematics concepts and properties related to logarithmic functions. From the perspective of pedagogical content knowledge, she used this network to guide both her own teaching and her students' exploration activities. Nadine also introduced the focal concepts and properties more from real-world contexts, and provided students with more opportunities to discuss and explain their ideas and thinking during learning processes. During this process, she took student learning into her considerations, and made stronger connections between prior and current concepts and skills.

From the perspective of teacher's beliefs, Nadine recognized that student leaning was a step-by-step process. The purpose of classroom teaching was not to transfer all knowledge to students like force-feeding ducks in a lesson, but to teach students how to learn through classroom interactions.

DISCUSSION AND CONCLUSION

Discussion

Field observations, microteaching and student teaching are three commonly used types of pedagogical training for prospective teachers in China. In this chapter, we described general purposes and requirements for these clinical experiences. Meanwhile, we also examined their impact on prospective teachers' development of mathematics knowledge for teaching and their conception and beliefs about mathematics teaching. It was found that these three experiences are interconnected and function differently. The filed observation experience helped prospective teachers' get a taste of being a teacher: a challenging job, needs to be well prepared in content knowledge, pedagogical content knowledge, specific teaching skills (such as board-writing, selection and use of problems), and needs to pay great attention to students' learning. During microteaching, prospective teachers got much more concrete experience with developing specific teaching skills, and developing an entire lesson through the SCDD approach. They further developed their conception of mathematics teaching, their knowledge in pedagogical content and pedagogy in general. During student teaching, prospective teachers were deeply involved in observing lessons, grading homework, participating in school-based teaching research group activities, and developing public lessons. These experiences provided multiple pathways to learning to teach. The case studies revealed that the prospective teachers developed a more interconnected knowledge structure, shifted their teaching toward reform-oriented approaches by focusing on core mathematics content, developing students' conceptual understanding, and paying attention to student learning progression.

This study demonstrated that limited clinical experiences could provide prospective teachers with valuable experience in familiarizing with a teacher environment (field observations), teaching skills (microteaching) and live classroom teaching (student teaching). When these three experiences are put together, prospective teachers can develop a vision of teachers (the nature of teachers and concept of a good teaching practices). They can also develop relevant knowledge needed for effective teaching including knowledge about curriculum, pedagogy, content, student learning through microteaching and actual teaching, and relevant effective teaching repertoires for further testing, refining and enriching. Thus, based on the framework by Hammerness and his colleagues (2005), the three types of Chinese pedagogical training practices can provide valuable experiences by prompting prospective teachers' learning to teach. In particular, the specific SCDD approach to prospective teachers' learning should be highlighted.

In addition to a strong content preparation in Chinese mathematics teacher preparation programs (Huang, 2014; Li et al., 2008), this study reveals that the limited experiences in pedagogical training adds valuable pedagogical knowledge and skills for prospective teachers by making them ready to become teachers. As many studies have shown, there is a systematic and multiple-tiered teacher professional development system helping practicing teachers develop from beginners, to competent teachers and to master teachers (e.g., Han & Paine, 2010; Huang, Peng, Wong, & Li, 2010; Huang et al., 2011).

More and more teacher preparation programs pay much attention to these pedagogical trainings. However, several problems still exist, such as prospective teachers' lack of opportunities to observe the classes and teach in the classroom, the supervisors' incompetence for coaching prospective teachers, and prospective teachers' lack of preparation before student teaching (Tong, 2013). We argued that a detailed and systematic schedule is necessary for dealing with these problems. For field observations, we need to provide a more detailed checklist to train prospective teachers' ability to observe the classroom teaching. For microteaching and student teaching, we should provide prospective teachers more preparation work. The SCDD activity may be an effective way to do so.

Conclusion

This chapter examined three major pedagogical training practices of field observations, microteaching and students teaching with regard to their procedures, requirements and impacts on prospective teachers' learning. The pedagogical trainings provide limited but valuable experiences for prospective teachers to learn how to teach. They could develop a vision of what it means to be a teacher, their knowledge for teaching, and a disposition toward teaching and student learning. Systematic quality control over these practices and qualified mentors for student teaching are two big issues that are needed to be further addressed.

REFERENCES

Allen, D., & Wang, W. (1996). *Microteaching*. Beijing: Xinhua Press.
Hammerness, K., Darling-Hammond, L., Bransford, J., Berliner, D. Cochran-Smith, M., Mcdonald, M., & Zeichner, K. (2005). How teachers learn and develop. In L. Darling-Hammond & J. Bransford (Eds.), *Preparing teachers for a changing world: What teachers should learn and be able to do* (pp. 358–389). San Francisco, CA: Jossey-Bass.
Han, X. (2012). Improving classroom instruction with apprenticeship practices and public lesson development as contexts. In Y. Li & R. Huang (Eds.), *How Chinese teach mathematics and improve teaching* (pp. 171–185). New York, NY: Routledge.
Han, X., & Paine, L. (2010). Teaching mathematics as deliberate practice through public lessons. *Elementary School Journal, 110*(4), 519–541.
Hart, L. C., Alston, A. S., & Murata, A. (2011). *Lesson study research and practice in mathematics education: Learning together*. New York, NY: Springer.
Hiebert, J., Gallimore, R., Garnier, H., Givvin, K. B., Hollingsworth, H., Jacobs, J., Chui, A. M.-Y., Wearne, D., Smith, M., Kersting, N., Manaster, A., Tseng, E., Etterbeek, W., Manaster, C., Gonzales, P., & Stigler, J. W. (2003). *Teaching mathematics in seven countries: Results from the TIMSS 1999 video study*. Washington, DC: National Center for Education Statistics.
Hsieh, F., Lin, P., & Wang, T. (2012). Mathematics-related teaching competence of Taiwanese primary future teachers: Evidence from TEDS-M. *ZDM: The International Journal on Mathematics Education, 44*, 277–292.
Huang, R. (2014). *Prospective mathematics teachers' knowledge of algebra: A comparative study in China and the Untied States of America*. Wiesbaden: Springer Spektrum.
Huang, R., & Li, Y. (2009). Pursuing excellence in mathematics classroom instruction through exemplary lesson development in China: A case study. *ZDM: The International Journal on Mathematics Education, 41*, 297–309.
Huang, R., Li, Y., Zhang, J., & Li, X. (2011). Improving teachers' expertise in mathematics instruction through exemplary lesson development. *ZDM-The International Journal on Mathematics Education, 43*, 805–817.

Huang, R., Peng, S., Wang, L., & Li, Y. (2010). Secondary mathematics teacher professional development in China. In F. K. S. Leung & Y. Li (Eds.), *Reforms and issues in school mathematics in East Asia* (pp. 129–152). Rotterdam, The Netherlands: Sense Publishers.

Huang, R., Su, H., & Xu, S. (2014). Developing teachers' and teaching researchers' professional competence in mathematics through Chinese lesson study. *ZDM: The International Journal on Mathematics Education, 46*, 239–251.

Kieran, C., Krainer, K., & Shaughnessy, J. M. (2013). Linking research to practice: Teachers as key stakeholders in mathematics education research. In M. A. Clements, A. J. Bishop, C. Keitel, J. Kilpatrick, & F. K. S. Leung (Eds.), *Third international handbook of mathematics education* (pp. 361–392). New York, NY: Springer.

Lewis, C. C., Perry, R., & Hurd, J. (2009). Improving mathematic instruction through lesson study: A theoretical model and North American case. *Journal of Mathematics Teacher Education, 12*, 285–304.

Li, S., Huang, R., & Shin, H. (2008). Discipline knowledge preparation for prospective secondary mathematics teachers: An East Asian perspective. In P. Sullivan & T. Wood (Eds.), *Knowledge and beliefs in mathematics teaching and teaching development* (pp. 63–86). Rotterdam, The Netherlands: Sense Publishers.

Li, X., & Li, B. (2008). Consideration on some problems in mathematics student teaching [关于数学教育实习中几个问题的思考——兼谈高师数学教育类课程与教法改革]. *Journal of Mathematics Education, 17*(3), 41–44.

Li, Y., & Li, J. (2009). Mathematics classroom instruction excellence through the platform of teaching contests. *ZDM: The International Journal on Mathematics Education, 41*, 263–277.

Li, Y., & Qi, C. (2011). Online study collaboration to improve teachers' expertise in instructional design in mathematics. *ZDM: The International Journal on Mathematics Education, 43*, 833–845.

Li, Y., Ma, Y., & Pang, J. (2008). Mathematical preparation of prospective elementary teachers: Practices in selected education systems in Asia. In P. Sullivan & T. Wood (Eds.), *Knowledge and beliefs in mathematics teaching and teaching development* (pp. 37–62). Rotterdam, The Netherlands: Sense Publishers.

Li, Y., Tang, C., & Gong, Z. (2011). Improving teacher expertise through master teacher work stations: A case study. *ZDM: The International Journal on Mathematics Education, 32*, 763–776.

Li, Y., Zhao, D., Huang, R., & Ma, Y. (2008). Mathematical preparation of elementary teachers in China: Changes and issues. *Journal of Mathematics Teacher Education, 11*, 417–430.

Lin, P. (2005). Using research-based video-cases to help pre-service primary teachers conceptualize a contemporary view of mathematics the teaching. *International Journal of Science and Mathematics Education, 3*, 351–377.

Ministry of Education (MOE). (1994). *An outline for training prospective teachers' professional skills in normal universities* (trial version) [高等师范学校学生的教师职业技能训练大纲 (试行)]. Retrieved June 9, 2014, from http://www.jsxl.hue.edu.cn/c010.htm

Ministry of Education (MOE). (1999). *Regulations on continuing education for primary and secondary school teachers* [中小学教师继续教育规定]. Retrieved April 2, 2014, from http://www.moe.edu.cn/publicfiles/business/htmlfiles/moe/moe_621/201005/88484.html

Ministry of Education (MOE). (2011). *Curriculum standards for teacher education* (trial version) [教师教育课程标准(试行)]. Retrieved April 3, 2014, from http://www.moe.edu.cn/ewebeditor/uploadfile/2011/10/19/20111019100845630.doc

Mishra, P., & Koehler, M. J. (2006). Technological pedagogical content knowledge: A framework for teacher knowledge. *Teachers College Record, 108*(6), 1017–1054.

Peng, A. (2007). Knowledge growth of mathematics teachers during professional activity based on the task of lesson explaining. *Journal of Mathematics Teacher Education, 10*, 289–299.

Ricks, T. (2007). *The mathematics class as a complex system* (Unpublished doctoral dissertation). University of Georgia, Athens, GA.

Schmidt, W. H., Tatto, M. T., Bankov, K., Blo¨meke, S., Cedillo, T., Cogan, L., Hun, S., Houang, R., Hsieh, F., Paine, L., Santillan, M., & Schwille, J. (2007). *The preparation gap: Teacher education for middle school mathematics in six countries.* East Lansing, MI: Michigan State University.

Stevenson, H. W., & Stigler, J. W. (1992). *The learning gap: Why our schools are failing and what we can learning from Japanese and Chinese education*. New York, NY: Summit Books.

Tatto, M. T., Schwille, J., Senk, S., Ingvarson, L., Peck, R., & Rowley, G. (2008). *Teacher Education and Development Study in Mathematics (TEDS-M): Policy, practice, and readiness to teach primary and secondary mathematics conceptual framework*. East Lansing, MI: Teacher Education and Development International Study Center, College of Education, Michigan State University.

Tong, L. (2013). A survey on pedagogical training of pure and applied mathematics program [关于"数学与应用数学"专业实践性教学的调查与分析]. *Journal of Chongqing Normal University, 30*(3), 130–133.

Yang, Y. (2009). How a Chinese teacher improve classroom teaching in teaching research group: A case study on Pythagoras theorem teaching in Shanghai. *ZDM: The International Journal of Mathematics Education, 41*(3), 279–296.

Yuan, Z., & Li, S. (2012, July 8–15). *Developing prospective mathematics teachers' Technological Pedagogical Content Knowledge (TPACK): A case of normal distribution*. Paper presented at 12th International Conference on Mathematics Education, Seoul.

Yuan, Z., & Li, X. (2015). "Same content different designs" activities and their impact on prospective mathematics teachers' professional development: The case of Nadine. In L. Fan, N. Y. Wong, J. Cai, & S. Li (Eds.), *How Chinese teach mathematics: Perspectives from insiders* (pp. 567–590). Hackensack, NJ: World Scientific.

Zhiqiang Yuan
School of Mathematics and Computer Science
Hunan Normal University
Changsha, Hunan, China

Rongjin Huang
Department of Mathematical Science
Middle Tennessee State University
Murfreesboro, Tennessee, USA

DESPINA POTARI

8. MATHEMATICS TEACHER PREPARATION IN CHINA

What Do We Learn?

INTRODUCTION

Reporting Mathematics Teacher Preparation around the world has been the focus of a number of studies and surveys. For example, the ICMI Study 15 survey in 17 countries reported by Tatto, Lerman, and Novotna (2010) used a framework to study the process of becoming a teacher for teaching mathematics at school by considering both systemic dimensions (e.g. institutional arrangements and regulations, teacher's recruitment structure of the programs) as well as content specific ones (structure and content of the curriculum). The study showed a great diversity among the participating countries and pointed out the need for comparative studies to consider (a) the impact of teacher education systems on teacher knowledge, teaching practice and pupil learning, and (b) the conditions for conceptualizing and improving mathematics preparation of teachers. Similar findings are reported by the Teacher Education and Development Study in Mathematics (TEDS-M) (see Li, 2012 in the special issue in ZDM "Measuring Teacher Knowledge – Approaches and Results from a Cross-National Perspective"). Li highlighted that teachers' mathematical knowledge for teaching is related to the educational system offering as an example the centralized education system and the important role of the textbook in developing knowledge needed for teaching in China in comparison to the US. He also commented that further discussion is needed about mathematics teacher preparation and its quality improvement at the international context. The chapters in this section that I read and comment on contribute to our understanding of these teacher education practices in China.

China is one of the countries where attention has been given on the mathematics teaching in school since the TIMSS video study (Stigler & Hiebert, 2009) and lately in the teacher education and professional development structures that exist. A search in the *Journal of Mathematics Teacher Education* leads to a number of studies that present the preparation of mathematics teachers for teaching at primary (Li, Zhao, Huang, & Ma, 2008) and secondary (Liang, Glaz, DeFranco, Vinsonhaler, Grenier, & Cardetti, 2013) level. Moreover, through comparative studies, we see various challenging questions to emerge about the preparation of mathematics teachers. For example, the study of Norton and Zhang (2017) questions the nature of mathematical content knowledge that is developed in the preparation of primary school teachers. Is strong content knowledge and mathematical fluency developed in China teacher education more important than mathematical processes such as problem solving and reasoning developed in Australia and other Western countries? Is mathematical fluency and basic processes a prerequisite for high levels of problem solving and mathematical creativity? Expect from these ontological and epistemological considerations, cultural and educational issues are also addressed. For example, in China the structural educational system with entry examinations at the University and in the profession, the specialist primary teachers and teacher-centered pedagogies are different from the more open system in Australia with less testing, generalist teachers and more student-centered teaching approaches. Research in mathematics education offers rather conflicting responses to these dichotomies and working for a synthesis of them is possibly a way of transforming the teaching and the teacher education activity.

The chapters 5, 6 and 7 in this section report current reforms in mathematics teacher preparation at elementary and secondary level in China and offer us insights about the quality of teacher education at the international level. Teacher knowledge is an important aspect of the impact of these programs covering in this way the gap that seems to exist in the research literature about the process of developing prospective mathematics teacher knowledge (Potari & da Ponte, 2017). I present briefly each chapter discussing important issues addressed in relation to the teacher education practices and to their impact on the development of prospective teacher knowledge.

ELEMENTARY MATHEMATICS TEACHER PREPARATION

Chapter 5 describes the teacher education approaches in sixteen programs in China related to the development of prospective elementary mathematics teachers' knowledge. The study focuses on the Mathematical Knowledge for Teaching (MKT) of prospective elementary mathematics teachers, the sources for acquiring MKT and the ways of improving it in the context of teacher education. From content analysis of documents and materials used in teacher education and from interviews with the teacher educators the underlying philosophy and principles of teacher education related to the development of MKT, the content of the teacher education curriculum and the sources of MKT in two different models of programs (discipline-based model for specialist teachers and comprehensive models for general teachers) are presented. Finally, from prospective teachers' responses on a designed questionnaire the level of MKT of prospective teachers and their views about the teacher education program are identified.

The Principles and the Content of Teacher Education

The development of mathematics content knowledge (MCK) is an important goal in the curriculum in both discipline and comprehensive models. This complements with other aspects of teacher knowledge (pedagogical content knowledge (PCK) or general pedagogical knowledge (GPK)) through courses related to mathematics teaching, educational theory and field experiences. The mathematics courses (especially in the discipline model) are both content (e.g., mathematical analysis, 3D analytical geometry, theory of probability, non-Euclidean Geometry) and process oriented (e.g., mathematical modelling, mathematical thinking method) and about mathematics (history of mathematics). The overall aim of the educational courses and especially of the field experiences is to support prospective teachers to make links between the theoretical knowledge promoted in the courses with the practice of mathematics teaching in classroom and school context. The theory-practice relation and the development of strong mathematical knowledge in relation to mathematics teaching, are principles of the teacher education curriculum discussed in this study. However, as the authors claimed this is difficult to be achieved as the traditional teacher education approaches do not allow prospective teachers to make links between different sources of knowledge with mathematics teaching in schools in a more active way.

Teacher Education from the Prospective Teachers' Perspective

The prospective teachers seemed to value practice related courses and especially those from comprehensive models of teacher education seemed not to appreciate university mathematics courses especially when they were taught by mathematicians. They found the university mathematics courses often too difficult and disconnected from mathematics teaching in elementary schools. The relevance of mathematics knowledge to teaching for prospective mathematics teachers is an issue that has also been reported in other studies (Adler, Hossain, Stevenson, Clarke, Archer, & Grantham, 2014), and is an important issue in mathematics teacher education. Different teacher education approaches have been reported that aim to show this relevance and this is related to the rather difficult goal of teacher education to develop PCK (Aguirre, Zaval, & Katanoutanant, 2012). The blending of mathematics and pedagogy is a vision that prospective teachers appear to value most also in the context of the study reported in this chapter.

Prospective Elementary Teachers' Mathematics Knowledge for Teaching

The study also indicates that prospective teachers' MCK and especially PCK remain weak during teacher education. From the teacher educators' perspectives MCK also includes understanding of the nature of mathematics while the development of PCK is also an important goal in teacher education. They also valued the integration of different aspects of teacher knowledge to the practice of mathematics teaching as a way of contributing to the prospective teachers' professional growth by linking theoretical and practical issues in the courses. The development of teacher education resources that integrate research, the engagement of prospective teachers in inquiry based learning and task-based activities and the promotion of reflection on classroom events are seen as ways that support the development of MKT. These have also been considered as important elements of teacher education internationally (Ponte & Chapman, 2016).

Institutional and Policy Issues on Teacher Preparation

The teacher education programs for prospective elementary teachers developed in universities are framed by the Teacher Education Curriculum Standards that have been enacted by the Ministry of Education in China. The standards give the following directions: (a) improving subject matter courses to ensure that prospective teachers grasp a good understanding of mathematics essential to effective teaching in classroom, (b) PCK is a key factor for effective teaching and teacher educators need to consider how to teach PCK effectively, (c) practice-oriented approaches are important through the synthesis of theoretical and practical knowledge, (d) cooperation with expert elementary mathematics teachers and sharing of teaching resources among universities (massive open online courses), (e) focusing on primary students' learning and its development, and (f) lifelong learning and reflection abilities – educate prospective teachers as reflective practitioners who possess theoretical based and research competencies.

Challenges

The following challenges related to the preparation of elementary mathematics teachers in the Chinese context are also challenges at the international level and the focus of research: How can teacher education programs for elementary teachers balance different aspects of teacher knowledge and their relation to teaching? In what ways the teacher education practices can meet prospective teachers' professional needs and expectations? How are the teacher education standards and educational policy implemented and in what ways they transform the mathematics teacher education?

SECONDARY MATHEMATICS TEACHER PREPARATION

Chapter 6 addresses issues of educational policy related to secondary mathematics teacher preparation and recruitment through the analysis of official policy documents in China. The authors highlight that research has shown that prospective secondary mathematics teachers in China have a solid subject matter knowledge but rather poor knowledge related to teaching and learning of mathematics referring to national and international research studies. Prospective mathematics teachers' knowledge is also considered and linked to the teacher education policy through a survey on algebra for teaching.

The Principles and the Content of Teacher Education

The analysis of the secondary mathematics teacher education program of a prestigious university in China also shows that mathematics content knowledge is highly emphasized. From the analysis of the curriculum standards for Teacher Education (mentioned also in chapter 5) it emerged that the basic educational principles for teacher education are whole person development, practice oriented and life-long learning. Under these principles, the emphasis in teacher education is on the development of beliefs and positive attitudes towards students, teachers and education, appropriate knowledge and skills to educate students, and self-development. The development of strong mathematical knowledge is also a central expectation in secondary mathematics teacher education. The standards also describe structures on which the teacher education programs can be based where general pedagogy is emphasized more than subject specific pedagogy undervaluing however the development of pedagogical content knowledge. Moreover, field experiences receive an increasing attention in the teacher education programs (expected to be 18 weeks).

The Recruitment of Teachers

Secondary mathematics teachers get admission to teach in secondary schools through a rather competitive national teacher qualification examination. The examination consists of three written papers and one oral examination. One paper investigates teachers' knowledge of educational laws and ethics as well as the ability to use language in written communication. The second paper focuses on the assessment of teachers' knowledge on education and pedagogy as well as on classroom management. The third written paper concerns essential knowledge of college and school mathematics as well as how to apply this knowledge in mathematics teaching through analysis of teaching scenarios and lesson planning. The oral examination is made for the applicants who have passed the written tests and focuses on professional dispositions and ethics as well as on communication and teaching skills. In the oral test, the teachers also conduct teaching on a given topic for ten minutes and answer questions asked from a panel. From the description of the different phases of the teacher qualification, the multiple aspects of teacher knowledge and their link to lesson planning and classroom teaching appears to be central in the recruitment of teachers.

Prospective Secondary Teachers' Mathematics Knowledge for Teaching

From a survey on the characteristics of knowledge of algebra teaching it appears that the prospective mathematics teachers have a strong knowledge and skills in algebra for teaching especially in the rather traditional area such as algebraic expressions, equations and inequalities and quadratic functions. They also provide interpretations of their solutions, fluency in dealing with problems of quadratic equations and functions and flexibility in using multiple representations. However, they have difficulties in making connections between algebraic and geometrical representation, modelling real life situations and in conceptualizing different aspects of algebraic notions.

Challenges

Overall, the study shows that in teacher education programs in China content knowledge is emphasized and prospective secondary mathematics teachers have a strong mathematics content knowledge. The reform aims to integrate practice-based approaches into mathematics teacher education and there is a challenge of how to develop PCK. On the other hand, the professional development structure in China provides teachers rich opportunities for professional growth and content knowledge is certainly a prerequisite. The dilemma that seems to exist in teacher education is how to develop PCK and at the same time keep the strong mathematical knowledge that teachers have when leaving the teacher education programs. Finally, implementing mathematics curriculum standards in teacher education in the spirit of the principles and standards posed by the state is a challenge for the universities and the mathematics teacher educators.

PEDAGOGICAL DEVELOPMENT OF PROSPECTIVE MATHEMATICS TEACHERS

Chapter 7 focuses on the structure and content of the pedagogical activities related to field experiences for prospective secondary mathematics teachers in China. In particular, it addresses: (a) the way that three main activities (field observations, microteaching and student teaching) have been conducted in a teacher education institute in the southeast coast of China, and (b) their impact on the development of secondary prospective teachers' knowledge, teaching practice and beliefs.

Pedagogical Activities

Field observations consist of observing teaching, grading students' homework, writing lesson plans, helping students as an everyday assistant class supervisor, participating in teaching research group activities, writing a summary of the observations. Microteaching is based on the Same Content Different Designs (SCDD) activities where the same lesson topic was taught in a simulation classroom with all prospective teachers. The prospective teachers were engaged in (a) preparing for and writing lesson plans individually, (b) teaching simulation and peer observation, (c) explaining and evaluating lessons, (d) observing self-videotapes and writing reflective journals and (e) revisiting lesson plans. Microteaching provided prospective teachers the opportunity to compare designs of the same topic. Student teaching lasts 12 weeks and is situated between one round of microteaching and 4 weeks of collective discussion on the experience. In student teaching the prospective teachers also participate in the teaching research group activities where they plan collectively lessons, they observe and discuss open lessons while few of them also teach the open lessons. Overall, the pedagogical activities reported in this chapter follow lesson study and Keli principles (Huang & Bao, 2006) and appear to balance content and teaching. A question that appears to remain is to what extend issues related to students' learning are brought to prospective teachers' attention and whether it can be an important parameter in the lesson designs and reflection reports of the prospective teachers.

Prospective Mathematics Teachers' Knowledge for Teaching and Professional Practice

The chapter reports three case studies that show a positive impact of the pedagogical part of teacher education on prospective teachers' knowledge. In particular, prospective teachers developed knowledge more interrelated to teaching and adopted more reform oriented teaching practices where the focus is on students' learning processes. The authors argue that the three aspects or field experiences contribute significantly to the professional development of prospective teachers. The continuous professional development experiences that these teachers will have later in their professional life as teachers will complement and extend these initial teaching experiences of the teachers. The SCDD structure of the Chinese system of teacher education provides experiences to prospective teachers of noticing and reflecting. Overall, improving prospective teachers' noticing of others and their own teaching by identifying critical moments of the teaching and learning process would be one way to see teaching at its complexity. This is an extending issue in the research literature about the structures that mathematics teacher education can provide to prospective teachers to promote deeper levels of noticing and development of their knowledge (McDuffie et al., 2014).

Challenges

The authors show that the suggested structure in teacher education (field observations, microteaching and student teaching) through the cycle of observing, planning, teaching and reflecting has a positive impact on teacher knowledge and teaching. However, a number of challenges that the authors also address include the systematic quality control of these teacher education practices and the connection between school and university especially concerning the quality of the mentors. Other challenges include the link between research/theory and teaching in the context of field experiences so that the prospective teachers learn to question existing teaching practices and align critically to their new teaching experiences (Jaworski, 2006). Another challenge is the connection of these pedagogical experiences to the professional development activities that are organized through the teachers' professional life. The fact that continuous professional development is an important goal in China offers the opportunity for a beginning teacher to make connections to his/her teacher education experiences and see this as an initial part of the process of becoming teacher.

DISCUSSION

Mathematics Teacher Education for prospective primary and secondary teachers in China has some characteristics that are unique in the Chinese culture as they have also been reported in different studies mentioned in the introduction. The three chapters of this book report the development of strong mathematics knowledge for prospective primary and secondary school teachers but also show the current teacher education reform in China where the development of pedagogical competence of teachers has become an important issue. Individual (e.g., personal portfolio) and collective activities (e.g., co-planning of teaching and joint reflection of research lessons) are initiatives in teacher preparation that promote prospective teachers to make connections between different experiences from the university and the school. Moreover, we see how the teacher education is related to systemic structures that are promoted by the state such as the national teacher qualification examination, the teacher education curriculum standards and the professional development opportunities offered to the teachers during their professional life. The authors of the chapters show especially for the prospective secondary teachers that the teacher education has positive effects on their mathematics knowledge for teaching. Research during the last ten years on initial secondary mathematics teacher education (Potari & da Ponte, 2017) shows that prospective teachers' active mathematical and pedagogical engagement has a positive impact on their learning:

> The studies on the impact of teacher education practices and the processes of how PSMT knowledge develops in teacher education programs suggest that the active engagement of participants in doing mathematics and discussing strategies and results has a positive influence in their mathematics learning. In addition, PSMT active engagement in preparing tasks, analyzing students' work, giving feedback to students, and discussing with colleagues and teacher educators are also positive influences on their knowledge about mathematics teaching. (p. 15)

The Chinese mathematics teacher education offers also examples of similar prospective teachers' engagement. However, linking micro and macro issues related to teacher education and school reality (see Jaworski & Potari, 2009) will offer researchers a better understanding of what is actually going on in the actual teacher education interactions (micro-level) and in what way this is framed by the educational policy and Chinese culture (macro-level). This connection may offer to the researchers and practitioners in other countries specific actions and interactions in the context of mathematics teacher education in relation to the existing Chinese educational system and culture. I do not mean that these teacher education practices can be "transferred" to other contexts and traditions but these links between teacher education practices, educational policy structures and prospective teachers' learning may provide answers about what we can consider "effective" in other countries and cultures.

REFERENCES

Adler, J., Hossain, S., Stevenson, M., Clarke, J., Archer, R., & Grantham, B. (2014). Mathematics for teaching and deep subject knowledge: Voices of mathematics enhancement course students in England. *Journal of Mathematics Teacher Education, 17*, 129–148.

Aguirre, J. M., Zaval, M. R., & Katanoutanant, T. (2012). Developing robust forms of pre-service teachers' pedagogical content knowledge through culturally responsive mathematics teaching analysis. *Mathematics Teacher Education and Development, 14*, 113–136.

Huang, R., & Bao, J. (2006). Towards a model for teacher professional development in China: Introducing keli. *Journal of Mathematics Teacher Education, 9*, 279–298.

Jaworski, B. (2006). Theory and practice in mathematics teaching development: Critical inquiry as a mode of learning in teaching. *Journal of Mathematics Teacher Education, 9*, 187–211.

Jaworski, B., & Potari, D. (2009). Bridging the macro and micro divide: Using an activity theory model to capture sociocultural complexity in mathematics teaching and its development. *Educational Studied in Mathematics, 72*, 219–236.

Li, Y. (2012). Mathematics teacher preparation examined in an international context: Learning from the Teacher Education and Development Study in Mathematics (TEDS-M) and beyond. *ZDM: The International Journal on Mathematics Education, 44*, 367–370.

Li, Y., Zhao, D., Huang, R., & Ma, Y. (2008). Mathematical preparation of elementary teachers in China: Changes and issues. *Journal of Mathematics Teacher Education, 11*, 417–430.

Liang, S., Glaz, S., DeFranco, T., Vinsonhaler, C., Grenier, R., & Cardetti, F. (2013). An examination of the preparation and practice of grades 7–12 mathematics teachers from the Shandong Province in China. *Journal of Mathematics Teacher Education, 16*, 149–160.

McDuffie, A. R., Foote, M. Q., Bolson, C., Turner, E. E., Aguirre, J. M., Bartell, T. G., Drake, C., & Land, T. (2014). Using video analysis to support prospective K-8 teachers' noticing of students' multiple mathematical knowledge bases. *Journal of Mathematics Teacher Education, 17*, 245–270.

Norton, S., & Zhang, Q. (2017). Primary mathematics teacher education in Australia and China: What might we learn from each other? *Journal of Mathematics Teacher Education*. doi:10.1007/s10857-016-9359-6 (online first)

Ponte, J. P., & Chapman, O. (2016). Prospective mathematics teachers' learning and knowledge for teaching. In L. English & D. Kirshner (Eds.), *Handbook of international research in mathematics education* (3rd ed.). New York, NY: Taylor & Francis.

Potari, D., & da Ponte, J. P. (2017). Current research on prospective secondary mathematics teachers' knowledge. In *The mathematics education of prospective secondary teachers around the world* (pp. 3–15). Cham: Springer International Publishing.

Stigler, J. W., & Hiebert, J. (2009). *The teaching gap: Best ideas from the world's teachers for improving education in the classroom*. New York, NY: Simon and Schuster.

Tatto, M. T., Lerman, S., & Novotna, J. (2010). The organization of the mathematics preparation and development of teachers: A report from the ICMI study 15. *Journal of Mathematics Teacher Education, 13*, 313–324.

Despina Potari
Department of Mathematics
National and Kapodistrian University of Athens
Athens, Greece

PART III

ACQUIRING AND IMPROVING MATHEMATICS KNOWLEDGE FOR TEACHING THROUGH TEACHING AND PROFESSIONAL DEVELOPMENT

SHUPING PU, XUHUA SUN AND YEPING LI

9. HOW DO CHINESE TEACHERS ACQUIRE AND IMPROVE THEIR KNOWLEDGE THROUGH INTENSIVE TEXTBOOK STUDIES?

INTRODUCTION

Textbooks are readily available and commonly used in school education around the world. However, their formats and roles in school education can vary dramatically across education systems and school contexts (Li & Lappan, 2014). Nevertheless, their importance is generally acknowledged as textbooks are viewed as connecting the intended curriculum and the implemented curriculum in classrooms (Howson, 1995; Schmidt, Wang, & McKnight, 2005). They are main materials that reflect curriculum standards (e.g., China Ministry of Education, 2011; NCTM, 2000). Textbooks are commonly taken as content resources for both students and teachers (Li, 2007, 2008; Reys, Reys, & Chavez, 2004; Sun, 2011), and are often used as a guidance for structuring teaching and learning activities (Li, 2007, 2008; Reys, Reys, & Chavez, 2004; Shield & Dole, 2013). They play an important role in transforming knowledge from the form of content disciplines to school subjects, as one of the main sources to delineate the content to be taught and the pedagogical approaches to be used in classrooms (Pepin & Haggarty, 2001; Sun, 2011). At the same time, textbooks embody crucial interfaces among culture, policy and teacher curricular practices (Pepin, Gueudet, & Trouche, 2013). Ma (2013) also argued that mathematics textbooks should play a role to meet the developmental needs of children, society, and mathematics.

How do teachers use textbooks in and for teaching? Existing studies also documented dramatic variations across system contexts, schools, and individual teachers (e.g., Li & Lappan, 2014; Remillard, Herbel-Eisenmann, & Lloyd, 2009; Stein, Remillard, & Smith, 2007). In general, textbooks are used to a limited extent in many western countries and mainly perceived as a source of tasks (e.g., Randahl, 2012). In contrast to many education systems in the West, textbooks and related teacher manuals in Chinese education system have been valued as an authoritative teaching resource (Ma, Zhao, & Tuo, 2004). They have played a crucial role in school education to guide and structure classroom instruction as the most fundamental instructional materials (e.g., Li, Chen, & Kulm, 2009; Li, Zhang, & Ma, 2009). They also play a much more important role for teacher professional development in Chinese culture than others (Ma, 2010).

Some researchers also explored how pre-service teachers may interact with textbooks in the process of reading and using textbooks. For example, Nicol and Crespo (2006) found that prospective teachers have various approaches of using textbooks ranging from adherence, elaboration, and creation. It is not easy for a prospective teacher to use textbooks flexibly. Their beliefs and instructional practices are likely influenced directly by textbook presentation (Nathan, Long, & Alibali, 2002; Vincent & Stacey, 2008). Davis (2009) examined the influence of reading mathematics textbooks and planning instruction on prospective teachers' pedagogical content knowledge and content knowledge of exponential functions, and found that textbook quality is one important factor. Indeed, it is often suggested that textbooks should contain both mathematical knowledge and basic pedagogical ideas and present them in meaningful and coherent ways (e.g., China Ministry of Education, 2011; Schmidt, Wang, & McKnight, 2005).

Teachers' Use and Study of Mathematics Textbooks in China

Mathematics textbooks are used in several important ways in the Chinese school education, including, as tools for teachers' professional development through studying textbooks (e.g., Ding, Li, Li, & Gu, 2013; Ma, 2010); as self-paced learning materials for out-of-school children; and as the main resource for teaching and learning in the classroom (Sun, 2011; Sun, Neto, & Ordóñez, 2013). Both in urban and rural schools, mathematics textbooks and related teaching manuals are important resources for teachers' instructional and lesson planning (Ma, Zhao, & Tuo, 2004). In general, Chinese teachers spend much more time on instructional planning and reflection than teachers in many other countries (Marton, 2008).

Ma (2010) further pointed out:

> In China, textbooks are considered to be not only for students, but also for teachers' learning of the mathematics they are teaching. Teachers study textbooks very carefully; they investigate them individually and in groups, they talk about what textbooks mean, they do the problems together, and they have conversations about them. Teacher's manuals provide information about content and pedagogy, student thinking and longitudinal coherence. (pp. 148–149)

Textbooks and teachers' manuals in China are different from American textbook manuals that offer teachers little guidance (Armstrong & Bezuk, 1995; Schmidt, 1996), possibly because teachers are not expected to read them (Ma, 2010). Burkhardt (cited in Ma, 2010) noted:

> The math textbook provides a script (with stage directions) for the teacher to use in explaining the topic and guiding the lesson; the students are only expected to read and do the exercises at the end of the chapter. Nobody reads "teachers' guides" except on master's courses. (p. 148)

Learning to study and analyze textbooks is regarded as a fundamental requirement of professional development for both pre-service and in-service mathematics teachers in China (Ma, Zhao, & Tuo, 2004). In the curriculum for pre-service mathematics teacher preparation, there is a course called "Mathematics Curriculum Theory and Didactics." In this course, prospective teachers are expected to learn how to analyze and use mathematics textbooks for instructional purposes to enhance practical knowledge and to develop skills (Wu, 2006; Xu, 2006). For veteran in-service teachers, "studying textbooks deeply" is an important activity to gradually improve their subject content knowledge and pedagogical content knowledge (Ding, Li, Li, & Gu, 2013; Ma, 2010).

"Group lesson preparation" (集体备课), a basic form of lesson preparation activity broadly used in China, refers to such group activities where several teachers in the same subject content discipline meet to discuss and plan lesson instruction together, including analyzing possible difficulties in students' learning and challenges in teaching the lesson, and sharing ideas and suggestions to each other (Ma, Zhao, & Tuo, 2004). Group lesson preparation is often taken as a regular activity of Teaching Research Group (TRG), which is a basic unit in the school structure in China. TRG's main responsibility is to facilitate curriculum and instructional planning, to collaborate on professional activities that help each other in addressing teachers' practical problems in instruction, and to collaborate on public lesson development and instruction (e.g., Yang & Ricks, 2013). Mathematical TRG exists in every school in mainland China. The three-level teaching research network (i.e., province-level Teaching Research Office (TRO), city or county-level TRO, and school-level TRO) has been put in place for more than 50 years (Yang, 2009). Textbook study activity is mainly organized by the school-based teaching research group as a routine activity, which has played a role in furthering teachers' understanding of textbooks over time in China.

Given the importance of TRG in school system in China, some researchers have examined how Chinese teachers worked together to develop and conduct public lesson through TRG (Hua, 2009; Sun, 2015; Yang, 2009). However, much remains to be examined how teachers may study textbooks together in TRG for instructional planning and knowledge improvement. In this study, we aim to examine "how" and "what" teachers may improve in terms of their knowledge and instructional design capabilities through studying textbooks intensively within a teaching research group. Specifically, we take a case study approach to focus on teachers' activities on the topic of "knowing and understanding circles" (认识圆) in a school-based TRG. It should be noted that existing studies on textbooks typically focused on the content, role and use of textbooks, there are very limited number of studies to investigate how in-service teachers improve their knowledge and skills through intensive textbook studies. Textbooks can be easily perceived as learning materials for students, but not for teachers. How a teacher may improve knowledge through intensive textbook studies has been neglected. With this study, we intent to share a meaningful way of utilizing textbooks for teacher professional development. At the school level, this study can provide evidence of the intensive textbook study approach for improving teacher knowledge and demonstrate its concrete application within an actual teaching context in a teaching research group. The results of this study should have the potential to inform textbook research and teacher training and knowledge enrichment practices, which are also important topics for teachers and teacher educators in China as well as in many other education systems.

RESEARCH QUESTIONS

In this study, we focused on two main research questions:

1. How do Chinese mathematics teachers use textbooks for their instructional planning in a teaching research group?
2. What may mathematics teachers learn from textbooks through intensive studies in a teaching research group?

METHOD

Research Design

A case study approach was adopted in this study. By focusing on the topic of "knowing and understanding circles", we intended to figure out how Chinese mathematics teachers in a TRG use textbooks to prepare for their lesson instruction, and what these teachers may learn through intensive study of textbooks.

Study Setting and Participants

The study was carried out at an ordinary primary school located in Western China in May, 2014. The teaching and research activity lasted two weeks. Every Tuesday and Friday during these two weeks, all teacher participants gathered and worked together to study textbooks and classroom instruction. The whole process was divided into three phases: the first phase is the preparation for studying textbooks, including reading and studying curriculum standards and teacher's manual; the second phase is textbook study; the third phase is classroom implementation and evaluation.

A total of 16 people participated in this teaching and research activities: in-depth study of textbooks. They include: ten mathematics teachers at this primary school, three master teachers,[1] one university teacher and two teaching research staffs[2] outside of this school.

In our study, we focused on Teachers A, B and C, as they are mathematics teachers of Grade 6, the other 7 teachers are mathematics teachers for other grade levels. Among teachers A, B and C, teacher C is the leader of mathematics teachers for Grade 6. Table 1 below lists the basic information of these three teachers. These three teachers A, B, and C with their teaching experience varying from early career stage, relatively experienced. All three teachers were heavily involved in and throughout the teaching and research activities.

All three teachers had more than 3 years' teaching experience. They not only participated in the whole process of textbook study, but also carried out classroom lesson instruction that was designed.

Table 1. Information of the three mathematics teachers

Teacher	Age	Entry Time(year)	Educational background	Remarks
A	29	2008	Bachelor (本科)	
B	31	2005	Bachelor (本科)	
C	34	2002	Associate (专科)	teacher leader of the group

The Study Topic: "Knowing and Understanding Circles"

The topic is "knowing and understanding circles." This is one of the topics in the first semester of sixth grade mathematics textbooks. It is in the fourth unit, "graphics and geometry", of the eleventh volume of a primary mathematics textbook series published by People's Education Press (PEP) (Lu & Yang, 2006), which is used in the first semester of Grade 6. The unit of "graphics and geometry" is one of the four main fields[3] in Chinese compulsory mathematics curriculum. Table 2 below lists the basic information of "Circles" in PEP textbooks.

Table 2. "Circles" in PEP textbooks

Version	PEP
Titles	Knowing and understanding Circles (认 识 圆)
Main contents	Drawing a circle by any possible methods and instruments; Definition, Center, Radius and Diameter of a circle; Drawing a circle by compasses; Symmetry property of a circle.
Instructional objectives	Exploring the components and properties of a circle
Activity	Operation and inquiry-based way

The content of this topic is presented and organized in PEP textbook as follows: chapter-head-diagram (see Figure 1)→ drawing a circle with a variety of tools or methods (see the upper part of Figure 2)→ having hands-on activities to find the components of a circle: folding the circle several times, then letting students find the circle center, diameter, radius, their definitions and corresponding expressions in letters and their relationships (see the lower part of Figure 2 & the upper part of Figure 3)→ exploring how to draw a circle accurately and easily: using a compass → practicing (see Figure 3).

Figure 1. Chapter-head diagram

Figure 2 "Drawing a circle with a variety of tools or methods" in the upper part of the figure; "finding the circle center, diameter, radius, their definitions and corresponding expressions in letters and their relationships" in the lower part of the figure.

Figure 3. "*Finding the circle center, diameter, radius, their definitions and corresponding expressions in letters and their relationships*" *in the upper part of the figure; "exploring how to draw a circle accurately and easily: using a compass" in the lower part of the figure.*

Data Collection & Analysis

Qualitative research methods were used to explore the ways in which mathematics teachers study textbooks and how textbooks improve their mathematical knowledge for teaching. Through the process we conducted field observations, video recording and interviews, to gather data on specific activities and related procedures of their textbook study.

Interviews were also used to collect data before the beginning of the "in-depth study of textbooks" activities. We interviewed the targeted three teachers with the following questions:

"How do you usually analyze textbooks? What aspects do you focus when studying textbooks? When making instructional design, how do you use the textbooks?"

Lesson explanation (Shuo Ke, 说课 in Chinese) is a special kind of instructional tasks in China. It requires a teacher to explain how the content is presented and organized in a lesson, and the nature of the mathematical challenges it deals with (Peng, 2007). Lesson explanation can thus provide a window for us to learn about teachers' initial understanding and design for the teaching content and how their understanding and design may change. In fact, lesson explanation is used not only for teacher evaluation but also as an effective form of teacher professional development. Lesson explanation serves as a teaching and learning activity in which teachers can develop their general pedagogical knowledge, pedagogical content knowledge and subject matter knowledge individually and collectively. In this study, lesson explanation was used to collect data about teachers' knowledge improvement and to examine what they learned after participating in the textbook study about a lesson.

The Analysis Framework of Teachers' Knowledge

In this study, we focused on both the mathematics subject knowledge and pedagogical content knowledge. For mathematical subject knowledge, it mainly includes basic concepts, and their relationships, multiple representations and longitudinal coherence among them (Ma, 1999). It also includes the history of specific mathematical concepts and topics, such as the knowledge of a specific topic's origin and evolutionary process (Kats, 1998). Mathematics subject knowledge is the important cornerstone of mathematics teaching. Shulman (1986) emphasized that effective teachers blend both content and pedagogical knowledge, transform this knowledge into knowledge specific to teaching. Shulman referred to this transformed knowledge for teaching as Pedagogical Content Knowledge (PCK). According to Shulman, PCK is "teachers' cognitive understanding of subject matter content and the relationship between such understanding and the instruction teachers provide for students" (p. 25). Grossman (1990) further elaborated on Shulman's PCK model and built on these ideas to define PCK as drawing on three knowledge domains: subject matter knowledge, pedagogical knowledge and contextual knowledge. Grossman's PCK model included the following components: (1) conceptions of purposes for teaching subject matter, (2) knowledge of students' understanding, (3) curricular knowledge, and (4) knowledge of instructional strategies.

Table 3. Framework for analyzing teachers' knowledge

Teachers' knowledge	Specific aspects	Teacher A	Teacher B	Teacher C
Mathematics subject knowledge	Basic definitions			
	Relevance			
	Multiple representations			
	Longitudinal coherence			
	History of mathematics			
Pedagogical content knowledge	conceptions of purposes for teaching subject matter			
	knowledge of students' understanding			
	curricular knowledge			
	knowledge of instructional strategies			

Based on Shulman (1986) and Grossman's (1990) work, we provided a framework for analyzing teachers' knowledge before and after their intensive textbook study (see Table 3).

RESULTS

In the following sections, we organize and present results as corresponding to each research question.

Research Question 1: How Do Chinese Mathematics Teachers Use Textbook for Their Instructional Planning in a Teaching Research Group?

The whole process of "in-depth study of textbooks" activity is identified as containing three closely linked phases: preparation, study, implementation & evaluation.

Phase 1: Preparation
1. Reading and studying curriculum standards
In China, the national mathematics curriculum standards for compulsory education (Ministry of Education, 2011) is the most authoritative document for the textbook compilation, mathematics teaching and learning. Thus, the first and most important thing that teachers should do is to analyze curriculum standards. In general, it involves two steps. The first step is to find a section's[4] general requirements from "the second part of the curriculum objectives" provided in curriculum standards.

The second step is to find the specific requirements about circles from "the third part of the curriculum objectives" in curriculum standards. In Chinese curriculum standards, there is only one sentence for the learning objectives on circle as stated below:

> *Through knowing and understanding circles, students can draw circles with compasses.*

Generally, the learning objectives need to contain three dimensions: knowledge and ability; process and methods; emotions, attitudes and values. How to balance the general and specific requirements and to formulate the three dimensions of learning objectives for this content? Teachers need an in-depth analysis of the teacher's manual and the content of textbooks.

2. Reading and discussing teacher's manual
For the content of "knowing and understanding circles" in textbooks, master teachers worked together with the primary school mathematics teachers, they studied the teacher's manual through reading, discussing and understanding related content guided by a strand of questions: (a) What are the main content of this topic? What teaching content and information are presented in the textbook? What may be the intention of including each content or problem? (b) what is the relevant knowledge

basis for this content? what is the specific role of such content for learning new content topic? (c) What is the value of studying this content topic? Specific questions include: where is it often used? what kind of cognitive basis or learning experience does it provide for the follow-up knowledge learning, and what subtle influence does it have on the development of students' thinking and emotions, attitudes and values?

Phase 2: Studying textbooks. One of the two teaching research staffs introduced the method and steps of studying contents in textbook intensively: studying the textbooks including its structure, content and exercises, etc. In addition, teachers can also obtain relevant pedagogical information from textbooks, such as instructional designs.

1. Understanding Chapter-Head-Diagram

The Chapter-Head-Diagram (CHD) in the Chinese textbook plays an important role by providing a real-world context, the direction of learning, and a realistic context for the knowledge to be learned, etc. In the CHD of "knowing and understanding circles" (see Figure 1), there are some real-world like drawings of the city square containing a lot of round objects, such as fountains, flower beds, wheels, etc. By studying and analyzing the textbook and its teacher manual, the teachers realized that the purpose and role of the CHD is to illustrate circles existing in everywhere and widely used in life. Inspired by these thoughts, the CHD is regarded as a starting point and a clue for knowing and understanding circles in their instructional design.

2. Understanding the prompting words and sentences

The role of prompting words or sentences usually includes three aspects: revealing concepts, hinting at methods, and providing suggestions for the classroom activities. There are 15 prompting words or sentences in the content of "knowing and understanding circles" in the PEP textbooks (see Figures 2 & 3), such as, "What objects are round?", "Why wheels are all circular?", "After folding a few times, what will you find?", etc. These sentences are usually brief and suggestive. They provide questions, which aim to elicit a train of thoughts, imply mathematics methods and thoughts, and guide teachers' teaching and students' learning. By reading and thinking, the primary school teachers begin to realize the importance of the prompting words and sentences to promote the development of students' thinking and problem-solving ability.

3. Analyzing the key and difficult points of content and drawing concept maps

Guided by the experts, the TRG identified the key and difficult points of the teaching content. The key points include: the circle definition, properties, radius, diameter; while, the radius, diameter of a circle and drawing a circle with compasses, particularly, how to guide students to find the center of the circle, are difficult points of this topic.

Combined with the analysis of key and difficult points, they also analyzed students' learning situation. This analysis is mainly based on whether the students

IMPROVING KNOWLEDGE THROUGH TEXTBOOK STUDIES

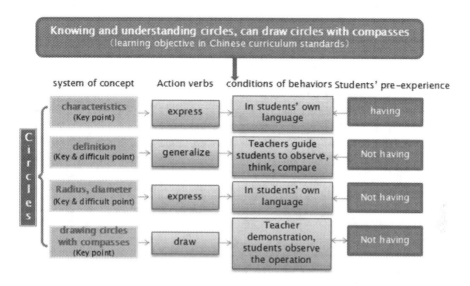

Figure 4. Analysis of content composition and instructional requirements

have necessary experience or knowledge (from daily life or previous mathematics learning) to learn these contents.

Then, these teachers were guided by the experts to build the concept maps. They firstly decomposed the study behavior verbs; secondly, they discussed and determined the conditions of the study behaviors; thirdly, they consulted to determine the degree of learning behavior (see Figure 4).

4. From analyzing textbooks to drafting learning objectives

Based on the above steps, the school teachers need to summarize the learning objectives, (1) Students can accurately articulate the characteristics of a circle; (2) Students can accurately describe the characters of radius and diameter; (3) Under the guidance of the teacher, and through the observation, thinking, comparison, students can systematically sum up the definition of circle; (4) Based on the teacher's demonstration, together with observation and practice, students can accurately draw circles with compasses.

At the end, these teachers formulate the learning objectives based on three dimensions as follows:

- Knowledge and skills: knowing the names of the component parts of a circle, understanding and mastering basic features, can draw circles with compasses, and can use the basic features of circles to solve practical problems in real life.
- Methods and processes: exploring the features of a circle by carrying out math activities (such as observe, discover, create, etc.), developing students' skills and ability of drawing circles, and developing their spatial sense.

175

- Attitudes and values: cultivating students' curiosity and innovative mindset so that students can enjoy mathematical thinking.

5. Studying ready-made sample lesson plan and instructional design

"Knowing and understanding circles" is an important topic in the Chinese primary school mathematics curriculum. It has some typical characteristics, such as being a representative content topic in geometric figures. It also contains a variety of thoughts and methods of mathematics in the learning process, etc. It is also a topic that every primary school mathematics teacher needs to teach. Over the years, a lot of great instructional designs have been developed. Such as Mr. Zhang's instructional design with a cultural perspective (Zhang, 2010), and Mr. Hua's instructional design based on students' real-life experience (Hua, 2009), etc. Those *ready-made* instructional designs could be good reference of lesson plan and sample design for school teacher.

Primary school teachers in this study had already studied even more, including the two sample lesson designs by Zhang (2010) and Hua (2009), before their own instructional design on the same topic. From those lesson examples, they also realized that the designs are not entirely following specific content materials and their orders as presented in the textbooks. By doing this, they realized the importance of studying textbooks intensively and thoroughly.

Phase 3: Classroom Implementation & Evaluation

1. Classroom instruction

With a solid foundation from studying textbook intensively, the subsequent activities involve designing and carrying out their classroom instruction. By the teaching evaluation, teachers could present what they have learned from the textbook and justify the rationale of their teaching plans. According to the content materials and methods provided by the PEP textbooks, classroom instruction should adopt an inquiry teaching method and explore hands-on approaches. However, the flexible selection and application of specific instructional method and content materials depend on an important foundation: thoroughly studying the textbooks.

2. Evaluation and Reflection by Shuo Ke

Lesson explanation (Shuo Ke) was used to collect data to examine what they learned after participating in the textbook study about a lesson. Through the process of pre- "in-depth study of textbooks" activity interviews and after-activity "Shuo Ke", the six experts – three master teachers, one university teacher and two teaching research staffs respectively gave their evaluation scores for the teachers' knowledge. The results in Table 4 are the average performance (in the 'before' column, the blanks indicate that the teachers do not have or do not show such knowledge). The scores of "A, B, C" in this table present the scores ranged from 90 ~ 100, 75 ~ 89, 60 ~ 74 respectively. Specifically, A is better than B, B is better than C.

Research Question 2: What May Mathematics Teachers Learn from Textbooks through Intensive Studies in a Teaching Research Group?

From the changes of the teachers' knowledge before and after the teaching research activities, we can learn what in-service teachers learn from textbooks, mainly through the observations, interviews, and Shuo Ke. The data were analyzed for the different stages and the results are provided below.

1. Before "in-depth study of textbooks" activities

Teacher A is a junior teacher. He began his teaching career about three years ago, and now teaches mathematics for Grade 6. He is relatively familiar with the requirements for mathematics teaching. He answered the interview questions as follows,

> Usually I firstly read to become familiar with content in the textbooks. The textbooks are an authoritative material, written by educational experts, but also align with mathematics curriculum standards, etc., so in most cases I generally develop and implement instruction in accordance with the textbooks. I mostly use textbooks to make instructional design, but sometimes I also learn from others' practices, such as the teaching design and their PPT downloaded from the Internet or others.

Teacher B was in her sixth year of teaching after started as a teacher of Grade 1. It was the first time that she teaches the topic of "knowing and understanding circles." She was close to complete the whole process of teaching from Grades 1 to 6. In more than five years of her professional teaching career, she had basically learned how to analyze, study textbooks, and instructional design. Her answers to the interview questions are as follows,

> Usually I read the textbooks, and look for the content material and didactical methods used or implied in the textbooks. When I make instructional design, I generally comply with the order of content, problems, methods, etc. in the textbooks, but sometimes I will arrange them flexibly to meet the actual situation of the content and the needs of students.

Teacher C is a relatively experienced teacher. This was the thirteenth year of her teaching. It was the second time that she teaches the topic of "knowing and understanding circles." She, as the leader of all mathematics teachers for the sixth grade, was responsible for leading all mathematics teachers of Grade 6 to carry out the teaching and research activities. She had already completed two cycles of the primary school mathematics teaching. In more than 12 years of her teaching experience, she is proficient in all kinds of mathematics instruction. She answered the interview questions as follows,

> Studying textbooks need to consider several aspects, such as students' learning situation, the difficulty of the topic, etc. Textbook presents us with teaching philosophy, resource materials, or even teaching approach. However, specific

to teaching, we need to have a comprehensive consideration. A basic approach is to grasp the essence of the content discipline and to study students' learning and cognition.

Based on the above answers from the three teachers, we could gain information about their methods of using textbooks and their knowledge of the textbook and its content. The junior teacher A takes the textbook as the "authority"; teacher B regards the textbook as the "database (信息库)" or the warehouse of teaching materials that are to be followed, and he could use textbooks "flexibly"; The textbook use for teacher C is more advanced in regards to the students' cognitive development and content features. All three teachers could study textbooks in different ways.

Based on the above information from the interviews, it can be concluded that different teaching experience brings teachers different understanding and practices to study and use textbooks. All three teachers did not give a specific textbook study methods and procedures, but using "reading" and other words to give a rough description, they have not yet mastered a systematic and comprehensive methods of analyzing and studying textbooks.

2. After "in-depth study of textbooks" activities

After two weeks of teaching research activities, how might three school teachers' knowledge be changed through "in-depth study of textbooks"? Through analyzing these teachers' performance before, during and after the activities of "in-depth study of textbooks", including the interviews with teachers, their performance in the identification and development of learning objectives, etc. we got some results about the changes of the teachers' knowledge through the activities.

a. Concerns and discussion that facilitated knowledge change

From textbooks to instructional designs at the beginning of the activities, some concerns in instructional considerations emerged mainly in the following three aspects:

First, the instructional design was initially focused more on organizing students to fold, measure, compare and other activities to discover the characteristics of a circle, than reasoning, imagination, and other thinking activities to generalize the characteristics of the circle. The reason for having such consideration perhaps is closely related to the way of how contents are presented in the textbook. However, if the teachers are just confined to the contents in the textbook, they would likely fail to find the methods and ideas hidden in the text and would not connect the materials, methods and thoughts with students' cognitive characteristics. The instructional design as a resulting product would just be a kind of formality. Such design would then fail to develop student's knowledge and skills, mathematical thinking, problem solving, and their attitudes.

Second, if the instruction focused on helping students to learn how to draw circles with compasses, it would fail to guide students to think about the question "why can

Figure 5. "Circles in (from) squares" in Chinese 《周髀算经》

we draw a circle with compasses?" This concern prompted the teachers to further adjust and improve their instructional design.

Third, if the instruction paid attention to mathematics teaching from a historical and cultural perspective, it would likely focus on the history but not necessarily on the historical and cultural thoughts and their mathematical essence about the topic. This concern led these teachers to think about taking advantage of the historical and cultural thoughts and their mathematical essence beyond the history itself.

The circle has been an important mathematical topic in Chinese history of mathematics, such as the thoughts of "circles in(from) squares" in ancient well-known book of Zhou Bi Suan Jing (《周髀算经》; Zhao, 1980, p. 58) (see Figure 5). The Ferris wheel in modern society are in fact excellent materials for studying and exploring the circles.

These three concerns as stated above furthered this teaching research group's instructional consideration in depth. What exactly should teachers do in the instruction of "knowing and understanding circles?" After discussions, teachers in this teaching research group developed a consensus on the following two aspects:

The first aspect is to study textbooks not only from the explicit ingredient, such as the content materials and their orders in textbooks and classroom teaching, but also to delve into a deeper level to find out the methods and ideas contained in the textbooks. We need to creatively use textbooks. Consequently, our primary mathematics teaching should not only focus on "what" and "how to do", but also guide students to explore the "why" and "why to do so." Such teaching would be able to reflect and highlight mathematical characteristics

179

as thinking tool, and allow students to develop in-depth understanding of the nature of mathematics.

The second aspect is that teachers should guide students to experience the essence of the phenomenon from the inquiry process, encourage students to develop good consciousness of studying the problem. "Questions are the heart of mathematics", mathematics teachers should provide students a procedure that can guide them to "know how to think", and students can master a way of thinking as a "non-verbal procedural knowledge" ("非言语程序性知识").

b. Changes of teachers' subject content and pedagogical content knowledge

Now we focus on the evaluation of teachers' mathematical knowledge development through the activities, to illustrate the changes in their subject content and pedagogical content knowledge.

With the data obtained through the method of "concentrated discussion and comprehensive evaluation", the six experts (i.e., three master teachers, one university faculty and two teaching research staffs) independently assessed the teachers' mathematics subject knowledge and their pedagogical content knowledge. Finally, by averaging the scores or taking the majority results given by the six experts, we obtained the results in Table 4 (the 'after' column).

For the teachers' subject content knowledge, the six assessment experts consistently agreed that Teacher A, B, C all have the basic definitions about circles, and this type of knowledge is not appreciably changed. But to the knowledge of "relevance", there are apparently difference and modification among the three teachers. As an example, at the outset, Teacher A only knew the components of

Table 4. Changes of three teachers' knowledge

Teachers' knowledge	Specific items	Teacher A Before	Teacher A After	Teacher B Before	Teacher B After	Teacher C Before	Teacher C After
Mathematics subject knowledge	Basic definitions	B	B	B	B	A	A
	Relevance		B	B	A	A	A
	Multiple representations	B	A	B	A	B	A
	Longitudinal coherence		B	C	B	B	A
	History of mathematics		B		B		A
Pedagogical content knowledge	conceptions of purposes for teaching subject matter	C	B	B	A	A	A
	knowledge of students' Recognizing	C	B	B	A	A	A
	curricular knowledge	B	A	B	A	B	A
	knowledge of instructional strategies	C	B	B	A	B	A

the circle, but not the relationship between these components. All three teachers' longitudinal coherence knowledge on the circle has gained a greater degree of improvement. Throughout the process of this study, the most obvious changes for mathematics subject content occur in the knowledge of history of mathematics. Before the teaching and research activity, all three teachers had little mathematics history knowledge, but after the textbooks study activities, they know more or less about the history of circles and its role and use for teaching.

For teachers' pedagogical content knowledge, from the process of identifying and developing learning objectives and classroom instruction, the six assessment experts gave their evaluation records for the three teachers by the horizontal (comparison among the three teachers) and vertical (before/after the 'in-depth study of textbooks' activities) compares. As shown in Table 4, it is clear that the three teachers' PCK have had considerable development and improvement.

DISCUSSION AND CONCLUSION

Based on the results and the whole process of the research activities as reported above, we can get the following conclusions:

Developing Profound Understanding of Mathematics through Teaching Research Activities

Ma (2010) found that Chinese mathematics teachers had profound understanding of elementary mathematics and they had better pedagogical content knowledge on mathematics. Yang (2009) pointed out one possible reason that all mathematics teachers were involved in various teaching research activities conducted in the school-based teaching research network. The results from this study suggest that studying textbook is one important reason, as argued by Ma (2010). It is easy to note all teachers participating in the activities of teaching and research have different extents of improvement in mathematics subject knowledge and pedagogical content knowledge. This activity provides support for teacher subject knowledge enhancement and pedagogy, as well as guidance on effective use of the textbook and supporting materials.

How Can Teachers' Knowledge be Improved through an in-Depth Study of Textbooks?

In many Chinese teachers' minds, textbook study is an important process of learning about curriculum, content and pedagogy. As illustrated in this case study, the specific process of studying and learning of textbooks mainly includes "study → design → teach→ evaluation and improvement." This process of "in-depth study of textbooks" is in fact commonly practiced because the intensive study of textbooks and the TRG network are normal in existence in mainland China. This study demonstrates the process of an organized teaching research activity, through which an in-depth study of textbook can further teacher's knowledge step by step in a specific case.

The results of this study shall also help inform the school-based textbook study approach for teacher training and knowledge enrichment practiced in China, and contribute to the thinking about professional development of mathematics teachers and teacher educators in general in other education systems. Indeed, at the school level, the intensive study of textbook approach in this study can provide a glimpse about this approach's feasibility and practice for improving teachers' knowledge and to demonstrate its concrete application in an actual school context.

NOTES

[1] In China, the "Master Teacher" is a special title possessing both advanced and professional qualities, in recognition of special outstanding school teachers, the teachers with these titles should set examples of moral, education and classroom instruction for others. http://baike.baidu.com/view/282390.htm?fr=aladdin
[2] In China, "teaching research staffs" are the backbones of the content discipline instruction and research in schools, they are also facilitators for other average teachers. They guide teachers to research the instructional objectives and to implement national curriculum standards in classrooms. http://baike.baidu.com/view/174988.htm
[3] The other three are "Number and Algebra", "Probability and Statistics", "Comprehensives and Practice."
[4] In China, the nine years of compulsory education is divided into three sections, each section including three years, among which the primary school stage contains the first and second sections.

REFERENCES

Armstrong, B., & Bezuk, N. (1995). Multiplication and division of fractions: The search for meaning. In J. Sowder & B. Schappelle (Eds.), *Providing a foundation for teaching mathematics in the middle grades* (pp. 85–119). Albany, NY: State University of New York Press.

Davis, J. D. (2009). Understanding the influence of two mathematics textbooks on prospective secondary teachers' knowledge. *Journal of Mathematics Teacher Education, 12*(5), 365–389.

Ding, M., Li, Y., Li, X., & Gu, J. (2013). Knowing and understanding instructional mathematics content through intensive studies of textbooks. In Y. Li & R. Huang (Eds.), *How Chinese teach mathematics and improve teaching* (pp. 66–82). New York, NY: Routledge.

Grossman, P. (1990). *Making of a teacher: Teacher knowledge and teacher education.* New York, NY: Teachers College Press.

Gu, L., & Wang, J. (2003). Teachers' development in the education actions. *Global Education, 3*(1): 44–49.

Howson, G. (1995). *Mathematics textbooks: A comparative study of grade 8 textbooks.* Vancouver: Pacific Educational Press.

Hua, Y. (2009). *I teach math in this way- Hua Ying-long and his classroom records* (in Chinese). Shanghai: East China Normal University Press.

Kats, V. (1998). *A history of mathematics: An introduction* (2nd ed.). Hong Kong: Pearson Education Asia Limited and Higher Education Press.

Li, J. (2004). Thorough understanding of the textbook: A significant feature of Chinese teacher manuals. In L. Fan, N. Wong, J. Cai, & S. Li (Eds.), *How Chinese learn mathematics: Perspectives from insiders* (pp. 262–281). Singapore: World Scientific Publishing.

Li, Y. (2007). Curriculum and culture: An exploratory examination of mathematics curriculum materials in their system and cultural contexts. *The Mathematics Educator, 10*(1), 21–38.

Li, Y. (2008). Transforming curriculum from intended to implemented: What teachers need to do and what they learned in the United States and China. In Z. Usiskin & E. Willmore (Eds.), *Mathematics curriculum in Pacific rim countries: China, Japan, Korea, and Singapore* (pp. 183–195). Charlotte, NC: Information Age.

Li, Y., & Lappan, G. (Eds.). (2014). *Mathematics curriculum in school education*. Dordrecht: Springer.

Li, Y., Chen, X., & Kulm, G. (2009). Mathematics teachers' practices and thinking in lesson plan development: A case of teaching fraction division. *ZDM: The International Journal on Mathematics Education, 41*, 717–731.

Li, Y., Zhang, J., & Ma, T. (2009). Approaches and practices in developing mathematics textbooks in China. *ZDM: The International Journal on Mathematics Education, 41*, 733–748.

Lu, J., & Yang, G. (2006). *Primary mathematics (Grade 6)*. Beijing: People's Education Press.

Ma, L. (2010). *Knowing and teaching elementary mathematics* (2nd version). New York, NY: Taylor & Francis.

Ma, Y. (2013). *Primary school teaching theory of mathematics* (In Chinese). Beijing: People's Education press.

Ma, Y., Zhao, D., & Tuo, Z. (2004). Differences within communalities: How is mathematics taught in rural and urban regions in Mainland China? In L. Fan, N. Wong, J. Cai, & S. Li (Eds.), *How Chinese learn mathematics: Perspectives from insiders* (pp. 413–442). Singapore: World Scientific Publishing.

Ministry of Education. (2011). *National mathematics curriculum standards for compulsory education*. Beijing: Beijing Normal University Press.

Nathan, M., Long, S., & Alibali, M. (2002). The symbol precedence view of mathematical development: A corpus analysis of the rhetorical structure of textbooks. *Discourse Processes, 33*, 1–21.

Nicol, C. C., & Crespo, S. M. (2006). Learning to teach with mathematics textbooks: How preservice teachers interpret and use curriculum material. *Educational Studies in Mathematics, 62*, 331–355.

Peng, A. (2007). Knowledge growth of mathematics teacher during the professional activity based on the task of lesson explaining. *Journal of Mathematics Teacher Education, 3*, 289–299.

Pepin, B., & Haggarty, L. (2001). Mathematics textbooks and their use in English, French and German classrooms: A way to understand teaching and learning cultures. *ZDM: The International Journal on Mathematics Education, 33*(5), 158–175.

Pepin, B., Gueudet, G., & Trouche, L. (2013). Investigating textbooks as crucial interfaces between culture, policy and teacher curricular practice: Two contrasted case studies in France and Norway. *ZDM: The International Journal on Mathematics Education, 45*, 685–698.

Remillard, J., Herbel-Eisenmann, B., & Lloyd, G. (Eds.). (2009). *Mathematics teachers at work: Connecting curriculum materials and classroom instruction*. New York, NY: Routledge.

Reys, B. J., Reys, R. E., & Chavez, O. (2004). Why mathematics textbooks matter. *Educational Leadership, 61*(5), 61–66.

Schmidt, W. (Ed.). (1996). *Characterizing pedagogical flow: An investigation of mathematics and science teaching in six countries*. Boston, MA: Kluwer Academic Publishers.

Schmidt, W. H., Wang, H. C., & McKnight, C. C. (2005). Curriculum coherence: An examination of mathematics and science content standards from an international perspective. *Journal of Curriculum Studies, 37*, 525–559.

Schoenfeld, A. (1988). When good teaching leads to bad results: The disasters of "welltaught" mathematics courses. *Educational Psychologists, 23*(2), 145–166.

Shield, M., & Dole, S. (2013). Assessing the potential of mathematics textbooks to promote deep learning. *Educational Studies in Mathematics, 82*, 183–199.

Shulman, L. S. (1986). Those who understand: Knowledge growth in teaching. *Educational Researcher, 15*(2), 4–14.

Stein, M. K., Remillard, J., & Smith, M. S. (2007). How curriculum influences student learning. In F. K. Lester (Ed.), *Second handbook of research on mathematics teaching and learning* (pp. 319–369). Charlotte, NC: Information Age.

Sun, X. (2011). Variation problems and their roles in the topic of fraction division in Chinese mathematics textbook examples. *Educational Studies in Mathematics, 76*(1), 65–85.

Sun, X., Neto, T., & Ordóñez, L. (2013). Different features of task design associated with goals and pedagogies in Chinese and Portuguese textbooks: The case of addition and subtraction. In C. Margolinas (Ed.), *Task design in mathematics education* (Proceedings of ICMI Study 22) (pp. 409–418). Oxford: Retrieved February 10, 2013, from https://www.hal.archives-ouvertes.fr/hal-00834054v3

Vincent, J., & Stacey, K. (2008). Do mathematics textbooks cultivate shallow teaching? Applying the TIMSS video study criteria to Australian Eighth-grade mathematics textbooks. *Mathematics Education Research Journal, 20*(1), 82–107.

Wu, Z. (2006). *Wu Zhengxian and her primary school mathematics* (in Chinese). Beijing: Beijing Normal University Press.

Xu, B. (2006). *Going into Xu Bin's mathematics education world* (in Chinese). Fuzhou: Fujian Educational press.

Yang, Y. (2009). How a Chinese teacher improved classroom teaching in teaching research group: A case study on Pythagoras theorem. *ZDM: The International Journal on Mathematics Education, 41*(3), 279–296.

Yang, Y., & Ricks, T. (2013). Chinese lesson study. In Y. Li & R. Huang (Eds.), *How Chinese teach mathematics and improve teaching* (pp. 51–65). New York, NY: Routledge.

Zhang, Q. H. (2010). *Examine the classroom: Zhang Qihua and primary mathematics culture* (in Chinese). Beijing: Beijing Normal University Press.

Zhao, J. Q. (1980). *Zhou Bi Suan Jing* (周髀算经) (p. 58) (in Chinese). Beijing: Cultural Relics Publishing House.

Shuping Pu
Chongqing Normal University
China

Xuhua Sun
University of Macau
Macau, China

Yeping Li
College of Education and Human Development
Texas A&M University, USA
and
Shanghai Normal University
Shanghai, China

XINGFENG HUANG AND RONGJIN HUANG

10. EXPERIENCED TEACHER'S LEARNING THROUGH A MASTER TEACHER WORKSTATION PROGRAM

A Case Study

INTRODUCTION

In China, as the school curriculum reforms progress, teacher professional development has been a critical issue. To meet the challenge, various innovative teacher professional development programs have been established, such as extensive teacher training workshops, teaching research activities, and novice teachers' mentoring (Huang, Ye, & Prince, 2016). The ultimate goals of teacher professional development programs are to improve classroom teaching, and eventually result in the improvement of student learning (Even & Ball, 2009). Ten years ago, Gu and Wang (2003a, 2003b) initiated the Keli study model, in which peer teachers studied a lesson collaboratively with the support of knowledgeable experts (Huang & Bao, 2006). In fact, investigation of lessons is the core component of various teaching research activities in China (Huang, Peng, Wang, & Li, 2010). For example, during the process of public lesson development (Han & Paine, 2010), teaching competition (Li & Li, 2011), or teaching research activity (Yang & Ricks, 2012) in general, participating teachers collaboratively work together to plan a lesson, teach, observe, reflect on and improve the lesson. Through this kind of collaboration, teachers can improve their teaching and develop their instructional expertise (Han & Paine, 2010; Huang, Zhang, Li, & Li, 2011; Wang & Paine, 2003; Yang & Ricks, 2012). Within the one-on-one mentor-mentee practice, experienced teachers help novice teachers to develop their competence in teaching through school based teaching research activities (Huang et al., 2010; Li & Huang, 2008). In order to help experienced teachers to become expert teachers, a great number of master teacher workstations (MTWs) have been established around the nation in the past decade. However, little is known about what and how experienced teachers could learn from participating in a MTW (Li, Tang, & Gong, 2011). To this end, we adopt a case study approach to examine how Mr. Wan, an experienced mathematics teacher, learned from Mr. Zhang, the mentor of the workstation located in East China. In particular, we will address how the teacher improved his teaching through the process of lesson study under the supervision of the mentor and how the experienced teacher continuously developed his expertise through extending the learning experience in his various teaching adventures.

Y. Li & R. Huang (Eds.), *How Chinese Acquire and Improve Mathematics Knowledge for Teaching*, 185–208.
© 2018 Koninklijke Brill NV. All rights reserved.

BACKGROUND AND THEORETICAL CONSIDERATION

Master Teacher Workstation (MTW) in China

Master Teacher Workstation (MTW) is a professional learning community that consists of a master teacher leader who is officially conferred and some key teachers (young and promising teachers). A MTW is officially recognized and financially supported by the local education department or government. The leader of the workstation, a master teacher (with an advanced, or exceptional title of teacher), is expected to play an essential role in the development of local experienced teachers' teaching expertise (Quan, 2009) and set their own specific goal and plan. Typically, the participating teachers are scheduled to engage in activities such as seminars, project research, and public lessons (Liu, 2010; Yang, 2013). The formats and contents of each MTW vary depending on local education policies and leader teacher styles (Chen & Wang, 2013; Shao & Zhu, 2013; Zhang, 2013). Although MTWs are quite diverse, studies on classroom teaching are the common and core tasks for all MTWs.

Zhang (2012) pointed out that the MTW "creates a bottom-up, teacher-focused, teacher-initiated teacher professional development, and creates a new form of teaching research so that doing teaching research becomes a common practice, and some teachers could develop toward outstanding teachers" (p. 52). However, many scholars have found numerous difficulties in administering MTWs including the lack of effective supervision of master teachers (Fang, 2012), and the lack of equal dynamic between the leading master teacher and participating experienced teachers in the community (Zhang, 2012). Thus, there is a need to explore how master teachers and practicing teachers work together to develop their professional competence.

Improvement of Classroom Teaching within a School-Based Teaching Research System in China

The core task in teacher professional development is the study of classroom teaching. Gu (2003) found that teachers appreciated experts' professional comments and suggestions on classroom teaching, including the process of preparing lessons and debriefing of post-lessons. Yang and Ricks (2012) revealed that teaching research activities could help teachers focus on students' conceptual understanding of mathematics learning through selection of tasks and exploration of big ideas. Han and Paine (2010) also demonstrated that teaching research activities could develop teachers' competences in designing mathematical tasks design, dealing with students' learning difficulties and their use of mathematical language. On the other hand, mentoring is also important for teachers to improve their teaching. Wang and Paine (2001) described how a novice teacher in Shanghai improved her classroom teaching under the guidance of her mentor.

Li and colleagues (2011) also presented two cases where experienced teachers improved their lesson plans under supervision of a master teacher. These studies show that teachers under the guidance of experienced or expert teachers can effectively improve their classroom teaching. However, it is not clear whether the improvement of a particular lesson is transferrable to a capacity of improving the teacher teaching in general.

Practicing Teacher Learning

There are mainly two perspectives regarding teachers' learning: constructivist and sociocultural. Fundamentally, constructivist perspectives focus on learning as the cognitive development of the individual while the sociocultural perspectives emphasize learning as the process of participation in sociocultural settings. However, many researchers treat these two perspectives as complementary (Borko, 2004; Lewis, Perry, & Hurd, 2009; Sfard, 1998). For example, Borko (2004) proposed a perspective of situated learning of teachers while Lewis et al. (2009) investigated teachers' learning focusing on three aspects: teachers' knowledge and beliefs, instructional practice and professional learning communities.

With regard to the process of teacher growth, the interconnected model illustrated by Clarke and Hollingsworth (2002) has been highly acclaimed (Bakkenes, Vermunt, & Wubbels, 2010; Goldsmith, Doerr, & Lewis, 2014; Opfer & Pedder, 2011). According to Clarke and Hollingsworht (2002), teacher professional growth occurs through iterative processes of enactment and reflection between four domains. The External domain represents the systems and policies that stimulate and shape teachers' learning; the Personal domain represents teachers' characteristics such as attitudes, beliefs, and knowledge; the domain of Practice represents teachers' instructional practice; and the domain of Consequence represents students' learning and other outcomes interpreted by teachers as a consequence of their professional actions. Based on a systematic review, Goldsmith et al. (2014) identified eight major areas related to teacher changes: teachers' identity, beliefs, and dispositions; mathematics content knowledge; teachers' instructional practices; teachers' collaboration/community; attention to student thinking; curriculum; and characteristics of professional development. Specifically, Bakkenes et al. (2010) highlighted six types of learning activities such as experimenting, considering own practice, getting ideas from others, and four associated learning outcomes such as changes in knowledge and beliefs, changes in intentions for practice, and changes in actual teaching practices that focus on experienced teachers' learning.

In summary, practicing teachers learn from formal as well as informal settings and may include changes in (1) knowledge for teaching, beliefs, and attitude, (2) instructional (intended and actual) practices; and (3) professional learning community.

Research Question

Since this study explores experienced teachers' learning within a specific master teacher workstation mechanism, we focus on teacher changes in their beliefs concerning mathematics teaching and their instructional practices. A case study was adopted to address the following research question: What does an experienced teacher learn from participating in a MTW?

METHODOLOGY

Participants

Mr. Zhang is a master teacher of mathematics in junior high school. He has taught mathematics nearly 20 years, and has led a MTW at local district since 2009. He is now the principal of a rural middle school and a member of the Education Expert Committee in Wuxi city. He won the national, provincial and municipal teaching competitions, and published multiple journal articles. In 2012, Mr. Wan become a member of Zhang-MTW by following a strict selection process including individual application – school recommendation – expert review. Mr. Wan is a senior secondary school teacher in another rural school, Xishan District, and has taught school mathematics for more than 15 years. He has been awarded several honors by the local education department and will study at Zhang-MTW with six other members for three years.

Mr. Zhang tailored a three-year plan for each member of the MTW. Mr. Wan's development objectives at the MTW included reinforcing theoretical foundation, forming an individual teaching style, learning teaching research, and leading young teachers. Zhang-MTW established an official web site, and created a QQ (the most popular social media service system in China) group initially for sharing and discussion among the members of the MTW. Later many teachers outside the MTW joined the group and now it has around one hundred members. Every year, the MTW would organize a series of activities to promote teachers professional development such as participating in public lesson activities, writing teaching materials, and attending academic meetings.

Data Collection and Analysis

In this study, we collected four types of data. The first part consisted of the MTW background information including planning, assessment criteria, and work schedule. These data were primarily obtained from the MTW website and QQ group. At the same time, we collected lesson-cases and articles published by Mr. Zhang so that we could understand his teaching belief and instructional style.

The second part was form public lessons organized by the MTW. In May 2014, Zhang-MTW held public lessons of *Fuxi Ke* (a type of lesson which reviews content from a textbook unit learned in the classroom) with other two members of MTWs from Jintan, a county near to Xishan. The public lesson development was similar to normal teaching research activity. First, three teachers implemented public lessons and twenty members of the three MTWs observed the lessons. One of the three teachers was a novice teacher from Mr. Wan's school and the others were experienced teachers from Jintan. They all taught the same topic of *Fuxi Ke of Congruent Triangles*. In a post-lesson meeting, the three teachers explained lesson plans and implementations briefly and then the three leaders of the MTWs commented on the lessons. Materials were collected including lesson plans, lesson videos (each lesson about 45 minutes), and a post-less debriefing audio (about 120 minutes). We mainly focused on Mr. Wan's experience therefore only the materials related to him were analyzed.

The third part was from Mr. Wan lesson preparation. Mr. Wan had designed his lesson plan one week prior to actual teaching. In fact, a few years ago he taught a public lesson on the same topic in another rural school where he had worked as a volunteer teacher for one year. Mr. Wan sent his plan through the QQ group to Mr. Zhang. They made brief communication in the group that led Mr. Wan to modify his lesson plan. Three days later, Mr. Wan taught the lesson at his school as a rehearsal for the public lesson. Mr. Zhang observed the rehearsal lesson and discussed with him after lesson. Mr. Wan then modified his lesson plan further. Throughout this process, we collected the materials including the lesson plan Mr. Wan designed when he was a volunteer teacher, and the lesson plan and video of the public lesson (about 45 minutes), and audio of a post-lesson debriefing meeting (about 60 minutes) in his school.

The fourth part included interviews with Mr. Wan. Three phone interviews were conducted having the first lasting approximately 120 minutes, the second about 50 minutes, and the third about 90 minutes. As supplementary to the interviews, Mr. Wan sent us additional documents including one lesson plan for a district teaching competition as well as a public lesson plan prepared for a district teacher training program. The two lessons were taught after the public lesson that we focused on.

Our purpose is to understand what and how Mr. Wan learned through MTW program, therefore particular attention was given to the changes in teacher's beliefs and practices in his teaching if any. We focused on teacher learning as changes in teaching practice and belief concerning about teaching. We carefully examined the two videotaped rehearsal and public lessons, and then analyzed the other related data: the discussion with Mr. Zhang and the meeting with members of the three MTWs, as well as the lesson plans and interviews with the teacher. Four themes emerged from the data analysis that included lesson structure design, open and autonomous classroom environment, dealing with learning difficulties, and understanding student reality. Our study revealed that all changes to the teacher's belief and practice were related to a specific classroom activity of students' problem posing. We analyzed Mr. Zhang' lesson plans and his published essays on mathematics teaching in order to understand what and how the experienced teacher learned from him.

RESULTS

Before presenting our major findings, we will describe the mentor's beliefs about mathematics teaching so as to understand to what extent the mentors' beliefs may have impacted the mentee's change of beliefs. The major findings include the salient changes between the two public lessons, the factors that caused these changes, and the extending efforts to further explore a teaching model.

Mr. Zhang Beliefs on Teaching Mathematics

Metaphor: pearls and a necklace. Mr. Zhang believed that a *Fuxi Ke (review lesson)* should have a main thread (i.e., focus and progression of knowledge development), along which scattered pearls (concepts and skills) that could be strung to make a beautiful necklace (knowledge network and structure). When he discussed this principle of teaching of Fuxi Ke with Mr. Wan, he used his previous lesson plan to illustrate his ideas.

The case provided by Mr. Zhang is a *Fuxi Ke* of *Geometry Transformation* (Zhang & Xu, 2013). In his lesson plan, he made use of two rotated squares as an agency to promote students' deep understanding of invariant features within various rotations. In order to achieve the learning goal of the lesson, he designed the first task (Figure 1). In this task, Mr. Zhang would initiate students to observe equal segments and inspire them to give necessary explanations or proofs. Then, based on this exploration, he would add more conditions (for instance, given square side length, or given rotation angle), and require students to calculate the shaded area.

A square ABCD is rotated α degree around the point A ($0° \leq \alpha \leq 90°$) to generate a square AEFG, where BC and FG intersect at H.

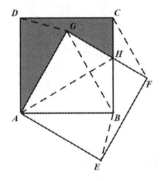

Figure 1. Task 1 in the lesson plan provided by Mr. Zhang

Mr. Zhang argued that each task in a lesson could be seen as a pearl, and we should find string to connect them together. Following this line of reasoning, he changed the center of rotation in task 1 as a main thread to generate different figures (Figure 2), so that he could use this line of reasoning to string all the pearls in his lesson.

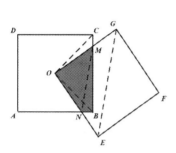 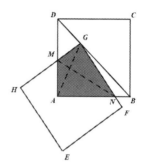

(1) Translate A to O and then rotate the square ABCD around O

(2) Translate A to G on BD and then rotate the square ABCD around G

Figure 2. Variations of Task 1 in the lesson plan provided by Mr. Zhang

Mr. Zhang believed that by having fewer tasks that were purposefully designed, rather than many isolated exercises, could provide more space for students to explore the nature of mathematics and experience mathematics thinking. If a teacher does this consistently, students may become smarter.

From this case, we see that the main flow of this lesson is a series of deliberately selected tasks: from a stereotype task (Task 1) to several interconnected variation tasks. The ultimate goal however, is to review the theorems and properties of congruent triangles. Research shows that systematic and purposeful variations of tasks help learners to develop rich experiences and various strategies in problem solving (Bao, Huang, Yi, & Gu, 2003).

Openness and autonomy in classroom activities. Mr. Zhang emphasized that a teacher should be concerned about students' learning in their classes. He believed that student learning in the classroom should be autonomous, lively, and intelligent. In other words, a teacher should pay attention to designing and organizing classroom activities so as to create a relaxed and lively classroom environment within which the students are engaged in problem solving. In a reflection report on his lesson of *Solid Figure World* in seventh grade, he wrote:

> In this lesson, I designed some activities for students. Let students touch physical models of solid shapes with hands, and match the solid shapes with given figure names. These activities should meet the students' characteristics at that age. ... So I think that students should generate their knowledge through engaging in various activities, rather than being told by the teacher.

Mr. Zhang also said: "In *Fuxi Ke*, I often ask students to pose problems by themselves, and then encourage them to solve these problems. After that, I will initiate students to extend these problems step by step. Moreover, students could design a lot of unexpected problems, and some of them are difficult. Therefore, a lesson is open enough to inspire students' learning" (Personal conversation, April 11, 2014).

Mr. Zhang's beliefs about mathematics teaching and his classroom practice reflect the notions presented in curriculum standards (MOE, 2011). It is recommended that effective mathematics teaching should focus on both the teachers' teaching and the students' learning as well as promote students' comprehensive development. The standard also highlights that students are the subject of mathematics learning and that they develop their capacities through actively participating in learning activities. A teacher should be an organizer, a guide, and a collaborator of student learning aiming to establish a productive environment and conditions for all students' development.

Making a balance between teachers' intention and classroom instances of sense making [生成性]. Curriculum standards (MOE, 2011) state that a lesson plan is a teacher's intended teaching process, which depends on the teacher's understandings and interpretations of teaching materials. Implementation of a lesson plan is the process of transforming the teacher's intentions into actual classroom practice where classroom instances of sense making such as students thinking and ideas (including incomplete ideas or errors) should be considered when making instructional decision. Thus, the teacher should make use of students' authentic thinking, give appropriate guidance to students, and adjust their plans in a timely manner in order to achieve better results from classroom teaching. Mr. Zhang had a profound understanding of the relationship between teaching intention and classroom instances of sense making, and provided an unique perspective on balancing them as illustrated in his article (Zhang, 2009, pp. 50–51) as follows:

> According to the new curriculum, making use of instances of sense making should become a significant feature of classroom teaching. However, it's a new challenge for many teachers. They are not able to appropriately use the classroom instances from complex classroom situations. So, some teachers deliver lessons by following their lesson plans step by step. They control the process of classroom teaching. While others even misinterpret the new curriculum. They see instance of sense making equivalent to ramble, respect to indulgence, and autonomy to freedom. Therefore, their classrooms become chaos. …New curriculum values the instances of sense making in classroom. But it does not mean to excessively use instances of sense making deviating from key mathematical contents. We should not interpret term superficially while missing the essence…

In fact, not all of instances of sense making are valuable. Therefore, teachers should purposefully elicit and make use of students thinking and ideas. To irrelevant or less important student ideas, teachers should pass it quickly; to important and valuable students' learning resource, they should encourage students to explore it further.

Brief Introduction of Mr. Wan's First Lesson

Review and introduction. At the beginning of the lesson, Mr. Wan reviewed how to determine triangle congruence. He asked the whole class: What are the conditions for determining if two triangles are congruent or not? Four students presented for basic facts – SSS (if three corresponding sides are equal, then these two triangles are congruent), ASA (if two corresponding angles and the included angle are equal, SAS, and AAS. Mr. Wan responded: "If this is a right triangle, are there other criteria for determining triangle congruence?" All students answered HL theorem (if corresponding hypotenuses and legs are equal, then the two right triangles are congruent). Mr. Wan finally reminded students to avoid using SSA to prove triangle congruence because it is a false proposition.

Patterns constructed with a model of two congruent triangles. Mr. Wan instructed: "Please cut two congruent triangles from a paper, and construct patterns via translation, reflection, and rotation using these two triangle models. How many different patterns can you construct?" He anticipated students would find diverse patterns with two congruent triangles in the activity, and discover some basic schemes to lay a foundation for further exploring congruent triangles in complex figures. Students began constructing different patterns using the two congruent triangles, and Mr. Wan picked several students to demonstrate their work (Figure 3). He then presented eight basic schemes using overhead slides.

Instruction on mathematics tasks. The first mathematics task was expanded through rotation of a Rt$\triangle$ABC with a 30° angle. $\triangle$DEC is generated from $\triangle$ABC rotated 90° around point C and then translated left along BC to obtain $\triangle$DEF (Figure 4). Mr. Wan required students to observe this translation of $\triangle$DEC in order for them to make and justify their conjectures about special positional relationships between AC and DE. He then asked students to add specific conditions such that $\triangle$APN$\cong\triangle$NCD. Finally, if given PB = BC, Mr. Wan posed two problems: (1) prove $\triangle$APN$\cong\triangle$NCD, and (2) prove $\triangle$EMP$\cong\triangle$BMF.

Mr. Wan continued to translate the triangle so that D and B overlapped (Figure 5), and then create the second mathematics task that was prepared for his lesson. He first connected AE, and asked students to make conjectures about the properties of $\triangle$ABE as well as to justify their conjectures. He then took the midpoint G of AE, connected GF and GC, and then asked students to prove that $\triangle$GFC is an isosceles right triangle.

Figure 3. Some students and Mr. Wan demonstrated work

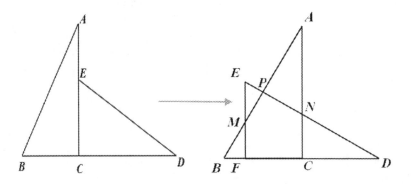

Figure 4. △ABC rotated 90° around point C to generate △DEC, then translated left along BC to obtain △DEF

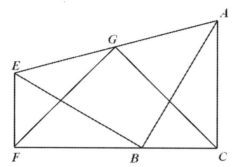

Figure 5. △EFB translated to make D and B overlapping

In the first lesson, Mr. Wan reviewed how to determine two congruent triangles, and then made use of models of two congruent triangles to generate basic patterns by translation, reflection, and rotation. In this way, congruent triangles and geometric transformations were interconnected. Finally, he created complex figures consisting of composite transformations including translation and rotation, and asked students apply their knowledge of congruent triangles to prove relevant conjectures. Through these teaching steps, Mr. Wan not only reviewed the properties of congruent triangles, but also connected it to geometric transformations. Thus, *Fuxi Ke* was no longer a simple repetition of what students learned (Zhang, 2012) but was a process that helped build connections between different types of knowledge that helps students gain new insights through reviewing.

Meanwhile, we can also find that the way Mr. Wan designed tasks was consistent with the ideas advocated by Mr. Zhang. Mr. Wan translated a right triangle, and rotated it in different positions to generate different complex figures, so that all mathematics tasks in his lesson were interconnected systematically. This strategy employed by Mr. Wan was similar to what Mr. Zhang used in *Fuxi Ke* of *geometry transformation* we described before.

Post-Lesson Discussion between Mr. Wan and Mr. Zhang

After watching the first lesson, there was a post-lesson debriefing. During the meeting, Mr. Zhang pointed out shortcomings found in the first lesson. He said that too much content was covered and the teacher went through the lesson in hurry in order to complete his lesson plan. Thus there were too few opportunities for students to actually think. On the other hand, Mr. Zhang stated that the activity of constructing patterns using triangle models was good because congruent triangles and geometric transformations are connected, and the manipulative activity engaged students and motivated them to explore. He did however feel that the openness of this activity could be widened a bit further. He suggested that the teacher post the patterns constructed by students on the blackboard and further encourage them to pose problems. The following excerpts demonstrated their conversations:

Episode 1

Mr. Zhang: Manipulation tools are very important. Papers used are too thin. Manipulating with the models was inconvenient. Change it with another. ... Also, the triangle should be a right triangle with a 30° so that it can be consistent with triangles included in mathematics tasks that followed. Specially, HL theorem can be used to prove that two right triangles are congruent.

Mr. Wan: In the lesson, we used the two congruent triangle models to construct patterns. Therefore, how can we elicit students to design some tasks to determine whether two triangles are congruent? Here, we just use the properties of congruent triangles in the lesson. But, determining congruence of triangle is also an important component of this unit. How to deal with this issue?

Mr. Zhang: OK! But, you will look at some new intersections, segments, and triangles when triangles are translated, reflected, or rotated. In these cases, we should determine triangle congruence firstly, and then prove that corresponding angeles and segments are equal.

Mr. Wan: Oh! Let me try it tomorrow.

In the episode, we can see that the suggestion given by Mr. Zhang was quite general. It was interesting to see how Mr. Wan adopted these ideas in his second teaching of the lesson.

Improvement of Student Problem Posing in the Second Lesson

At the beginning of the lesson, Mr. Wan asked students to use two pre-made congruent right triangle (with a 30° angle) models to construct patterns. Several students were invited to draw several typical figures of their patterns on the blackboard, which included ten more types of basic figures than the ones created in the first lesson. Subsequently, Mr. Wan invited students to pose problems based on these basic figures. In order to ensure that the lesson was coherent, he designated three figures to students and asked them to pose problems by adding or changing one or more of the conditions.

The first selected figure was composed two right triangles translating one to another (Figure 6). Mr. Wan asked students to give conditions so as to ensure $\triangle ABC \cong \triangle DEF$. Students provided different conditions as follows: (1) AD=CF, $\angle B=\angle E$, $\angle A=\angle EDF$; (2) $\angle B=90°$, BC=EF, AC=DF, $\angle BCA=\angle F$; (3) $\angle B=\angle E$, $\angle C=\angle F$, AB=DE; and (4) $\angle B=\angle E=90°$, BC=EF, AD=CF. In fact, when students figured out the conditions for ensuring that two triangles were congruent, they would have to use relevant concepts and theorems, resulting in reviewing relevant knowledge.

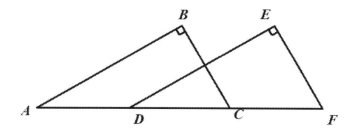

Figure 6. A translated figure in the second lesson by Mr. Wan

The second one is a symmetrical figure (Figure 7). The following episode shows the interactions surrounding the discussion of the figure.

Episode 2

T: In this figure, could you pose any problems? ... We know △ABC≅△ADC.
...
S1: Connect BD. BD is perpendicular to AC.
T: Why? We can connect AC, BD. They intersect at point O.
S1: Because △ABC≅△ADC, ∠BAC =∠DAC, AB = AD, and BD is perpendicular to AC.
T: Why?
S1: Isosceles triangle.
T: Very good. We also can prove it with congruent triangles. Any else...?
Ss: AC is a perpendicular to and bisects BD.
T: Ok! If we pick another point O at AC, what will you find? ...
S2: BO = DO.
T: Why?
S2: Because △ABO≅△ADO.
T: We all know this, all right? We have said. In fact, it is a symmetric figure. Folding, it will overlap. ... If we pick two points E and F on BC and DC, then there is...
Ss: AE = AF.
T: In order to AE = AF. What conditions have to be given?
S3: BE = DF.
T: Why?
S3: SAS.
T: Very good! ...

In the above episode, a student first conjectured that "BD is perpendicular to AC." Under the guidance of the teacher, students then made several conjectures. In fact, these conclusions were simple properties of symmetrical figures. Students made most of these conjectures under the guidance of the teacher; however they were able to eventually justify them. Most of their proofs needed to show that two

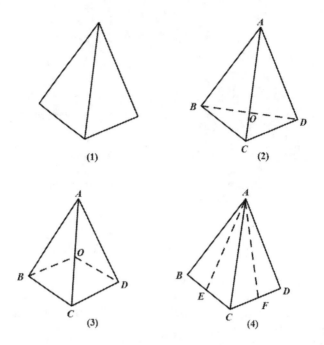

Figure 7. A symmetric figure in the second lesson by Mr. Wan

triangles were congruent in the figures, and then to find the corresponding sides that were equal using the properties of two congruent triangles.

The last one is a rotated figure (Figure 8). Mr. Wan and students discussed as the following episode.

Episode 3

    T: In the figure, please connect some segments. What do you find?
    S4:   Construct a line through point C. The line intersects AB and DE at F and G. There is CF = CG.
    T: Great! The segments are equal, why?
    Ss: The two triangles are congruent.
    T: (Mr. Wan marked the two triangles mentioned by students.) why are they congruent?
    S4:   Vertical angles.
    T: Vertical angles. (Mr. Wan marked them) ... others?
    S4:   ∠B =∠E.
    T: Ah. The two original triangles are congruent. Anymore?
    S4:   EC = BC.
    T: Very good! Do you understand? Are there any more findings?
    S5:   AE = BD.

T: If connect AE, BD, there AE = BD. What is the quadrilateral?
Ss: a rhombus.
T: Okay. Now, I make some changes, erasing BE, and then taking G and F on AD. What do you find? If you add some conditions, what conclusions can you make?
S6: If AF = DG, then EF = BG.

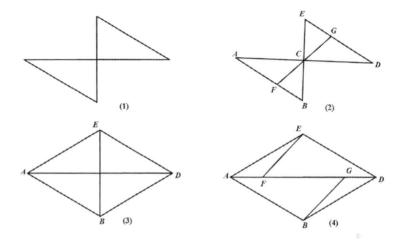

Figure 8. A rotated figure in the second lesson by Mr. Wan

As discussed above, students posed the first two problems independently while they posed the third one with the support of the teacher. It is worth recalling that in Episode 2, students proved their conclusions independently. It can be said that the teacher initiated students thinking and built a teacher-student interactive environment in the classroom so that students actively participated in classroom activities under the guidance of the teacher.

In the second lesson, we found that Mr. Wan employed problem posing as a strategy to build connection between congruent triangles and geometric transformations. In each figure generated through transformations, students added or changed some conditions to form different variations. In this way, students not only reviewed how to determine if two triangles are congruent, but also explored congruent triangles using these basic figures.

Developing a Model of Problem Posing-Based Fuxi Ke

After analyzing the interviews with Mr. Wan, we identified four phases of his journey in exploring how to use problem posing and problem solving in *Fuxi Ke*: (1) reflection and activating previous experience; (2) experiment under the supervision of Mr. Zhang and learning from others in the workstation; (3) further exploration and establishing a mode of teaching *Fuxi Ke*; and (4) transfer to teaching other types of lessons.

Reflection and activating previous experience. In the second lesson, problems posed by students played an important role for achieving the lesson goals. How did Mr. Wan consider this teaching approach in practice? Mr. Wan told us his story in an interview:

> Mr. Wan: Mr. Zhang said that the scattered pearls are strung to make a beautiful necklace. I think that his idea probably refers to build a problem context, which as a basis could be added or changed conditions by students. So that the problem is extended step by step, and put the scattered problems together…When I heard what he said about "main thread", I rethink its meaning. In fact, a *Fuxi Ke* should be carried out in this way. Students should generate important problems. Students should pose problems and solve them by themselves.

Mr. Wan connected the "main thread" idea with students' problem posing. He believed that teachers should create a problematic context throughout their lessons and then change the conditions for students to create problems of variation for their learning.

In fact, several years ago, Mr. Wan required students to design test papers for *Fuxi Ke* by themselves, or to select mathematics problems according to different classifications as stated in an interview.

> Mr. Wan: several years ago, I tried to let students to design a test paper. I found it was a good way. Later, when I was a volunteer teacher, I taught a lesson on the same topic [as the public lesson]. At that time, before the lesson, I assigned students homework to select mathematics problems and classify them according to different applications of congruent triangle. In the lesson, students discussed the problems they prepared in groups. To those difficult or valuable problems, each group can recommend them for whole class discussion.

That lesson was so impressive that he believed that this approach could not only initiate students' autonomy of learning, but also facilitate them to form a generalized method for solving a set of problems.

> Mr. Wan: Students prepare so well. The classroom environment is very lively. Every student was concerned with peers' questioning and demonstration. They became the masters of the classroom entirely. Once students' autonomy is initiated, their harvest is no longer how to solve one problem, or two problems, but methods for solving a group of problems.

Experiment with the supervision of Mr. Zhang and learning from others in the workstation. Based on our interviews, we know that Mr. Wan adopted student problem posing and problem solving in *Fuxi Ke* of *congruent triangles* several years ago. Why did he give up this approach at the first lesson? Mr. Wan shared his consideration as follows:

> Mr. Wan: Because in the public lesson activity, a young colleague would teach the same topic, and Mr. Zhang said teaching should embody the idea of same topic but different design structures so that this time I made an adjustment and change. ... I had this idea before, but very superficial. I suddenly gain insights when listening to Mr. Zhang's explanation.

Although Mr. Wan realized that students' problem posing in classroom teaching was very beneficial, he did not initially apply it in his classroom. However, after discussing with Mr. Zhang, Mr. Wan was aware of applying this strategy in *Fuxi Ke* purposefully. Building on his previous experience, Mr. Wan provided basic figures to students, and then initiated them to pose new problems by varying the conditions. He said: "This design makes the subject content more focused and teaching phases more coherent than before." During the meeting after the second lesson, Mr. Wu, another member of the MTW in Jintan district, made a positive comment.

> Mr. Wu: Mr. Wan made use of basic figures. Students changed conditions at the scene, then generated problems and solved problems. It facilitated students to think mathematically. To adopt his teaching approach, a teacher has to master some specific basic skills.

However, Mr. Wan also told us what he worried about.

> Mr. Wan: Because students get used to listening to teacher, they wanted to solve problems given by the teacher. Students may not be able to adapt to this teaching approach, and do not know how to express their thoughts. So we need to support them with patience. ... In addition, they cannot understand the difficulty of the problems they posed. Some problems may be too difficult, while others too easy.

Mr. Wan realized the necessity of giving guidance when students are asked to pose problems in his classrooms. However, if a teacher gives students too much guidance during their explorations, the autonomy of classroom learning will be discouraged. He also worried that the difficulty in carrying out problem posing in classroom may impact the smooth unfolding of a lesson. Mr. Lu, another member of MTW from Jintan, pointed out some related problems.

> Mr. Lu: You got a lot of basic figures with two congruent triangles' translation, reflection and rotation. This approach is very good, but in classroom discourses and explanations, simply repetition was

too much. For example, a symmetrical pattern, which is a simple operation, students must know that. It took too much time. Later, the time to solve other difficult problems was limited.

As Mr. Lu stated that student' problem posing with regard to basic figures is good practice, however, when the posed problems are easy or repeated, teacher should make appropriate choices. Some inappropriate problems should be ignored and enough time should be spent on mathematical worthwhile problems. This seemed to be a big challenge for Mr. Wan when students posed problems in his classroom. Mr. Wan discussed the following questions with Mr. Zhang: If a student poses a simple or a repetition problem, how does a teacher deal with? In order to ensure a lesson achieves the learning goals, how does a teacher guide students to generate variation problems ranging from the simple to the sophisticated? Mr. Zhang commented that although there are no simple answers to these questions, but the key is to make a balance between predetermined learning goals and students generated learning resources in classrooms.

Further exploration problems posing and establishing a model of teaching Fuxi Ke. Mr. Wan had consistently explored how to use student problem posing in his *Fuxi Ke*, and developed some effective strategies to make that balance. He had finally formed a style of problem posing-based teaching *Fuxi Ke*.

> Mr. Wan: Later, I taught a *Fuxi Ke* of *parallelogram* in a teaching competition in local district. I won the champion. I initiated students to try adding conditions on a quadrilateral to generate a parallelogram. Then let students to pick points and generate segments on the basis of the parallelogram so that the problem is extended.

We were interested in how Mr. Wan dealt with the relationship between learning goals and students generated learning resources in the class. Mr. Wan explained his approach using a lesson segment.

> Mr. Wan: I showed students a parallelogram ABCD (Figure 9), and initiated them to add conditions to pose problems. A student said that if picking up points E and F on the side of AB and CD respectively, when AE = CF, $\triangle ADF \cong \triangle CBE$ can be proved, as well as quadrilateral AECF is a parallelogram. I think this problem was relatively easy, so I did not give any explanation. I required students to pose problems further. Another student said that if AF and CE are bisectors of $\angle A$ and $\angle C$ respectively, the above conclusion is also true. I think this problem was similar to the previous one. I wanted to prove one of the conclusions that a generated quadrilateral is a parallelogram, because I was just a little worried. Later, the third student said that if picking up points on the diagonal to let AE = CF, then there are some conclusions, such as, $\triangle ADE \cong \triangle CBF$,

△ABE≅△CDF, and quadrilateral DEBF is a parallelogram. In this problem, the figure is more complex than the previous one, and different methods could be used to prove them. Therefore, I spent some time to discuss it in class. The fourth student selected a point on each of the four sides. He said if AE = CF and AH = CG, then there is a new parallelogram generated inside of ABCD. Based on the previous discussion, student should understand this problem easily, so there was no need to further prove. Through these problems, all knowledge points-how to determine a parallelogram are connected together. So that classroom teaching achieved the lesson goals.

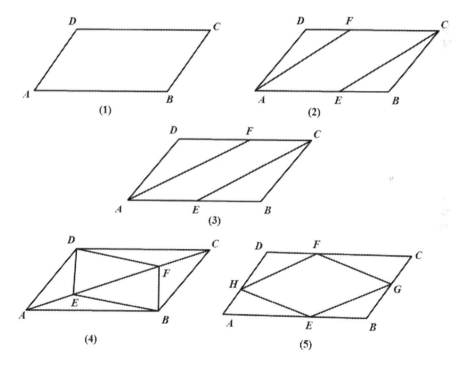

Figure 9. Problems posed by students in Mr. Wan classroom

Mr. Wan adjusted his strategy for using student problem posing in the classroom. He no longer gave excessive guidance for student problem posing, and broadened the space for student exploration. In his public lesson on parallelograms, how did Mr. Wan manage the time and moving rhythm? His strategy was adjusted such that the problems posed by students were deliberately selected to discuss further in class. Based on Mr. Wan, we can find that his decision-making depended on the following criteria: (1) difficulty of posed problems; (2) multiple solutions to the problem; and

(3) multiple knowledge content points covered. These considerations guided Mr. Wan to determine whether student-posed problems should be chosen to discuss in classroom. If a problem meets these criteria, he may further discuss in the class. If not, he may deal with it briefly, and move on. Therefore, Mr. Wan has explored the strategies of selection and use of student-posed problems to ensure the openness of the classroom and the encouragement of student engagement. This was to ensure that the class was moving forward smoothly and in accordance with the lesson objectives.

Transfer to teaching other types of lessons. Teaching is an art with endless potential to improve. Mr. Wan has never stopped exploring how to improve teaching although he has won several highly respected awards in teaching. He began to transfer the approach of student problem posing-based teaching *Fuxi Ke* to other types of lessons. When he taught a lesson of *Introduction of Quadratic Equations* for the teacher training school in a local district, Mr. Wan implemented this particular approach in his practice. After the lesson, he explained his approach to teachers as follow.

> Mr. Wan: I believe that there is a desire to become a discoverer, a researcher, or an explorer in every child heart. Children have a strong curiosity and thirst for the unknown world and knowledge. As long as teacher constructs a rich problem context, and gives appropriate guidance, students will be interested in the process of exploration, and pose good problems. Maybe, just in a lesson, knowledge gained by the students is not too much, but their other harvest is perhaps beyond our measure and imagination. Problem posing is more important than knowledge itself, and it has benefits in students' lifelong development. … Students learn how to learn independently is ultimate goal of education.

Mr. Wan believed that problem posing-based teaching approach created a learning space of self-exploration for students, and could improve their learning ability. The previous data analysis showed that he has developed a profound understanding of problem posing-based teaching.

In an interview, Mr. Wan explained that the implementation of problem posing and solving based teaching approach in other types of lessons is more challenging than in a review lesson, yet he will continue to explore effective strategies for implementation because this is a worthy adventure in his teaching career.

CONCLUSION AND DISCUSSION

In this chapter, we examined how an experienced teacher made continued improvements in his teaching with the support from a master teacher workstation (MTW), as well as a multiple-tiered teaching research system in general. It was found that the experienced teacher was inspired by the master teacher's guidance (the general teaching principles and concrete teaching strategies), and benefited from team members' comments. Building on his previous experience and the insight gained from the MTW, the experienced teacher developed an awareness of designing a problem posing-based teaching approach in *Fuxi Ke*. He has continuously experimented with the teaching approach, and had finally succeeded in finding some effective strategies for implementing this teaching approach in *Fuxi Ke*. Beyond that, the experienced teacher has started his new adventure in implementing problem-posing based teaching approach in other types of lessons. Mr. Wan's accomplishments from participating in MTW include developing an innovative teaching approach in such an old type of *Fuxi Ke*, and developing an awareness of implementing this teaching approach in other types of lessons. In addition, he has also published teaching reference materials based on his exploration in teaching *Fuxi Ke*, and has won several teaching competitions and awards.

As indicated by other studies, the teacher changed his belief concerning student-learning autonomy (Goldsmith et al., 2014; Opfer & Pedder, 2011) through a particular strategy of problem-posing-based teaching approach. The teacher benefited from the master teacher's supervision and comments from his peers (Bakkenes et al., 2010). The teacher improved his teaching practice through continued efforts in enactment and reflection on practice (Clarke & Hollingsworth, 2002; Goldsmith et al., 2014).

Within the specific professional development system in China, teachers have a great deal of opportunities to learn through collaboratively lesson planning, lesson observation and post-lesson reflection and improvement, known as Chinese lesson study (Huang et al., 2010; Li & Li, 2011). This study makes a unique contribution to understanding experienced teacher learning in such a context. Experienced teachers normally have a rich teaching experience and grasp basic teaching strategies and skills. A master teacher is required to demonstrate an effective teaching style and possess professional competence in researching and improving their practice (Li, Huang, Bao, & Fan, 2011). This study shows how the mechanism of a MTW may provide a pathway for experienced teachers to further their development towards a master teacher. The strong mathematical knowledge for teaching, rich teaching experiences, and experienced skills that teachers normally possess (Huang, 2014; Li, Huang, & Shin, 2008; Ma, 1999) lay a strong foundation for their further development. The master teacher' supervision and peer communications in a MTW provide insights and inspiration for the experienced teachers to reflect on their experiences and find new directions for their further development. As illustrated in this case, Mr. Wan was inspired when developing problem posing – based teaching approach in *Fuxi Ke* became his new direction to explore. Beyond the MTW, the school-based teaching research system provides a broad and supportive platform for teachers to experiment and explore the new teaching style. During this process, the experienced teachers could continuously learn from their practice and develop their professional competence.

Within the systematic teacher professional development in China, the master teacher workstation provides an additional vehicle, which is effective for helping experienced teachers to continuously develop their competence for becoming master teachers. The key ideas related to this mechanism are to provide a platform for experienced teachers to discuss their teaching with more knowledgeable persons and gain insights from the sharing. As emphasized by Wong (2004), reflection forms the core of Chinese way of "pedagogy." The major task of the master is "to arouse a disciple's reflection by generating a state of discomfort and perplexity" (p. 519). By using a metaphor of learning Chinese martial art, Wong used "entering the way" to describe how the master could help novice teachers to grasp basic knowledge and skills, and used "exiting the way" to illustrate how competent teachers become masters. Beyond the importance of repeated practices and reflection in promoting "exiting the way", Wong (2006, 2009) further argued the using deliberate variation in teaching and learning might help to bridge the basic skills to high-order thinking (Huang & Li, 2017). With regard to experienced teachers' learning, we would argue that repeatedly teaching a lesson as well as reflection on the lesson with immediate feedback from more knowledgeable experts is an effective way to promote teachers' learning toward masters (Ericsson, 2008). Within the master teacher workstation mechanism, the Chinese lesson study provides experienced teachers with this kind of learning opportunity (Huang, Fang, & Chen, 2017; Huang, Li, & Su, 2013).

ACKNOWLEDGEMENTS

We thank anonymous reviewers for their invaluable feedback on the reversions of the paper. We appreciate Ms. Kristin S. Hartland from Middle Tennessee State University for her helpful edits. Our thanks go to participating teachers and master teachers for their commitment to the study and support of data collection.

REFERENCES

Bakkenes, I., Vermunt, J. D., & Wubbels, T. (2010). Teacher learning in the context of educational innovations: Learning activities and learning outcomes of experienced teachers. *Learning and Instruction, 20*, 533–548.

Ball, D. L., & Even, R. (2009). Strengthening practice in and research on the professional education and development of teachers of mathematics: Next steps. In R. Even & D. L. Ball (Eds.), *The professional education and development of teachers of mathematics: The 15th ICMI study* (pp. 255–260). New York, NY: Springer.

Bao, J., Huang, R., Yi, L., & Gu, L. (2003). Research on teaching with variation. *Mathematics Teaching, 2*, 2–6. [鲍建生，黄荣金，易凌峰，顾泠沅. 变式教学研究. 数学教学，2003(2), 2–6.]

Borko, H. (2004). Professional development and teacher learning: Mapping the terrain. *Educational Researcher, 33*(8), 3–15.

Chen, D., & Wang, W. (2013). Understanding of and practice of master teacher station. *Secondary Mathematics, 7*, 71–73. [陈德前，王文. 对名师工作室的认识和实践. 中学数学, 2013(7), 71–73.]

Clarke, D., & Hollingsworth, H. (2002). Elaborating a model of teacher professional growth. *Teaching and Teacher Education, 18*, 947–967.

Ericsson, K. A. (2008). Deliberate practice and acquisition of expert performance: A general overview. *Academic Emergency Medicine, 15*, 988–994.

Goldsmith, L. T., Doerr, H. M., & Lewis, C. C. (2014). Mathematics teachers' learning: A conceptual framework and synthesis of research. *Journal of Mathematics Teacher Education, 17*, 5–36.

Gu, L., & Wang, J. (2003a). Teachers' development through education action: The use of Keli as a means in the research of teacher education model (Part A). *Curriculum, Textbook & Pedagogy, 1,* 9–15. [顾泠沅，王洁. 教师在教育行动中成长 – 以课例为载体的教师教育模式研究(上).课程教材教法, 2003(1), 9–15.]

Gu, L., & Wang, J. (2003b). Teachers' development through education action—The use of Keli as a means in the research of teacher education model (Part B). *Curriculum, Textbook & Pedagogy, 2,* 14–19. [顾泠沅，王洁). 教师在教育行动中成长 – 以课例为载体的教师教育模式研究(下).课程教材教, 2003(2), 14–19.]

Han, X., & Paine, L. (2010). Teaching mathematics as deliberate practice through public lessons. *The Elementary School Journal, 110,* 519–541.

Huang, R. (2014). *Prospective mathematics teachers' knowledge of algebra: A comparative study in China and the Untied States of America.* Wiesbaden: Springer Spektrum.

Huang, R., & Bao, J. (2006). Towards a model for teacher professional development in China: Introducing Keli. *Journal of Mathematics Teacher Education, 9,* 279–298.

Huang, R., Fang, Y., & Chen, X. (2017). Chinese lesson study: An improvement science, a deliberate practice, and a research methodology. *International Journal for Lesson and Learning Studies, 6*(4), 270–282.

Huang, R., & Li, Y. (2017). *Teaching and learning mathematics through variations: Confucian heritage meets western theories.* Rotterdam: Sense Publishers.

Huang, R., Li, Y., & Su, H. (2013). Improving mathematics instruction through exemplary lesson development in China. In Y. Li & R. Huang (Eds.), *How Chinese teach mathematics and improve teaching* (pp. 186–203). New York, NY: Routledge.

Huang, R., Li, Y., Zhang, J., & Li, X. (2011). Developing teachers' expertise in teaching through exemplary lesson development and collaboration. *ZDM: The International Journal on Mathematics Education, 43*(6–7), 805–817.

Huang, R., Peng, S., Wang, L., & Li, Y. (2010). Secondary mathematics teacher professional development in China. In F. K. S. Leung & Y. Li (Eds.), *Reforms and issues in school mathematics in East Asia* (pp. 129–152). Rotterdam, The Netherlands: Sense Publishers.

Huang, R., Ye, L., & Prince, K. (2016). Professional development system and practices of mathematics teachers in Mainland China. In B. Kaur & K. O. Nam (Eds.), *Professional development of mathematics teachers: An Asian perspective* (pp. 17–32). New York, NY: Springer.

Lewis, C., Perry, R., & Hurd, J. (2009). Improving mathematics instruction through lesson study: A theoretical model and North American case. *Journal of Mathematics Teacher Education, 12,* 285–304.

Li, S., Huang, R., & Shin, Y. (2008). Mathematical discipline knowledge requirements for prospective secondary teachers from East Asian perspective. In P. Sullivan & T. Wood (Eds.), *Knowledge and beliefs in mathematics teaching and teaching development* (pp. 63–86). Rotterdam, The Netherlands: Sense Publishers.

Li, Y., & Huang, R. (2008, June 22–27). *Developing mathematics teachers' expertise with apprenticeship practices and professional promotion system as contexts.* Paper presented at US – Sino Workshop on Mathematics and Science Education: Common Priorities that Promote Collaborative Research, Murfreesboro, TN.

Li, Y., & Li, J. (2009). Mathematics classroom instruction excellence through the platform of teaching contests. *ZDM: International Journal on Mathematics Education, 41,* 263–277.

Li, Y., Huang, R., Bao, J., & Fan, Y. (2011). Facilitating mathematics teachers' professional development through ranking and promotion practices in the Chinese Mainland. In N. Bednarz, D. Fiorentini, & R. Huang (Eds.), *International approaches to professional development of mathematics teachers* (pp. 72–87). Ottawa: Ottawa University Press.

Li, Y., Tang, C., & Gong, Z. (2011). Improving teacher expertise through master teacher workstations: A case study. *ZDM: International Journal on Mathematics Education, 43,* 763–776.

Liu, C. (2010). Interpretation of and reflection on master teacher station. *Jiangsu Education Research, 10,* 4–7. [刘穿石（2010）名师工作室的解读与理性反思，江苏教育研究, 10, 4–7.]

Ma, L. (1999). *Knowing and teaching elementary mathematics: Teachers' understanding of fundamental mathematics in China and the United States*. Mahwah, NJ: Lawrence Erlbaum Associates.

MOE. (2011). *Mathematics curriculum standards*. Beijing: Beijing Normal University Publishing House. [教育部. 义务教育数学课程标准, 北京: 北京师范大学出版社, 2011.]

Opfer, V. D., & Pedder, D. (2011). Conceptualizing teacher professional learning. *Review of Educational Research, 81*(3), 376–407.

Quan, L. (2009). Teacher professional development in master teacher station: A perspective in professional community. *Contemporary Educational Science, 13*, 31–34. [全力.名师工作室环境中的教师专业成长 – 一种专业共同体的视角, 当代教育科学, 2009(13), 31–34.]

Shao, H., & Zhu, L. (2013). The mentor who we seek in these years: Training experience in Zhu Leping primary mathematics master teacher station. *Henan Education, 3*, 41–42. [邵虹, 朱蕾. 这些年我们追求的导师 – 记朱乐平小学数学名师工作室培训历程. 河南教, 2013(3), 41–42.]

Wang, F. (2012). Effective strategies of master teacher station: A case of master teacher stations in Zhongshan District, Dalian. *Liaoning Education, 7*, 14–16. [王芳.名师工作室的有效策略 – 以大连市中山区名师工作室为例. 辽宁教育, 2012(7), 14–16.]

Wang, J., & Paine, L. (2003). Learning to teach with mandated curriculum and public examination of teaching as contexts. *Teaching and Teacher Education, 19*, 75–94.

Wong, N. Y. (2004). The CHC learner's phenomenon: its implications on mathematics education. In L. Fan, N. Y. Wong, J. Cai, & S. Li (Eds.), *How Chinese learn mathematics: Perspectives from insiders* (pp. 503–534). Singapore: World Scientific.

Wong, N. Y. (2006). From "entering the way" to "exiting the way": In search of a bridge to span "basic skills" and "process abilities." In F. K. S. Leung, G.-D. Graf, & F. J. Lopez-Real (Eds.), *Mathematics education in different cultural traditions: The 13th ICMI study* (pp. 111–128). New York, NY: Springer.

Wong, N. Y. (2007). Confucian heritage culture learner's phenomenon: From "exploring the middle zone" to "constructing a bridge." *ZDM: International Journal on Mathematics Education, 7*(2), 363–382.

Yang, S. (2010). Sing climbing mountain: A review of master teacher stations in Nantong city, *Jiangsu Education Research, 10*, 13–16. [杨曙明. 如歌行板唱攀登 – 南通市名师工作室透视. 江苏教育研究, 2010(10), 3–16.]

Yang, Y., & Ricks, T. E. (2013). Chinese lesson study: Developing classroom instruction through collaborations in school-based teaching research group activities. In Y. Li & R. Huang (Eds.), *How Chinese teach mathematics and improve teaching* (pp. 51–65). New York, NY: Routledge.

Zhang, B. (2013). Discussions on strategies of constructing second mathematics master teacher station. *The Monthly Journal of High School Mathematics, 8*, 61–63. [张必华.中学数学名师工作室建设措施的探讨. 中学数学月, 2013(8), 61–63.]

Zhang, H. (2012). Master teacher station: Problems and solutions. *Jiangsu Education, 3*, 52–53. [张华.名师工作室：困境与出路. 江苏教育, 2012(3), 52–53.]

Zhang, X. (2009). Good teaching plan and practice for effectiveness. *Jiangxi Education, 1*, 50–51. [章晓东.在精心预设和驾驭生成中追求实效. 江西教育, 2009(1), 50–51.]

Zhang, X., & Xu, X. (2012). Geometry transformation: A case of rotation of double squares. *Middle School Mathematics Teaching Materials, 1*, 110–113. [章晓东,徐新. 图形的变换 -以"双正方形"的旋转为例.中学数学教学参考, 2012(1),10–113.]

Xingfeng Huang
College of Education
Shanghai Normal University
China

Rongjin Huang
Department of Mathematical Sciences
Middle Tennessee State University
USA

YUDONG YANG AND BO ZHANG

11. IN-SERVICE MATHEMATICS TEACHERS' PROFESSIONAL LEARNING IN TEACHING RESEARCH GROUP

A Case Study from China

CHINESE TRG AND ITS PRACTICAL ORIENTATION

Although Chinese mathematics teachers do not exhibit the quantity of formal higher education that their Western or Japanese counterparts do, some studies have suggested that Chinese mathematics teachers have more profound understanding of fundamental mathematics, have better PCK of mathematics, and use that knowledge more coherently during instruction (An et al., 2004; Li & Huang, 2008; Ma, 1999). Despite a lack of formal training, one possible reason for the strength of Chinese mathematics teachers' practice may be their involvement in various teaching research activities conducted by their school-based teaching research network. In TRG activities, Chinese teacher usually discuss how to improve classroom instruction. The emergence of knowledge is quite similar to PCK, which was explained by Shulman (1987). That is, when a teacher copes with a special topic, he\she will organize, adjust and present subject matter knowledge to do task design by considering a special group of students' interest and capacity. Though some studies have been done on what the PCK is (Leinhardt, 1989; Wineburg, 1991; Lampert, 1990; Ball, 1993; Grossman, 1995; et al), even Carter (1990) and Gudmundsdottir (1991) compared the different understandings on PCK in different studies. PCK is looked at as a kind of practical knowledge which is situated in a special topic of a subject. Unique PCK makes a subject teacher to be a teacher without requiring him/her to be a subject expert or an education expert. It is "different from content knowledge and knowledge of general pedagogy; rather it is consisting of representations of subject matter, student conceptions, and understanding of specific learning difficulties" (Appleton, 2003).

Chinese educators have engaged in Teaching Research Group (TRG) activities for several decades since TRG was set up in each school by administrative power (MOE, 1952, 1957). TRG activities are similar to a form of lesson study, but less well-known than Japanese lesson study (Stigler & Hiebert, 1999). Like the West, Chinese schools are organized by age into primary, junior high, and senior high schools. All grade levels study the same three core subjects: the Chinese and English languages and mathematics. Unlike the West, Chinese students form class cohorts that stay together in the same classroom throughout the day, visited by their various teachers. Because most of the Chinese teachers who teach Chinese, English, or mathematics only teach one subject two or three times a day, these core-subject teachers are easily organized into subject-specific TRGs. City-level and province-level teaching research officers (TRO) are responsible for guiding the subject-specific teaching and research activities in their domain. This multi-tiered teaching research system is a network where province-level TROs oversee city-level TROs, which oversee school-level TRGs (Yang, 2009b). The TRG is the basic unit in this network; its main responsibility is conducting research on teaching to solve the practical problems facing teachers.

A TRG provides the learning opportunities for the front line teachers. Usually there are experienced teachers and green hands in a TRG. Although they have diverse understanding about mathematics learning and teaching, and experienced teachers are generally superior to the green hands in terms of practice knowledge, every teacher has an equal right to decide how to organize his/her own classes. That means the opinions in TRG activities are not compulsory for the implementer of a lesson, no matter whether he/she is a novice or an experienced teacher. Each one does not need to follow others' opinions or imitate what experienced teachers do, but should rely on his/her own knowledge and judgment. So, our hypothesis in the paper is that, once a teacher changed his/her behaviors in a lesson or showed his/her different understanding in post-lesson TRG activities, there will be a significant signal that he/she has acquired or understood something where learning or knowledge exists.

Over the last 60 years, Chinese teachers have developed a very focused framework to think about lesson preparation, observation, and post-lesson reflection called *Three Points*. These three points are: (1) the lesson's key point, (2) the lesson's difficult point, and (3) and the lesson's critical point (Yang, 2009b). Sometimes the Three Points are not explicitly written into the lesson plan, but help to frame TRG discussion during the lesson's construction, influencing the subsequent post-lesson discussion. Current Western educational scholars often refer to Shulman's (1986, 1987) PCK to describe the peculiarities that a subject-specific teacher must master to effectively instruct students about a particular content topic. It is the knowledge that teachers need to transform their own content understanding into instruction that helps students understand. Educational scholars have struggled to describe, define, and study pedagogical content knowledge. In effect, the Chinese teachers are using a practical way to think about various sub-domains of PCK with the Three Points during their lesson-improving process. In the paper, we will show a case of TRG activities to explore how this kind of frame was used and what the teachers have learned.

METHOD

We chose the case study method to investigate how teachers' participation in TRG activities help them to acquire practical knowledge or PCK during the process of improving classroom instruction. The case study allows in-depth analysis of teacher collaboration and aids theory development. We decided to conduct a single case study of a specific Teaching Research Group in Hangzhou, capital city of Zhejiang Province, China. We chose a Hangzhou TRG as our research case for several reasons: (1) The researcher as a supportive expert was involved in a training program and it is easy to collect the data needed, (2) The mathematics TRG from an average school (not a key or model school) in the region of Hangzhou, has some representativeness to major urban schools in east China area, (3) There are 10 primary mathematics teachers in the TRG and the age structure of this TRG is typical, which means the distribution of teachers is almost equal in different teaching experience level: three of the teachers had less than five years' teaching experiences, three teachers had roughly ten years' teaching experiences, three teachers had over fifteen but no more than twenty years' experiences and one teacher, the only master teacher in the TRG, has nearly thirty years' experiences.

In this study, we refer to these teachers by a pseudonym to indicate their level of teaching experience (T1 to T10), from least (T1) experienced to most (T10) experienced. The Hangzhou TRG activity was organized around teaching the content topic of Position to parallel classes of the sixth-grade students, two separate iterations. The least experienced teacher (T1) taught the first and second iteration of the lesson to each of her two mathematics classes. All the other TRG members were involved in classroom observation and post-lesson discussion.

We collected three main categories of data that reflected the three main stages of the Hangzhou TRG activity: (1) lesson preparation, (2) lesson implementation, and (3) post-lesson discussions. T1 created and recreated her lesson plans several times based on feedback. Therefore, we collected the finalized lesson plan as the first set of data. Second, we collected information from the lesson videos in each iteration. Third, we collected field notes during the post-lesson discussion to capture the chaotic nature of the rapid-fire and disparate perspectives as the ten teachers debated the Three Points and learning effect of each lesson implementation. Fourth, we collected information by individual interview after the TRG activities. Because the core of the TRG activity was the development of the lesson, we considered the two implementations (iterations) of the lesson plan enactments (created by T1) to be of paramount import in our analysis.

RESULTS: HOW LESSONS WERE DISCUSSED IN TRG ACTIVITIES

T1 worked collaboratively on the development of the lesson plan with the more experienced T8, who was nominated as her tutor in teaching for the first three years. In TRG of mainland China, almost every novice will have a tutor who will assist or guide his/her ordinary teaching skills from the time he/she enters into a school as a teacher. This model reflects the popular Western approach for instructing students of "Think, Pair, Share" pedagogy because T1 first independently thought about the Position lesson and created the lesson plan alone and then paired with her tutor (T8) to further develop the other iteration amidst the shared feedback from the other members of the TRG. The content Position in Primary 6 is mainly about using a pair of numbers to represent the object's position, forming the foundation of learning coordinates in junior high school mathematics. The TRG used the textbooks from the People's Publishing House of China.

THE ITERATION

The First Lesson

T1 designed worksheets for the lesson based on the textbook content, and began the lesson with such a task:

"In this lesson, we will learn how to represent a position. If I told you that Zhang Liang was seated in the second column and the third row, could you know where he was? Please use the worksheet and point out Zhang Liang's seat" (as shown in Figure 1).

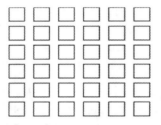

Figure 1. Seats table used by pupils on the worksheet

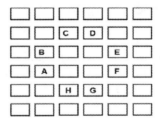

Figure 2. Zhang Liang's possible seats on the worksheet

After students' individual work, they found that the seat of Zhang Liang was marked in the worksheet differently. Then T1 represented all possible eight seats for Zhang Liang (Figure 2) and told students that these correspond with different ways to count rows or columns. Then the teacher emphasized the importance of clarifying the rule to represent a position and introduced how the row or column was defined here.

By defining the order of rows (from down to up as the 1st, 2nd…) and columns (from left to right as the 1st, 2nd…), T1 asked students how B/C/D/E/F would be represented if she used brackets and two numbers to represent position A as (2,3)? After students correctly represented B, …, H as (2,4), …, (4,2), T1 summarized that a paired number, like (2,3), could be used to represent a stable position.

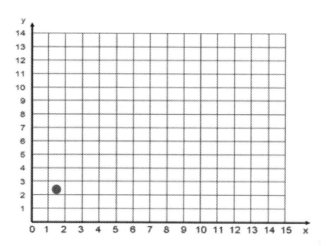

Figure 3. Students' representation of Zhang Liang's position

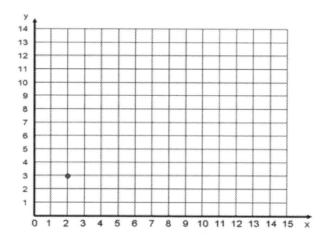

Figure 4. Teacher's correction in grid paper

Then T1 asked pupils to do another task.

T1: *If I know Zhang Liang's position can be represented as (2,3), could you marked where his seat is in the coordinates grid-paper?*
Ss: [some pupils raised their hands] *I know, I know!*
T1: *Who would like to point out Zhang Liang's position for me?* [She gave a gesture to invite one pupil to come up to the front of the classroom]
S: *It is here.* [He placed his finger on the inside of a rectangle in Figure 3]
T1: *Is this correct? Here in the grid, the vertical lines represent columns and horizontal lines represent rows, so where is the Zhang Liang's position? Who can point it out?*

By further explanation, the students eventually understood that the position of Zhang Liang should be on the crossover point of the 2nd vertical line and the 3rd horizontal line in Figure 4.

Later, T1 linked a classroom seats map with the coordinate grid-paper, asking students to apply paired numbers to represent position step by step: (1) T1 asked all the students whose first number is 5 in a paired number to stand up and tell others their paired numbers to represent the corresponding positions; (2) T1 drew the corresponding dots in the coordinate grid-paper on the blackboard, and asked students to look for the common characteristic of these dots on the coordinated grid paper; (3) T1 asked students to imagine a paired number (a,a), where the first number is the same as the second number, to represent some positions in the grid-paper and asked them to think about the characteristic of these dots if drew them as a line.

In the last section of the lesson, T1 introduced the history about how Decare invented paired numbers in coordinates system to represent positions and some living examples of using paired numbers to represent positions in daily life.

The lesson lasted about 44 minutes.

The First Post-Lesson Discussion

T9, as the leader of mathematics TRG, held the post-lesson discussion. He firstly invited T1 to state her design intention.

T1 said, *"This is my first time to teach the lesson Position. The teaching goal is to make students understand the meaning of using a paired number to represent a position, and to apply a paired number to represent position in grid-paper... The key point is to make students master the way to represent positions by an ordered paired number... The difficult point is to culture their spatial sense... From the result of students' learning during the lesson implementation, I think they have learned to use paired numbers and pay attention to the order of the numbers in a paired number... and I didn't expected the lesson to last so long and I had thought I could finish the teaching tasks in the normal lesson time* (35 minutes) ..."

Then T9 invited the other teachers to talk freely on the lesson and give some suggestions to T1 for her next iteration lesson in another parallel class.

T5, who has nine years' experience, positively stated that the strength of the lesson was building the connection between the grid-paper with students' seats map in the classroom, which gave pupils real experience from positions to paired numbers. But she pointed out there was difficulty for pupils to represent the seat as a crossover point because the seat and desk position was more like a rectangle while the paired number was a dot in the grid paper, which was the reason why a student represent Zhang Liang's seat as the area of a rectangle (as shown in Figure 3).

T3, T7 and T8 also talked about the difficulty for students from a rectangle area to a crossover point. T8 pointed out: "*The difficult point of the lesson, is not culturing a spatial sense as T1 mentioned, but the converting process from real life positions in a plane to algebraic paired numbers, which is abstract for students...*"

T2, who was teaching Primary 1 mathematics, shared her thought: "*I taught the lesson Position in grade one several weeks ago. In my lesson I just taught some concepts, like front and back, left and right, up and down... Pupils only need to understand that position is relative, that's my key point of the lesson. So what is the key point of the lesson Position in Primary 6?*" T4 was teaching the Primary 4 mathematics and put forward her similar questions, "*...in primary 4, pupils will learn east, south, west and north in the lesson of Position, the key point is that making the students understand at least two elements, that are direction and distance, could accurately represent a position in a plane... So what is the key point of the lesson Position in Primary 6? I think the students in Primary 6 who have understood two numbers could accurately represent a position in a plane, so the key point of the lesson should be focused on the order of the two numbers, which means the rule of paired numbers...*"

T10, who has nearly 30 years of experiences in primary mathematics, was also concerned about the coherency of mathematics content. She mentioned that, "*...the lesson of Position was taught spirally in grade 1, 4 and 6 in primary mathematics. In different grade, there is a different focus...And we need be aware that the Position lesson in primary stage is a basis for students' learning of Coordinates System in grade 8, in junior high school. (In junior high school) students will be asked to create geometric figures in the coordinates system or write out coordinates value of corresponding points in the Decare coordinates system. So in primary 6, we just asked students to represent concrete position as a point in grid paper and write down a paired order-number... the key point of the lesson in primary 6 is that applying paired numbers to represent an object's location in grid paper, and the difficult point for students is the obstacle of experiencing abstract process from a pictorial position to a dot in grid paper and then finding the corresponding paired order-numbers. In my opinion, the critical linkage of the lesson, is to help students transform the position of an object to the area of grid paper and then to a dot (crossover point) on grid paper where the label order of columns and rows are listed out...*"

The discussion on the lesson lasted about one hour and T9 summarized all above opinions on the lesson and generalized that mathematics is a language; that is, the teacher should help students to experience the process from natural life language of describing a position to mathematical symbol language of representing a position.

THE SECOND ITERATION

The Second Lesson

Based on post-lesson discussion, T1 redesigned her lesson plan with her guider T8, implementing the following main steps in another parallel class.

Firstly, T1 asked some students to introduce themselves where they seated and she presented a seats table by black dots arranged in rows and columns as Figure 5. The students found they had different ways to tell others where their seats were. So the teacher inspired the discussion on the importance of the common rules if they could accurately express positions and make others quickly know where their seats were.

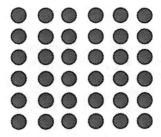

Figure 5. Seats table used by T1

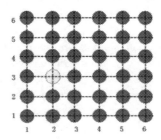

Figure 6. Zhang Liang's seats on the seats map

Secondly, when students recognized they must have common rules to express the location of seats, T1 suggested to tell others column number firstly and then row number. Naturally, T1 labeled the numbers on the seats map and use some dotted lines to allow students to easily count the columns and rows (Figure 6). When students felt it was a quick and easy way to represent their seats, T1 explained that paired numbers were used to represent positions in mathematics and the order rule was very important when a paired number was used. Then T1 asked, "*if I told you Zhang Liang's paired number was (2,3), where he was seated in?*" Students easily pointed out the corresponding position (the yellow dot) in the seats map (Figure 6).

Thirdly, T1 introduced the story on how Decare got the idea by observation of cobwebs to create coordinates system. She then transformed the seat maps into grid paper with coordinates, which was same as Figure 4 in the last lesson.

Fourthly, T1 asked students to find out the crossover points in grid paper which shared one number (and position) with (3,4). When Students found all the positions whose paired number was (3,...) or (...,4), T1 generalized them as (3,Y) and (X,4) and students found all the positions of (3,Y) or (X,4) were in a same line by observing the corresponding position in the grid paper.

Fifthly, T1 asked students to play a game called Closing Eyes and Imagining. She separately gave some groups paired numbers and asked them to close their eyes and imagine the corresponding positions in grid paper. She then asked them to raise one hands to show the possible figures in grid paper.

Lastly, T1 summarized the lesson and concluded that, a position in a plane could be represented as a paired number and a paired number could be converted into a crossover point on grid paper when the rules of representing was built.

The lesson lasted about 42 minutes.

The Second Post-Lesson Discussion

At the beginning of post-lesson discussion, T1 was invited to share her thoughts. She expressed her appreciation to T8 who assisted her new lesson design. She said, "*By T8's help, I used small black dots to represent the classroom seats map... I think this is a good way to help pupils overcome the obstacle and naturally understand a position could be simplified as a crossover point in mathematics... I emphasized the importance of a common rule to represent a position as a paired number before learning paired numbers in the lesson and the corresponding relationship between a position and a paired number was addressed, too. I think students have learned how to use paired numbers to represent positions in grid paper during their excises.*"

T2 and T3 thought the second lesson was more related to students' real life. T3 said:" *...in the former lesson, T1 just told students Zhang Liang's position and directly asked students to mark it on their worksheet. But in this lesson, she firstly asked students to introduce their own seats and which maked students feel they have different ways... when students confused about the chaos and then the importance of coincident rule came out... In real life, a position have a size but in mathematics it is only a point without caring about its size. I think the difficult point was overcome* *ingeniously...*" T6 further pointed out that column or row in students' real life had size too, but in mathematics it was a line without width. He then praised T1 for the dotted line used in Figure 6, stating that it was a very smart teaching strategy to help students to abstract mathematical objects from real life.

T8 shared her thoughts when she redesigned the lesson with T1: "*The key point of the lesson is applying a paired number to represent a position on grid paper. When we addressed the coincident rule to represent a position, not only the order of columns or rows, from left to right or from down to up, but also the order of a paired number, that is column number firstly and row number secondly, was emphasized. So applying a paired number to represent a position means to do it based on the rules in mathematics...*"

T7 commented positively on the game, Closing Eyes and Imagining. He compared the exercises used in the former lesson with the game in this lesson. He said, "*In the first lesson, the exercises mainly focused on representing positions as paired numbers... but in this lesson, the game gave out groups of paired numbers to make students imagine the corresponding positions. In addition to being the inverse of applying the knowledge, this game disciplined students' spatial ability...*" At the same time, he suggested to add some exercises on drawing figures according to groups of paired numbers on grid paper before the game, like a rectangle. In the process of drawing a rectangle, he thought students could profoundly understand that positions on the same line share one number in their paired numbers.

T10, the master teacher, from the angle of students' learning effect to express her opinions. She suggested that T1 should compare students' homework between the two classes and she believed that students in the second lesson would show better understanding than students in the first lesson. She addressed that, "*...in grade 1, when students learned the concepts of front and back, or left and right, or up and down, they grasped how to describe a relative position in a lineal situation. In grade 4, they began to understand at least two elements could represent a relative position in a plane, like an angle and distance, but the relativity of position in a planar situation was built on the position of somebody himself/herself... Only in grade 6, the relativity of a position was built on a common rule, the order of columns or rows, which was not depended on the position of anyone himself/herself. This means the common rule has been built for students to study the Positions of the planar world... pupils really get rid of the way to reference the self-centered position. This paves the way for Coordinates System to be learned formally in high school...*" T10 commented that the lesson Position of grade 6 had built stable basis for students' learning coordinates system in the future by reinforcing the importance of rule and order of a paired number.

Finally, T9 summarized the discussion and suggested that T1 write a teaching reflection note and put the one lesson of Position in grade 6 into the whole perspective of knowledge system within the compulsory education phase.

DISCUSSION: WHAT HAS BEEN DEVELOPED IN TRG ACTIVITIES

Though the dialog in TRG activities was excerpted based on the author's intention, we could find that what they talked about in the post-lesson meeting is closely related to PCK and could be described as practical knowledge. We will discuss what has been learned in TRG activities in this section. Here, a basic hypothesis should be recognized that behavior changes imply teaching or situated knowledge has been acquired and improved.

From Grossman's PCK Angle

Shulman (1986) described PCK, or pedagogical content knowledge, as a "specialized knowledge" that teachers have that separates them from the content expert. This kind of knowledge will not exist without any specific teaching topic as the supported background. Some researchers have further explored the nature of PCK. Grossman (1990) elaborated on PCK further by describing it as a "conceptual map for instructional decision-making, serving as a basis for judgments about textbooks, classroom objective, assignments, and evaluation of students" (p. 86). In explaining this notion of a conceptual map, she arranged and merged original categories of Shulman's PCK model. In the case of science subject teachers, she stated that there were four components contributing to PCK development: (a) knowledge and beliefs about purpose, (b) knowledge of students' conceptions, (c) curricular knowledge, and (d) knowledge of instructional strategies. If we review the teachers' discussion from the angle of Grossman's statement, we could find that teachers' PCK was developed in TRG activities.

Teacher's knowledge and beliefs about purpose. This is related to the teacher's main idea on mathematics teaching, which means teacher's knowledge about mathematics, the teaching goal of the lesson, and what is important for students to learn. T1, in her first post-lesson reflection, began her talk referring to "the teaching goal" and "the key point." T2 and T4, by reflection their own lessons of Position in Primary 1 and 4, aroused the question of what was most important for students' learning in Primary 6. T10 gave her thoughts on the question and stated her understanding of the "key point" of the lesson. T9, in his summary of the first post-lesson discussion, mentioned mathematics as a language, which is typically related to the belief on the nature of mathematics.

Teachers' knowledge of students' conceptions. This related to students' experience or pre-concepts, what is easier for students to learn based on their former learning or daily experience, and what is difficult for them; that is to say, the experience or pre-concepts will obstruct their learning of the new topic. From the first post-lesson discussion, we can see teachers have different understanding of students' experience or pre-concepts. Here, T1, T3, T7, T8, and T10 prefer using "difficult point" to name it. In the second post-lesson, they repeatedly talked about the "difficult point."

Teacher's curricular knowledge. This related to the knowledge on textbooks or other resources, including the teacher's understanding on the lesson topic; how it was connected with other topics in the subject matter. The typical statement could be found from T10's talking. In her two paragraphs of speech we could see her overview of the Position lesson from the primary mathematics to junior high school mathematics and she had a consistent knowing of the Position knowledge system. T1 and T4 also noticed the related content from textbooks and contributed their thoughts. T9 suggested that T1 should write a reflection note from "the whole perspective of knowledge system."

Teacher's knowledge of instructional strategies. This is hard to be discussed by teachers, because it is a kind of tacit and practical knowledge. But from the talking of T1 and T8 in the second post-lesson sharing, we could find why they designed small black dots to represent students' seats map and why the dotted-line was used when columns and rows were counted by students. Additionally, in T7's comments and suggestion on the game and exercises, he showed his knowledge of instructional strategies.

From the above discussion on what kind of knowledge was developed or learned in TRG activities from Grossman's PCK angle, it is easy to see how PCK was shared and how the lesson was improved by these TRG activities. The four components of PCK developed by Grossman (1990) better explained what kind of knowledge was easily acquired based on lesson studies in TRG activities. Of course, the *Three Points* could not be simply corresponded to Grossman's (1990) elaborated work on PCK. Sometimes they are intercrossed, but by the above analysis we tried to show two things: there exists a linkage between PCK and Three Points and what kind of knowledge was acquired in TRG activities.

From Grossman's Angle to Chinese Teacher's Three Points

In the two post-lesson discussions, we could find three terms mentioned by teachers frequently, called *Three Points,* which was a very focused framework to think about lesson preparation, observation, and reflection during the past 60 years in TRG activities. These three points are: (1) the lesson's key point, (2) the lesson's difficult point, and (3) and the lesson's critical point (Yang & Ricks, 2012).

The lesson's key point refers to the central objective of the lesson for which the lesson is constructed. The success of the lesson depends on how well students learn this core mathematical content topic. The key point describes the emphasis that the teacher must put on the topic and the essentials that the students must grasp. The key point is also a reflection upon the subject matter. Some in the US have described this as the Big Mathematical Idea. T1 and T10, in the first post-lesson discussion, explicitly talked about the key point of the lesson; and T8 explicitly expressed her thinking on the key point in the second post-lesson discussion.

The difficult point is the cognitive difficulty that the students might encounter as they try to learn the mathematical key point. If the key point is the goal of the lesson, the difficult point is the metaphorical mental stumbling block. Being able to clearly state and anticipate this difficult point helps Chinese teachers to plan lessons that go beyond just attempting to transmit content knowledge to students in a univocal fashion, but helps to proactively mold instruction to maximize student erudition. T1, T3, T5, T7, T8 and T10 had used the term difficult point in their talking during the first post-lesson discussion.

The critical point is the heart of the lesson that shapes the teaching methods. If the key point is what the lesson is about, the critical point emphasizes how to reach that objective. The critical point is the teacher's consideration of how to help students navigate the mathematical terrain, to eventually reach the instructional objectives while avoiding or overcoming the pitfalls that might arise (the difficult point). Though teachers in the two post-lesson discussions seldom directly used the term, their comments on strategy or methods used in lessons hinted such a perspective. This can be seen in the second post-lesson discussion comments by T2, T3 and T7. T10 talked about "critical linkage" in the first post-lesson discussion; this was a more clear expression of critical point.

If we have chances to look at teachers' lesson plans in China, we might find the key point and difficult point were frequently and explicitly written in their lesson plans. The critical point was not so clearly written in lesson plans but it often emerged in a spoken way in different names, like critical linkage, critical part, critical knot or sometimes critical point itself. The Three Points actually helped teachers to frame TRG discussion during the lesson's construction. They also provide a frame to help teachers to accumulate mathematics PCK in TRG activities.

We could find the linkage between Grossman's PCK angle and the Three Points. A teacher's knowledge and beliefs about purpose, and curricular knowledge are closely related to the key point of a lesson. A teacher's knowledge of students' conceptions directly corresponds with the difficult point of a lesson. Finally, a teacher's knowledge of instructional strategies is mainly about the critical point of a lesson. Chinese in-service teachers gradually learn to analyze teaching content and processes from the perspective of the Three Points in TRG activities. Their ability to approach lesson planning, implementation, and reflection through this structured manner is an important instrument for increasing professional knowledge.

CONCLUSION

By the case study of Hangzhou TRG activities, we wanted to show how Chinese mathematics teachers improve their lessons in TRG and what kind of knowledge they could learn in TRG. This case study may provide an understanding that teacher's PCK could be easily acquired in practice-orientation professional activities though PCK could also be learned in theoretical training. In the study by Park and Oliver (2008), they indicated that PCK was developed through reflection-in-action and reflection-on-action within given instructional contexts, and teacher efficacy emerged as an affective affiliate of PCK.

From Grossman's PCK angle and the Chinese *Three Points* angle, PCK is a specialized knowledge which makes a teacher to be a teacher. But this kind of knowledge can not be separated from specific context, such as specific topic of mathematics and specific students. To make teachers quickly accumulate mathematics PCK, practice-orientation professional activities must be addressed, and the community of teacher's professional learning should be constructed, like Chinese TRG or Japanese professor-teachers group.

We hope as mathematics education researchers that this article can shed light on the finer details inherent in TRG activities. We hope that making this kind of tacit knowledge more recognizable will help more teachers improve their lesson analysis skills. We anticipate further research in this area in order to test hypotheses and also inductively study excellent practice.

REFERENCES

An, S., Kulm, G., & Wu, Z. (2004). The pedagogical content knowledge of middle school mathematics teachers in China and the US. *Journal of Mathematics Teacher Education, 7*, 145–172.

Appleton, K. (2003). How do beginning primary school teachers cope with science? Toward an understanding of science teaching practice. *Research in Science Education, 33*, 1–25.

Argyris, C., & Schon, D. A. (1974). *Theory in practice: Increasing professional effectiveness*. San Francisco, CA: Jossey-Bass.

Ball, D. L. (1993). With an eye on the mathematical horizon: Dilemmas of teaching elementary school mathematical. *Elementary School Journal, 93*(4), 373–397.

Carter, K. (1990). Teachers' knowledge and learning to teach. In W. R. Houston, M. Haberman, & J. Kikula (Eds.), *Handbook of research on teacher education* (pp. 192–310). New York, NY: Macmillan.

Grossman, P. L. (1990). *The making of a teacher: Teacher knowledge and teacher education*. New York, NY: Teachers College Press.

Grossman, P. L. (1991). Overcoming the apprenticeship of observation in teacher education coursework. *Teaching and Teacher Education, 7*, 345–357.

Gudmundsdottir, S. (1991). *The narrative nature of pedagogical content knowledge*. Paper presented at the annual meeting of the American Educational Research Association, Chicago, IL.

Kagan, D. M. (1990). Ways of evaluation teacher cognition: Inferences concerning the Goldilocks principle. *Review of Educational Research, 60*, 419–469.

Lampert, M. (1990). Connecting inventions with conventions. In L. P. Steffe & T. Wood (Eds.), *Transforming children's mathematics education: International perspectives* (pp. 253–265). Hillsdale, NJ: Lawrence Erlbaum Associates.

Leinhardt, G. (1989). Math lessons: A contrast of novice and expert competence. *Journal for Research in Mathematics Education, 20*(1), 52–75.

Li, Y., & Huang, R. (2008), Chinese elementary mathematics teachers' knowledge in mathematics and pedagogy for teaching: The case of fraction division. *ZDM: Mathematics Education, 40*, 845–859.

Ma, L. (1999). *Knowing and teaching elementary mathematics*. Mahwah, NJ: Lawrence Erlbaum Associates.

MOE. (1952). *Zhongxue zanxing zhangcheng* (Secondary School Provisional Regulation). Chinese Governmental Document (in Chinese).

MOE. (1957). *Zhongxue jiaoyanzu tiaoli* (Caoan) (Secondary school teaching research group rulebook draft). Chinese Governmental Document (in Chinese).
Park, S., & Oliver, J. S. (2008). Revisiting the conceptualisation of Pedagogical Content Knowledge (PCK): PCK as a conceptual tool to understand teachers as professionals. *Research in Science Education, 38*(3), 261–284.
Polanyi, M. (1966). *The tacit dimension*. Chicago, IL: University of Chicago Press.
Shulman, L. (1986). Those who understand: Knowledge growth in teaching. *Educational Researcher, 15*(2), 4–14.
Shulman, L. (1987). Knowledge and teaching: Foundations of the new reform. *Harvard Educational Review, 57*(1), 1–22.
Stigler, J., & Hiebert, J. (1999). *The teaching gap*. New York, NY: The Free Press.
Wineburg, S. S. (1991). On the reading of historical texts: Notes on the breach between school and academy. *American Educational Research Journal, 28*, 495–519.
Yang, Y. (2009a). Capturing the critical incidents in teaching research (in Chinese). *Peoples' Education, 1*, 48–49.
Yang, Y. (2009b). How a Chinese teacher improved classroom teaching in Teaching research group. *ZDM: Mathematics Education, 41*(3), 279–296.
Yang, Y., & Ricks, T. (2012). How crucial incidents analysis support Chinese lesson study. *International Journal for Lesson and Learning Studies, 1*(1), 41–48.

Yudong Yang
Shanghai Academy of Educational Sciences
China

Bo Zhang
YangZhou University
China

SU LIANG

12. LEARNING FROM DEVELOPING AND OBSERVING PUBLIC LESSONS

INTRODUCTION

In the past thirty years, China has established a rigorous system of on-going professional development (PD) for grades K-12 teachers. Multi-dimensional PD activities have been utilized to help in-service teachers acquire and accumulate the knowledge for teaching. Once a teacher starts teaching career, s/he has to actively and constantly engage in a variety of PD activities which are mandatory, including pairing up with an experienced teacher of the same grade level for one-on-one mentoring, participating in teaching research groups in which teachers study textbooks and develop lessons together, and giving public lessons for colleagues and administrators to observe (Liang et al., 2012). The PD activities are continuous, practice-oriented, relevance-focused, and integrated as a whole to be an important part of a teacher's entire teaching professional life. A teacher never stops engaging in PD activities unless s/he leaves the teaching profession.

Characterized as a special Chinese way of teacher PD, developing public lessons is a very important component for grades K-12 teachers to acquire the knowledge for teaching, and it is commonly considered as a powerful tool of teacher PD in China. It originally started in middle of 1950s as a tool of pre-service training. Expert teachers from elementary or high schools were invited to conduct exemplary lessons to students in Teachers' Colleges or Normal Universities to model how to teach (Deng, 2010). In 1980s, the scope of public lessons was expanded to in-service training (Pei, 2006). Since then, public lessons became very important teacher PD activity in in-service teacher training. Starting from 2001, China published new mathematic curriculum standard corresponding to educational reform initiated in 1999. The new curriculum emphasizes connections between mathematics and real life and promotes fostering students' interest and creativity in mathematics (Liu & Li, 2010; Wang & Cai, 2007). Public lessons have been utilized to help teachers understand the new curriculum and implement them effectively in their classroom teaching (Liang, 2012). From the existing research documentation in mathematics education in China, public lessons have been recognized as the most effective way of promoting teachers' professional growth. In a research on the professional growing process of 36 master teachers, Hu found that public lessons were most often mentioned as a crucial impact on their professional development (2006). In a survey about public lesson, conducted by four websites all over China, there was a zero vote for abolishing public lesson in teacher PD (Yu, 2008).

Based on the data resource that includes current research papers published in Chinese educational journals, existing documentation posted online, and analysis of research data collected in 2009 and 2010, this chapter describes how Chinese K-12 mathematics teachers develop public lessons, how involving teachers learn from the process of developing public lessons, how this process affects the involving teachers' learning and classroom teaching, how pubic lessons has been utilized as teacher PD materials for teachers to learn and reflect upon, and what were teachers' reflective perceptions on developing public lessons. Evidence has indicated that engaging teachers in developing public lessons is indeed a productive way to help teachers acquire and accumulate useable knowledge for teaching mathematics (Kersting, Givvin, Sotelo, & Stigler, 2010). This way of conducting teacher PD adds a new dimension for teacher educators in other countries to consider when designing a teacher PD program that intends to help teachers systematically construct the knowledge for teaching mathematics and improve their teaching.

PUBLIC LESSONS

Public Lesson is also called Open Class which "is taught by one teacher and open for observation to a group of teachers and administrators from inside a school or outside a school" (Liang, 2011, p. 1). Different from regular daily lesson, public lesson is a special teaching activity that is offered to other teachers and related people to observe, listen, evaluate, and analyze (Gu, 1999). The purpose of conducting public lesson is to study teaching content, forms, and methods, to promote innovative teaching ideas and strategies, and to improve teachers' teaching quality (Liang, 2011; Yu, 2008).

Classification of Public Lessons

According to the objectives and participants, public lessons can be classified as evaluative lesson, visiting-exchange lesson, research lesson, exemplary lesson, and competitive lesson. The five types of public lessons are described in detail as follows.

Evaluative lesson. Evaluative Lesson is a public lesson utilized to evaluate teachers. Elementary and high schools in China require every teacher to present one or two public lesson each academic year to demonstrate teaching competency. Other teachers and administrators come to observe the lesson and evaluate if a teacher meets the criteria of teaching competency. There is a ranking system for teachers at elementary and high schools in China. The professional ranks include three levels – second level, first level, and high level from low to high. A beginning teacher with three years of teaching experience is entitled to apply for a promotion to the rank of second level; a teacher with the rank of second level for five years can apply for a promotion to the rank of first level; a teacher with the rank of first level for five years can apply for a promotion to the rank of high level (Liang, 2012). A teacher's performance in an Evaluative Lesson would be evaluated and the evaluation result will be recorded as one of the indicators for promotion. Salary would be increased correspondingly for higher rank. There is also an honour title for expert teachers, called master teacher which is not in the rank system, but is the highest honour title for teachers. Master teachers are usually invited to present exemplary lessons (see description in the section of exemplary lessons below) to model good teaching practice for teachers in in-service teacher training.

Visiting-exchange lesson. It is very common in China that teachers observe each other's class and then discuss teaching ideas and strategies on the purpose of learning knowledge for teaching from each other and growing professionally together (Liang et al., 2012; Ma, 1999). It could happen between a new teacher and an experienced teacher. Letting new teachers form a one-on-one mentoring group with an experienced teacher is another common practice in China. A new teacher would observe mentoring teacher's class as often as possible to learn how to teach; mentoring teacher would also observe new teacher's class to give suggestions or advices on problems revealed during class observation. Visiting-exchange lesson could also take place among the teachers in the same group of lesson preparation at the same grade level. Teachers teaching the same grade are required to form a lesson preparation group. They meet once or twice a week preparing next week's teaching plan. Working collaboratively, teachers at the same lesson preparation group seek knowledge for teaching and gain professional growth together.

Research lesson. Research Lesson is conducted to accomplish a teaching research project. Teachers are encouraged to involve in teaching research project at grades K-12 schools in China. In a research lesson, all the participating teachers work together to find the most effective and feasible strategy and method for implementing some teaching ideas, strategies, methods, or models, and one of the involving teachers would present the lesson to teachers at local schools, or schools in other cities if a research lesson is widely recognized.

Exemplary lesson. Exemplary lesson is also referred to model lessons (Wang & Cai, 2007). It is presented by master teachers or other recognized excellent teachers (eg. a good research lesson) demonstrating good teaching practice for many teachers to observe and learn. Exemplary lesson usually provides teachers new teaching ideas and strategies. For example, when new curriculum was adopted since 2001, experts who involved developing new curriculum provided exemplary lessons nationwide for teachers to understand the content and implement the new curriculum appropriately. Award-winning lessons in teaching competition are also offered to teachers as exemplary lessons (see the section of Competitive Lesson below). Award-winning teachers usually are invited to give exemplary lessons for other teachers to study.

Sometimes teachers at affluent area go to schools in remote area and offer exemplary lessons to the teachers there.

Competitive lesson. Competitive lesson is a lesson taught in a teaching competition. Usually the participants prepared several topics and then blindly take a draw to decide the topic to teach. Except competitions at school level, students in class are assigned to participating teachers who don't know the students before teaching the lesson. There are different levels of teaching competitions – school, district, city, province, and national levels. Starting with school level, winners will go to competition at district level; winners at district level will go to competition at city level; winners at city level will go to competition at the province level; finally province level winners will compete at national level (Liang et al., 2012; Li & Huang, 2013).

Young teachers (under 30 years old) are encouraged to participate in teaching competition. Some teachers describe teaching competition as "our Chinese way of training novice teachers", because preparing competitive lesson is a process of learning. Preparation of a competitive lesson involves not only the teacher who will present but also all other teachers teaching the same grade level at school. The whole process of preparation from planning to teaching, every teacher contributes ideas on discussions and trial teaching. Before a competition, a lesson has been "polished" repeatedly. Teachers reflected that the polishing process help them learn usable knowledge for teaching mathematics and make them grow fast (Liang et al., 2012).

Characteristics of Public Lessons

Public lessons are different from regular lessons and can be distinguished by the following characteristics:

- Carefully prepared: Teachers giving public lesson are informed in advance. From planning to actual teaching, a lesson is repeatedly refined through trial teaching before it is presented to observers publically.
- Collaborative: A public lesson is a joint effort of teachers and a product of collective wisdoms. The process of preparation involves a group of teachers working together.
- Sharing: Both preparation process and post-lesson discussions engage participants in sharing teaching ideas, knowledge, and strategies.
- Reflective: When preparing a public lesson, involving teachers have to practice reflective thinking on trial teaching in order to refine the lesson. In addition, post-lesson discussions involve the participants reflecting upon the lesson presented (Liang, 2012).

These characteristics make public lesson a unique teacher PD activity appreciated by teachers in China.

PUBLIC LESSONS AND TEACHERS' LEARNING

Analyzing ten award-winning mathematics teachers' growing path of professional development, Liang et al. (2012) find that these teachers actively involve in a variety of teacher PD activities including one-on-one mentoring for new teachers, public lesson, teaching research, and professional training (workshops). According to the teachers studied, developing pubic lessons was an irreplaceable teacher PD activity that significantly helped constructing their knowledge for teaching and formation of their good quality of teaching with their own style. It is commonly agreed in China that public lesson help teachers grow professionally in general and it is more helpful for novice teachers to acquire and accumulate useable knowledge for teaching (Deng, 2010; Hu, 2006; Liang, 2012; Yu, 2008). Existing research and documentation indicate that public lessons constantly provide opportunities for teachers to gain useful knowledge for teaching mathematics in the path of their professional growth. In this section, the five types of public lessons mentioned above will serve as main threads to help illustrate how teachers in China acquire knowledge for teaching mathematics through public lessons.

Evaluative Lesson's Meaningful Impact on Novice Teachers

A brand new teacher may be equipped with mathematics subject knowledge from colleges, but having mathematic knowledge is not enough for teaching mathematics. In a narrative study on teachers' professional growth, teacher Z recalled what she experienced at her beginning of teaching:

> I had thought I would have no problem to teacher middle school with my qualification of studying mathematics for four year in a normal university. However, when I started teaching, I realized what I learned from the university textbooks, was not helping me to teach. When I observed that other [experienced] teachers were able to easily teach well but my students' feedbacks were not good, I was deeply perplexed. Now I know that when just starting teaching, I didn't know how to prepare a lesson, how to design a teaching plan, and didn't know what teaching strategies should be used, not mentioning being sensitive to my students' needs. As a novice teacher, I didn't know how to organize my teaching. I basically followed the textbook teaching whatever written in there. (As cited in Yu, 2008, p. 16)

Usually a novice teacher is not able to distinct either important or difficult knowledge points from textbook materials and can't recognize connections between knowledge taught before and after. Lack of teachers' knowledge (Shulman, 1986) makes a novice teacher teach rigidly by telling "what it is" but not showing "why it is" or meaningful connections among mathematics ideas. In addition, textbooks in China don't include a lot of content. A large amount of materials used in classroom instruction are literally added by teachers based on their experience. New teachers don't have good idea(s) for what to add because of insufficient teaching experience. They have to learn it from the experienced teachers. Just as one teacher interviewed by the author in 2009 described:

> By observing experienced teachers' teaching, new teachers learn the basic procedures of teaching such as how to achieve the objectives of the lesson, ways of teaching, and the use of language, etc. By classroom observation and trial teaching, new teachers build up their confidence and ability to handle teaching. To be honest with you, when I taught the first class, my voice was trembling. I was nervous. I did not know what to do first and after. New teachers take notes and reflect while observing experienced teacher's classes. From imitating [experienced teacher's teaching] in the beginning to creating new ways of teaching, a new teacher would gradually pick up and gain confidence and know how to handle some unexpected incidents in class.

Evaluative Lesson motivates new teachers to work hard to learn and to improve teaching. In Yu's study (2008), one teacher recalled that the first time she gave an Evaluative Lesson about rational equations; she carefully prepared the lesson from multiple layers – teaching design, teaching process, technology use, and class outcome expectations, etc. Comparing to her regular class, this Evaluative Lesson took her much more time and prepared very carefully. After this lesson, she got to know what to do when teaching similar type of lessons. Conducting this evaluative lesson had a positive effect on her daily lessons. The same teacher described her second Evaluative Lessons as follows:

> In the second semester, I taught an Evaluative Lesson which was a geometry class. This lesson discuss (1) the areas of triangles that located between two parallel lines are equal, (2) using the fact of equal areas to prove that the lines are parallel. In the beginning, I believed that my teaching design was pretty good. However, our subject leader, teacher Z pointed out that my design was not focused, just showing one problem after another problem. He suggested me put all the problems in the same background and use one thread to connect them systematically to form a coherent sequence activities. In this way, the difficult level of the lesson was increased, but students were not just learning an isolated knowledge piece. Using one big idea, one main method throughout the lesson from the beginning to the end, students would be able to understand and apply what learned to solve different problems in similar situations. Taking his suggestions, I started the process of looking problems, solving them, and making comparisons. I searched and solved numerous problems in order to determine several problems to be used in the lesson. With teacher Z's help, I had dug the underlying mathematical principals and ideas hidden in these selected problems. As a result, when I taught the lesson, I was not only teaching the content superficially, but paid attention to disclose the relationship between

knowledge. ... In general, this Evaluative Lesson had improved another big step than the one I did in the first semester, and still have influence on my teaching. (Yu, 2008, p. 21)

Evaluative Lesson set up a starting point of expectation for a new teacher to reach. Because it is an Evaluative Lesson and the teaching evaluation is recorded for promotion purpose, teachers work very hard to prepare the lesson and receive helps from one-on-one mentoring teachers. Many face-to-face discussions take place between a new teacher and his/her mentor before the lesson is given. Mentoring teachers have rich-experience of teaching, in many cases they are very knowledgeable teaching experts in the school. These experienced teachers assist and guide new teachers to build up knowledge for teaching mathematics including principle knowledge (mathematics knowledge, general pedagogical knowledge), knowledge of special cases (individual experience), and strategic knowledge (strategies formed by using principle knowledge and special experience knowledge based on reflective thinking) (Zhang, 2005). With their help, new teachers go through the entire process from designing lesson plan to actual teaching. Additionally in post-lesson discussions, other experienced teachers would provide valuable comments and suggestions for novice teachers to realize his/her existing problem; the novice teacher would reflect his/her teaching for future improvement. This whole process itself provides a precious opportunity for new teachers to construct and enrich their teaching knowledge such as recognizing connections between mathematical ideas, noticing students' misconceptions, selecting worthwhile tasks (National Council of Teachers of Mathematics, 2000) and appropriate class activities, organizing teaching materials coherently (Chen & Li, 2009; Liang, 2013). Conducting Evaluative Lesson helps novice teachers meet basic requirement for becoming qualified mathematics teachers.

Visiting-Exchange Lesson's Role in Teachers' Learning

It is very common practice at schools in China that teachers observe each other's class. Novice teachers observe experienced teachers' classes; experienced teachers observe novice teachers classes; experienced teachers observe each other's classes. This kind of open lessons is not for evaluation purpose but for learning and sharing. One expert teacher recalled that when he just started teaching, he tried very hard to explain a mathematics concept several times but students still could not understand, then he watched that an experienced teacher had students understand the same concept just used a few sentences and one student was very excited, saying "so easy to understand! So easy to understand! So simple!" The teacher was impressed deeply and learned his insufficiency in teaching and was motivated to change for better (Lei, 2008, p. 210). Visiting-exchange lessons provide good opportunities for teachers to learn from their colleagues and share worthwhile resources for improvement.

After observation, there are post-lesson discussions which help teachers think reflectively about the lesson taught aiming at sharing their perspectives of improving. For novice teachers, they can observe his/her mentoring teacher or experienced teachers' classes as often as they want. Every time after observing an experienced teacher's lesson, a novice teacher could ask questions and discuss why the experienced teacher chose to teach a specific content the way s/he did. This kind of communication and sharing experience would help novice teacher keep adding up their knowledge for teaching. Usually a mentoring teacher and the mentored teacher teach the same grade level. It makes it easier for novice teachers relate what observed to their own classroom teaching and apply acquired knowledge into their own teaching.

Mentoring teachers are required by schools to observe new teachers' classes. For example, some schools required mentoring teachers observe new teachers' class at least once a week. After observing a new teacher's lesson, a mentoring teacher help the new teacher reflect on the lesson taught; pointing out both good traits and shortcomings of the lesson and providing suggestions to improve (Liang et al., 2012). An award-wining teacher interviewed by the author in 2009 commented: "Classroom observation and face-to-face discussion have huge effects on new teachers. They shorten the time of new teachers' improvement. I can see that new teachers have been improving every week, every month."

It is worthwhile to mention that experienced teachers also realized that they learn from new teachers too. Some expert mathematics teachers pointed out, "Young teachers also have their own ways and ideas worth learning from"; "New teachers have their own strengths. For example, they are usually good at new teaching technologies and bring some fresh ideas from college. I should say that we help each other to improve teaching" (liang, Glaz, & Defranco, 2012).

Visiting-exchange lessons allow teachers to form a professional learning community in which teachers support each other, discuss mathematics content taught, share teaching strategies and resources, and help each other to gain and accumulate knowledge for teaching in order to improve teaching. For new teachers, they don't necessarily struggle alone, feeling isolated and lost; they are guided to learn, to grow, to improve teaching constantly.

Research Lesson – Connecting Educational Theories to Teaching

Teachers are encouraged to conduct teaching research seeking professional growth. Teachers in China believe that a high quality of teacher should be a scholar not only a teacher. Usually teachers voluntarily form a team to work together applying a certain learning theory to classroom teaching. Teachers appreciated their experience of engaging in developing a research lesson. They recognized what had learned through this process significantly change their teaching philosophy and structure of teaching knowledge in general. An award-winning teacher realized that his participation in the research project – Mathematics Methodology, changed his view of teaching mathematics and had him envision how important to foster students' interest in learning and create a learning environment encouraging students to explore, investigate, and discover mathematical patterns, rules, and principles. Another award-winning teacher involved a research project of "Participation and Discovery" – having students actively engage in the process of teaching and learning, he was the one teaching the research lessons. Being recognized by their province, the research lessons developed by the team became exemplary lessons to demonstrate how to motivate students to participate and how to lead students to discover mathematics knowledge on their own. This teacher appreciated highly being given the opportunities to involve in the project and teach the research lessons; he gained confidence, knowledge, and experience which have remarkable influence on his daily teaching. Research indicated that award-winning teachers utilized teaching research to expand their professional opportunities (Liang, Glaz, & DeFranco, 2012). Research lesson indeed is a very efficient teacher PD activity for teachers to develop expertise and leadership in teaching (Zhang, 2005).

Research lesson usually is tied to a teaching research project that leads teachers to have a professional growth beyond minimum requirement. Developing a research lesson is an opportunity for participating teachers to learn educational theories, connect these theories to their teaching practice and learn how to implement innovative teaching approaches to their classroom teaching. When a teacher reflected his own involvement with a research lesson of mathematics, he used three words to describe his experience of the process: making sense, understanding, improving. The research lesson was taught three times. At the first time the teacher himself and another teacher taught the same topic (using bisection method to find the root of an equation) at different class, after observing the two lessons, all the participating teachers worked together analyzing the two different teaching approaches, provided their feedback, exchanged ideas and decided how to redesign the teaching plan. The redesigned lesson plan eliminated the part of recalling previous learned knowledge to avoid limiting students' thinking, emphasized knowledge application, and integrated technology (*Excel*) into teaching the lesson. Following the redesigned lesson plan, the teacher taught the lesson again. At this round of teaching, more application problems were added for students to think and Excel was used to help students study the process of finding the root of an equation. The teaching design became richer; the outcome of this lesson was good overall. After the second time of teaching, the teacher reflected that he had a breakthrough improvement in grasping the main ideas of the textbook and understanding big ideas of mathematics comparing to his shallow understanding to the content taught at the first time. He realized that his teaching should allow students to experience the concurrent development of constructing a mathematics method, applying techniques, and utilizing computation methods, and to understand the formation of mathematical ideas from a whole to a part, from quality to quantity, from accuracy to approximation, from computation to techniques, and from techniques to algorithms. Many expert teachers observed the second round of teaching and a university professor was also invited to observe the lesson, making the research lesson be refined again with teaching experts' reflective feedbacks. The third round of teaching set a good example of teaching bisection method under new curriculum standard in front of teachers from all over the province (Wang & Zhang, 2008).

Development of an exemplary lesson is a teaching research model focusing on teachers' reflection and developing process of a lesson. All participating teachers learn and benefit from this process at a certain level. In China, many K-12 teachers published numerous research papers or books through teaching research (Liang, Glaz, & DeFranco, 2012; Yu, 2008). They are proud of being teaching research scholar but not only good teachers.

Exemplary Lesson – Demonstrating Good Teaching Practice

An Exemplary Lesson is a public lesson modelling high quality of teaching. It could be taught by master teachers, experienced teachers, or teaching competition winners. Administrators at school, district, city, and province level often organize exemplary lessons for teachers to observe, discuss, and reflect on the purpose of demonstrating effective teaching practice and setting up a high standard of quality teaching. Teaching competition lessons are also considered as exemplary lessons which are usually very well prepared with teachers' joint effort through the process of repeated refinement. They are conducted to help teachers acquire teaching knowledge. Teachers involving the preparation of the lesson would learn deeply about a teaching design; teachers observing the lesson would see a good model of teaching and learn from it (Liang, 2011; Yu, 2010).

Usually in one district, schools take turns to offer exemplary lesson to teachers from all schools. The frequency could be once a week or once a month. Teachers, who are recognized as effective teachers, would be selected to present exemplary lesson, even though the preparation of the lesson could involve a group of teachers.

Teachers in affluent area are also sent to schools in remote area where education is weak, to present exemplary lessons to teachers there, intending to demonstrate good teaching practice that the remote area teachers don't have opportunity to observe (Liang, 2011).

Exemplary lessons allow teachers to actually see what an effective mathematic teaching would look like and how it is performed. In a study, a survey to 636 elementary teachers in Zhejiang province revealed that exemplary lessons and peer teachers' collaboration are the most common ways for their teaching knowledge acquisition (Wu et al., 2005). On the one hand, Exemplary lessons play a leading role functioning as model teaching for teachers to follow; On the other hand, the process of refining an exemplary lesson help participating teachers improve teaching and expand professional capability.

Usually exemplary lessons are videotaped and provided to teachers who could not observe in-person. Many schools organize teachers to watch the videos of exemplary lessons and discuss reflectively after watching. This is also a part of teacher PD activities.

Competitive Lesson – Professional Life Changing Experience

Participants of teaching competition are usually young teachers under 30 years old. Most expert teachers or master teachers were winners at different levels of teaching competition. They may have gone through several times of the process: (1) preparing lessons for teaching competition, (2) teaching the lesson in front of unfamiliar students, judges (expert teachers), and other participating teachers, (3) listening to the comments from the judges, and (4) reflecting the lessons taught. A national teaching competition winner has to pass all the way through school, district, city, province, and finally up to the national level and won over other award-winning teachers from other provinces. If a teacher won the national level of competition, s/he is really highly recognized and honoured as an outstanding teacher. This honour is a significant mark on her/his teaching achievement. Research documentations and published teachers' reflections agreed that participating teaching competition is a very important path for young teachers to grow and improve teaching (Deng, 2010; Liang et al., 2012; Wang & Zhang, 2008; Yu, 2008).

Competitive lessons were frequently mentioned as a huge influence on them in their successful story when expert teachers/master teacher reflect their professional growth, For example, a master teacher recalled that when he just started teaching mathematics, he had hard time to have his student understand. The principal was very supportive and sent him to observe the big event – national teaching competition. He had the opportunities to observe the best teaching from all over the country and ask the expert teachers many questions he had. After the event, he started preparing every lesson as if he was preparing a competitive lesson. Later on, he himself attended the teaching competitions; from school to district, from district to city, from city to province, from province to nation, he won the first place in all levels of competition in two years. He described his experience:

> That was unforgettable years in my life! Preparing lessons again and again, conducting trial teaching repeatedly, revising teaching design for numerous times, like a silkworm I had been experiencing life changes, painful but also happy! In countless nights, I practiced teaching in an empty classroom, thinking [the lesson] over and over; many times on the road riding the bicycle, I was very close to an accident when being puzzled by a question on my mind; my head was full of the teaching content, almost all circle shapes can make me related to my lesson – Circumference of a Circle. (Lei, 2008, pp. 211–212)

His path of professional development is inseparable to his active involvement with teaching competition. He had been invited to give exemplary lessons hundreds of time and published more than one hundreds of papers in provincial and national journals. Because of his high achievement, he was honoured as master teacher.

Many teaching competition winners value the process of preparing the competitive lessons and found themselves learned significantly from repeated lesson refinement and continuous reflections with other teachers' working together as a team. Just as another four-time winning teacher said, for him, in a teaching competition, what he learned most was not from performing the lesson or being evaluated in the end but was from the preparing process with a team of teachers who helped design the lesson and gave suggestions (*Reflection on participating national mathematics teaching competition*, 2009, March 24). Some winning teachers also pointed out that the teaching competition lesson came out of teachers' joint work and collective intelligence and all participating teachers learn from the process. As one teacher stated:

> The process of preparing a teaching competition improves all teachers' teaching. If I prepare for a regular class by myself, I may not think so much and not be able to cover every detail. When a lesson is studied by all the teachers, everyone contributes their good ideas to a high quality of lesson. Even though only one teacher won the prize, all of the teachers who participated in the preparation benefit from the process. (Liang, 2011, p. 19)

Teaching competition provides a learning platform for teachers, especially for young teachers, to grow from ordinary to outstanding. Many exemplary lessons were produced through the events of teaching competition.

CONCLUSION

In order to conduct effective mathematics teaching in classroom, a teacher needs to be equipped with mathematics content knowledge (procedural and conceptual knowledge), general pedagogy knowledge (eg. teaching goals, appropriate tasks, and activities, students' engagement), and pedagogical content knowledge (eg. helping student construct mathematics ideas, fixing students' misconceptions, foster students' mathematical thinking) (Hill, Schiing, & Ball, 2004; Huang, Kulm, Li, Smith, & Bao, 2011). Not all these useful knowledge for teaching mathematics can be learned from college or university pre-service teacher programs. Indeed a lot of the teaching knowledge has to be acquired through practicing teaching and constant reflection on teaching. It is commonly agreed in China that teachers' colleges or universities are responsible for equipping solid mathematics content and pedagogic knowledge, however teacher preparation program alone can't create effective teachers. A teacher's teaching expertise is grown out of the continuous process of learning how to teach while teaching. Teaching expertise is not formed in a few days or even a few years but is developed over many years of teaching, and only teaching experience itself is not enough for teachers to acquire teaching knowledge. Studies have found that Chinese K-12 mathematics teachers' knowledge for teaching is acquired and developed through their active involvement with public lesson activities which engage teachers in the process of working collaboratively from a lesson design to a lesson refinement and to post-lesson reflection.

Public lesson involves teachers with learning and studying how to teach effectively through various perspectives: (1) learning from observing other teachers' teaching; (2) learning from collaborating; (3) learning from repeatedly refining a lesson; (4) learning from expert teachers' suggestions; (5) learning from reflection; (6) learning from shared teaching resources. Unlike other teacher PD activities (eg., a workshop, a college course, teaching conference, etc.), public lesson allows teachers to be both observers and participants; teachers can visualize what exactly happened in other teachers' teaching, reflect on that, and directly relate to their own classroom teaching. Going through this process, teachers acquired knowledge for teaching. As public lesson experience increases, a teacher would accumulate more and more teaching knowledge.

Studies have revealed that even though what specifically learned was individualized depending on individual teachers, public lessons helped teachers in China gain mathematical knowledge for teaching which may be classified into two categories: mathematics content knowledge and mathematical pedagogical knowledge. Analysis of the interview data and documentation of teachers' reflections supported that Mathematics content knowledge included (1) deepening understanding of certain connected mathematical concepts, (2) learning a new way to solve a type of mathematics problems, (3) interpreting some mathematical ideas from different perspectives, (4) recognizing connections between certain mathematical ideas; and mathematics pedagogical knowledge included (1) lesson designing techniques, (2) effective ways to help students understand certain difficult concepts, (3) skills of posing questions, (4) problem selection and ordering, (5) effective classroom communication skills, (6) strategies of engaging students in doing mathematics, (7) effective ways to address students' common mistakes or misconceptions, (8) ways of designing a coherent mathematics lesson, (9) usage of technology in mathematics teaching. The list of teachers' learning from public lesson is not exclusive and could be added more as more research results come out.

Public lesson motivates teachers to constantly learn and improve their teaching. As an award-winning teacher had stated:

> Open classes can motivate me, inspire me, and force me to best use the textbook, to design the best examples, and to stimulate students' learning interests. Through the problems exposed in the open classes and the comments from other teachers, I can keep improving myself and move forward. (Liang et al., 2012, p. 9)

Public lesson helps build a healthy professional learning community at schools, districts, cities, provinces, and nation. Sharing teaching wisdoms has become a culture in which teachers work together to face difficulties and solve problems in teaching instead of being alone with limited resource to deal with problems, feeling isolated. Public lesson makes both the whole teaching community and individual teachers constantly grow and improve professionally.

REFERENCES

Chen, X., & Li, Y. (2009). Instructional coherence in Chinese mathematics classroom: A case study of lessons on fraction division. *International Journal of Science and Mathematics Education, 8*, 711–735.

Deng, J. (2010). 公开课对教师专业成长影响的叙事探究 [*A narrative study of public lesson's effects on teachers' professional growth*] (master's thesis). Nanjing Normal University, Nanjing. Retrieved from http://www.doc88.com/p-386438200179.html

Gu, M. (1999). 教育大辞典 (*Education dictionary*). Shanghai: Shanghai Education Press.

Hill, H. C., Schilling, S. G., & Ball, D. L. (2004). Developing measures of teachers' mathematics knowledge for teaching. *The Elementary School Journal, 105*, 12–30.

Hu, D. (2006). 影响优秀教师成长因素 – 对特级教师人生经历样本分析 [The impacting factors on teachers' growth: An analysis of master teachers' life experience]. 教师教育研究 [*Teacher Education Research*]

Huang, R., Kulm, G., Li, Y., Smith, D., & Bao, J. (2011). Impact of video case studies on elementary mathematics teachers' ways of evaluating lessons: An exploratory study. *The Mathematics Educator, 13*(1), 53–71.

Kersting, N. B., Givvin, K. B., Sotelo, F. L., & Stigler, J. W. (2010). Teachers' analyses of classroom video predict student learning of mathematics: Further explorations of a novel measure of teacher knowledge. *Journal of Teacher Education, 61*(1–2), 172–181.

Lei, L. (2008). 小学数学名师教学艺术 [*The teaching arts of elementary mathematics master teachers*]. Shanghai: East China Normal University Press.

Li, Y., & Huang, R. (2013). *How Chinese teach mathematics and improve teaching*. New York, NY: Routledge.

Liang, S. (2011). Open class: An important component of teachers' in-service training in China. *Education, 1*(1), 17–22.

Liang, S. (2013). An example of coherent mathematics lesson. *Universal Journal of Educational Research, 1*(2), 57–64.

Liang, S., Glaz, S., & DeFranco, T. (2012). Investigating characteristics of award-winning grades 7–12 mathematics teachers from the Shandong Province in China. *Current Issues in Education, 15*(3), 1–11.

Liang, S., Glaz, S., DeFranco, T., Vinsonhaler, C., Grenier, R., & Cardetti, F. (2012). An examination of the preparation and practice of grades 7–12 mathematics teachers from the Shandong Province in China. *Journal of Mathematics Teacher Education, 16*(2), 149–160.

Liu, J., & Li, Y. (2010). Mathematics curriculum reform in the Chinese Mainland: Changes and challenges. In F. K. S. Leung & Y. Li (Eds.), *Reforms and issues in school mathematics in East Asia: Sharing and understanding mathematics education policies and practices* (pp. 9–31). Rotterdam, The Netherlands: Sense Publishers.

National Council of Teachers of Mathematics. (2000). *Principles and standards for school mathematics*. Reston, VA: Author.

Pei, D. (2006). 在追问中把握公开课的现代意义 [Understanding the modern meanings of open lesson by questing]. 中小学教育 [*Elementary and High School Education*], *1*, 1–5.

Reflection on participating national mathematics Teaching competition. (2009, March 24). 参加全国初中数学优质课观摩与评比活动感受. Retrieved from http://mdjsznlm.blog.163.com/blog/static/11300475220092240181 7536/

Shulman, L. S. (1986). Those who understand: Knowledge growth in teaching. *Educational Researcher, 15*(2), 4–14.

Yu, X. (2008). 公开课在数学教师专业发展中作用的个案研究 [*A case study on the functions of open class in mathematics teachers' professional development*] (Master Thesis). Capital Normal University, Beijing. Retrieved from http://www.m.doc88.com/p-4913937605945.html

Wang, S., & Zhang, S. (2008). 走进高中数学新课程 [*Entering high school new mathematics curriculum*]. Shanghai: China East University Press.

Wang, T., & Cai, J. (2007). Chinese (Mainland) teachers' views of effective mathematics teaching and learning. *ZDM: Mathematics Education, 39*, 287–300.

Wu, W. et al. (2005). 小学教师教学知识现状及其影响因素的调查研究 [Investigation on elementary teachers' knowledge and influencing factors]. 教师研究 [*Teacher Education Research*], *4*, 61.

Zhang, Z. (2005). 中学数学教师职后教研培训模式的研究 [*An examination of high school mathematics teacher's professional development models*] (master thesis). Tianjin Normal University, Tianjin. Retrieved from http://www.doc88.com/p-90935595537

Su Liang
Department of Mathematics
California State University
San Bernardino, USA

YEPING LI AND RONGJIN HUANG

13. IMPROVING TEACHERS' EXPERTISE AND TEACHING THROUGH APPRENTICESHIP PRACTICE IN MAINLAND CHINA

Case Studies

INTRODUCTION

The search for the most effective approaches to achieve excellence in mathematics education in many education systems has led to the increased interest in teaching and teacher education practices in some high-achieving education systems, including Mainland China (e.g., Li, Ma, & Pang, 2008; Li & Shimizu, 2009). In particular, recent investigations revealed remarkable differences between the U.S. and Chinese teachers in their mathematics knowledge for teaching (e.g., An, Kulm, & Wu, 2004; Huang, 2014; Ma, 1999). Paradoxically, Chinese mathematics teachers often received less formal education than their U.S. counterparts (Ma, 1999). A possible explanation is that Chinese teachers continue to improve their knowledge and skills for teaching over the years (Huang, Ye, & Prince, 2016; Li & Huang, 2008). Although it is recognized that Mainland China has a coherent teacher professional development system (Stewart, 2006; Wang, 2009), it is rarely examined what may happen in Mainland China and how Chinese teachers may experience different professional growth as compared to their counterparts in other education systems such as the United States. In this chapter, we aim to focus on the apprenticeship practice developed and used in Mainland China for supporting teachers' professional development of their expertise and their pursuit of excellence in mathematics instruction.

BACKGROUND: TEACHER AND TEACHER PROFESSIONAL
DEVELOPMENT IN MAINLAND CHINA

Teacher as a Professional in Mainland China: A Brief Background

As a system with a long history in education, Mainland China has formed some deeply-rooted cultural value orientations about teachers and teaching. Chinese teachers are respected for their exceptional teaching and in-depth knowledge and for their exemplary moral traits: a conception that presents more differences from than similarities with what is viewed as a good teacher in the West for example in

England (see Beishuizen, Hof, van Putten, Bouwmeester, & Asscher, 2001; Jin & Cortazzi, 1998). It is common for Chinese mathematics teachers not only to sit in others' classrooms and discuss teaching with fellow teachers, but also to develop and polish lesson instruction together (Li & Li, 2009). Senior teachers who earned a good reputation in teaching are often assigned the responsibility of mentoring junior teachers for their professional development. However, being a teacher as a professional from the law perspective, specific regulations were established not long ago.

According to the regulations (MOE, 1995), there are different professional titles for secondary and primary teachers. For example, the positions within secondary teachers include senior-rank teacher, secondary level 1, and secondary level 2. For each, there are specifications with respect to political, moral, and academic standards. Local educational authorities further provide detailed and specific requirements for promotion at each level (MOE, 1999). In August 2015, the Ministry of Human Resources and Social Security [MHRSS] and Ministry of Education [MOE] of China (MHRSS & MOE, 2015) jointly released another document, *The Guidance for Deepening the Reform of Secondary and Primary School Teachers' Promotion System* which unifies the ranks for both secondary and primary teachers into three levels that are aligned with other professional ranking systems. The senior-rank level includes full-senior and senior teachers. The intermediate-rank level is called level 1 teacher, and the primary-rank level consists of level 2 and level 3 teachers. Building on the previous regulations (MOE, 1995), the recent guidance (MHRSS & MOE, 2015) stresses teacher morality, practical accomplishment and practical experience, and de-emphasizes academic articles and academic degrees. Teacher performance should be evaluated by expert panels using multiple ways including explaining and delivering lessons, interviews and expert reviews (Huang et al., 2016).

The Conception of Teacher Professional Development in Mainland China

The conception of teacher professional development in Mainland China shares some similarities but more differences from what teachers normally perceive in a Western context (e.g., Li, Kulm, & Smith, 2006). Although teachers' professional development may commonly be perceived as to improve teachers' knowledge and skills needed for teaching, there are big differences cross-culturally in terms of what needs teachers may perceive and how to improve as part of professional development. For example, when teachers' professional development in the United States are often provided in the form of workshops and recommendations, they are remote, inconsistent and can even be contradictory (e.g., Guskey, 2003). In contrast, teachers' professional development in Mainland China is taken as activities that are practical in nature. Rather than providing lengthy recommendations and workshops, professional development in Mainland China happens in teachers' daily life that is pertinent to the teachers' professional needs.

The cross-cultural difference in professional development relates to what is perceived as the professional nature of teaching. Teaching is taken as a professional, not private and personal, practice in Mainland China. Model teaching behavior is often established informally through frequent exchanges of ideas among teachers in and outside their own schools (e.g., Huang & Bao, 2006). In particular, master teachers[1] in Mainland China are often identified as ones who bear most of the culturally-valued moral characters and expertise for others to follow (Li & Huang, 2008). What teachers can learn from others is the type of knowledge and skills that are publically valued and often locally proven as effective. Teachers realize the needs to improve their expertise in mathematics instruction and know what they can learn from others through their daily teaching activities. Teachers' expertise, reflected as knowledge and skills needed for and in mathematics teaching, is real and teachers acknowledge expertise difference from one teacher to another.

Strategies for Teacher Professional Development in Mainland China

Taking professional development as an independent effort, U.S. teachers receive information provided in professional publications about what can be considered as good instructional practice and what is needed for achieving that. Many large-scale professional development programs and workshops are also made available to facilitate teachers' professional growth and changes. Thus, the U.S. teachers' knowledge and skills for teaching is often enhanced through the approach of "learning from being told and by doing" (Li et al., 2006). In contrast, Chinese teachers' knowledge and skills for teaching are often improved through the approach of "learning from example and by doing." Some strategies have been in place for quite many years and they include (1) school-based apprenticeship approach; (2) school-based teaching research group; (3) teaching research activity organized by municipality; (4) master teacher workstations; and (5) teaching contests (e.g., Huang et al., 2016).

As Mainland China has been undertaking curriculum reform over the past decade, teacher professional development becomes even more important. Given the fact that new content topics and innovative instructional ideas need to be implemented in classroom instruction, traditional strategies for professional development alone are not enough to meet these new challenges. Recent investigations have revealed that professional development is taking up some new approaches, while keeping valuable existing ones (Huang & Li, 2008). For example, Chinese lesson study (Huang & Han, 2015) has become a nationwide teacher professional approach. Nevertheless, one fundamental feature stays the same, that is, teacher professional development still stems from teachers' needs and bears direct connections with what teachers need to do in their own classrooms.

APPRENTICESHIP PRACTICE IN MAINLAND CHINA: FEATURES AND PATHWAYS OF LEARNING

Apprenticeship practice as a popular approach for professional development has been developed and used in Mainland China for a long time. The typical practice is the school-based coaching that is mainly for teacher induction. It is often adopted and used as a basic part of school-based teaching research activities, which focuses on novice teachers' adaption to culturally-valued daily teaching with specific requirements. In recent years, another form of apprenticeship practice has also been developed and used in Mainland China. It is a subject content-focused coaching practice for mid-career teachers from different schools, which aims to help them further develop teaching expertise and get promotion to senior rank. It is often structured and practiced as a part of district or city-level teaching research activities. It could be more flexible and is content based in general. In the following sections, we provide more information about these two types of apprenticeship practice in Mainland China.

School-Based Apprenticeship Practice

In general, new teachers are required to be mentored for three years. To carry out this practice, two teachers form a pair as matched and endorsed by the school administration. One of the members is a novice teacher (mentee) and the other is an experienced or master teacher (mentor). In most cases, there is a signed formal agreement between the mentor and mentee that outlines the responsibilities for each of the pair. For example, in one school, the agreement specifies that the mentor should (a) establish an effective measure for enhancing the development of the novice teacher; (b) share with the novice teacher about the experience of teaching and thinking of education twice per month; (c) share and discuss the thoughts, learning, life and teaching of the novice teacher, and help him/her solve some practical difficulties or put forward those problems to relevant authorities; (d) inspect the teaching schedule and lesson plans of the novice teacher and observing his/her classroom instruction at least 10 times per semester, and help him/her improve teaching skills through reflections and discussions and so on. Relevant requirements for the novice teacher (mentee) are also set in the agreement (Huang et al., 2010). Through such an apprenticeship practice, it is expected that novice teachers could familiarize themselves with teaching materials and effective teaching methods as soon as possible.

Recently, there are also some changes in this practice. In some schools, mentor and mentee may look for their partners from different grade levels or even different subject areas on the basis of mutual needs and satisfaction.

Content-focused Coaching Practice for Teachers from Different Schools

In addition to the school-based apprenticeship for induction teachers, there are content-focused coaching practices to support mid-career teachers' further development.

In some cases, local education institutes organize a team of master teachers to supervise a group of mid-career teachers selected from different schools. Those teachers should have demonstrated a potential to be promoted to higher levels, and the school principals or teacher leaders in specific subject content areas recommend them to join out-of-school coaching activities. In this case, there is a regular activity schedule for developing, observing and evaluating public lessons.

In other situations, the coaching relationship may be formed due to developing public lesson for different purposes, including research lesson for regular teaching research activities, contest lessons at different levels, and exemplary lessons developed through projects.

Studies on Perceived Effectiveness of Apprenticeship Practice in Mainland China

Based on interviews with 18 mathematics teachers, teacher educators, and Teaching research specialists on their perceived effects of apprenticeship, it was found that apprenticeship practice has helped teachers learn how to analyze and use mathematics teaching materials (e.g., important points, difficult points, and examples and exercises), improve teachers' basic teaching skills, form a classroom learning environment, and cultivate teachers' ethic and moral (Huang & Li, 2008). Through investigating the collaboration of a first-grade novice teacher and her mentor teacher in the context of the former's learning to teach mathematics over one year in Shanghai, China, Wang and Paine (2001) concluded that the mentor contributed to the novice teacher's progress as follows: (1) developing and implementing a clear and consistent focus that is aligned with reform-oriented curriculum; (2) modelling and analyzing and reflecting such mathematics teaching; (3) defining and refining the zones of the novice's proximal development in learning to teach and pushing the mentee to move ahead gradually. Lee and Feng (2006) investigated eight dyads of mentoring teacher and first-year secondary school teachers in Guangzhou, China with regard to the natures of novices' learning and associated factors affecting mentoring support. The findings revealed that mentors provide four types of support: provision of information, mutual lesson observation, collaborative lesson preparation and discussion in the office. Factors affecting mentoring support include teaching workload, grade and subject, style of mentor–protégé interactions, relationships between mentor and mentee, incentives for the mentors, and collegial culture in the case study schools. The salient foci of mentoring tend to be the teaching of specific content rather than curriculum and pedagogy in general.

SCHOOL-BASED APPRENTICESHIP PRACTICE: A CASE STUDY

Data Source and Analysis

Two pairs of mentor -mentee were conveniently selected from a city-level key high school in southern part of China. In the school, all the beginning teachers who just graduated from normal universities, and experienced teachers who transferred from middle schools are required to be mentored for three years. There are specific requirements for mentor and mentee. For example, mentees need to observe mentors' 20 lessons each semester while mentors should observe mentees 10 lessons per semester. In this study, one pair of mentor-mentee consists of a beginning teacher who had half year of teaching experience after graduation (Mentee A) and a mentor teacher who had six years of teaching experience (Mentor A). While the other pair of mentor-mentee includes a mentee who was a beginning teacher at high school but with years of teaching experiences at middle school level (Mentee B) and a mentor who was the head of mathematics department in the high school (Mentor B). One interview with each member of the two pairs was carried out and audio recorded. The interviews focused on the following three aspects: background of the interviewees, analysis of the lesson observed (lesson plan, lesson delivery, and lesson reflection), and mechanism of apprenticeship and its perceived effectiveness. Nevertheless, this case study only focuses on the analysis of the questions related to apprenticeship practice.

All the audio recorded interviews were transcribed into Chinese. We had read the transcriptions several times to capture the overall ideas emerged from the interviews. After that, we took each pair of mentor-mentee as a unit to analyze and identify patterns of interactions between the members of the pair. Some features of their interactions and their perceptions of the effects of the apprenticeship practice were identified and illustrated in the following sections.

Results

The basic activities of their apprenticeship practices include collaborative lesson planning, mutual lesson observation and informal post-lesson exchange. Through these activities, both mentors and mentees can learn from each other in general. In particular, mentees can learn from their mentors in the following aspects: (1) mastering teaching materials, dealing with difficult and important points, and designing lesson; (2) understanding student's knowledge readiness and learning difficulties, using appropriate teaching strategies and managing classroom. In the following parts, these aspects are illustrated.

Mastering teaching materials and dealing with difficult and important points. Both mentees and mentors appreciated the mutual benefits in developing their understanding of teaching materials and proper dealing difficult and important points through collaborative lesson planning. Usually, the grade-based lesson preparation sub-group within the mathematics group develops a prototype of lesson plan (including examples and exercises) for each teacher's adaptation. Each teacher can modify and further develop his/her own teaching plan based on individual class reality and personal teaching style. Mentors and mentees need to read the other's teaching plans, and discuss how to improve their own lesson designs. Through this process, the mentees can learn a lot from the mentors as stated below:

When preparing a lesson, although I understand the content itself, I am not quite sure how many topics I need to cover, and what instructional objectives I should set for the particular lesson. I should listen to the mentor's suggestions on these aspects first, and then develop my own lesson plan. (Mentee A)

Learning of teaching strategies and students' learning through lesson observations. The biggest challenges for mentees are their lack of teaching strategies and limited understanding of students' knowledge readiness and difficulties. Through frequent observations of mentors' lesson instruction, mentees can learn effective teaching strategies, and get to know students' typical responses to specific topics and in problem solving. Then, mentees will have pre-prepared strategies to deal with different situations. So they will have confidence in teaching, as one mentee explained as follows:

Through observing mentor's lessons, we can learn the procedures of lesson instruction, and students' responses in lessons. That really helps us understand students' real situation, and improve our lesson planning. (Mentee A)

On the other hand, one mentor explained ways to help the mentee as follows:

Firstly, collaborative lesson plan will identify the important and difficult points and find ways to tackle difficult mathematics problems. Then, through reading and discussing the lesson plan, some improvements could be made such as topics' arrangement and connection, summarizing the mathematics thinking after solving problems. It will benefit each other. Thirdly, through lesson observations, we are concerned with the implementation of the teaching plan and classroom management, such as dealing with improper students' behaviors (i.e. sleeping), activating student motivation and atmosphere. (Mentor A)

The mentor (Mentor B) further noticed the differences in supervising new teachers versus experienced ones as follows:

With regard to experienced teachers, the main task is to help them advance subject content knowledge and foster their problem-solving ability. Novice teachers often have sound subject content knowledge and strict mathematical language, but they need to improve their teaching skills, master teaching materials, and understand students' learning.

Establishing a learning community that benefits each other. Both mentors and mentees agreed that the team members are mutually benefited. Not only do the mentees learn something from mentors through collaborative lesson planning and classroom observation, but the mentors are also willing to learn from the mentees. Usually, after lesson observations, there is a discussion within the pair. They discuss the strength and weakness of the lesson and what can learn from the lesson. Since the mentee and mentor usually seat nearby in the same office, they often have pertinent and informal discussions. The mentors also know well about their responsibility in helping mentees to develop successful open lessons from design to rehearsal to final teaching. In general, the mentor-mentee practice benefits from "establishing a harmonious atmosphere in the subject content group and making the mentees to ask mentors more specific questions legitimately and freely" (Mentor B).

Summary of This Case Study

The basic activities of mentoring in that school include collaborative lesson planning, lesson observations, and informal post-lesson discussion. Overall, the mentoring practice benefits both mentees and the mentors. In particular, mentees can learn from mentors how to analyze teaching materials, identify and deal with difficult and important content points, design lessons, adopt appropriate teaching strategies, understand students' learning situation, and manage classrooms. In general, the mentoring practice can help establish a harmonious and collaborative subject-based learning community.

CONTENT-FOCUSED COACHING PRACTICE: A CASE STUDY

In this case study, we investigated one pair of mid-career teacher and master teacher (coach) who had jointly developed an open lesson as part of a nationwide research project. This case study aimed to explore the following three questions: (1) what and how are the main improvements of an open lesson design made? (2) What are the teacher's gains through participating in open lesson development with the support of the coach?

The Participants

In order to complete an open lesson design assigned by a nationwide teaching research project (see Huang & Li, 2009 for details), a mid-career teacher had developed an open lesson on "*sampling*" with the help of an expert who is working at middle-school teaching research department of a city educational bureau.

The practicing teacher. The teacher, Miss. Wang (hereafter called T1), held a bachelor degree in mathematics, had 6 years of teaching experience at a middle school in suburb of a big city located in Eastern part of China. The teacher had an entrance position title (level 2 of secondary teachers). Usually, after three years of a school-based mentor-mentee program, a teacher could be promoted to a position at level 2, and then after three to five years, the teacher may be promoted to a position at level 1 based on their performances. Through next three to five years of efforts, some excellent teachers can be promoted to the advanced position (e.g., Huang et al., 2010; Li et al., 2009). So, the teacher in focus was on her middle way to pursue higher position levels, which would require the teacher not only having an excellent teaching performance such as open lesson and students' performance, but also having some teaching research publications. For a self-motivated teacher, it is a valuable opportunity to actively participate in some open lesson development in order to improve her/his teaching competence and teaching research ability.

The teaching research specialist. Mrs. Zhang had a bachelor's degree in mathematics, and a master' degree in mathematics education, and currently a part-time Ph.D. candidate (hereafter called C1). She had taught mathematics at all grade levels in secondary school (grades 7 to 12) for about 10 years, and held an advanced rank of secondary school teachers. She has been supervising middle school teachers as a teaching research specialist since 2002. She has won several teaching excellent awards at municipal and provincial levels. She also has supervised five master degree students in a nationally recognized Normal University. She was the local coordinator of the aforementioned nationwide project.

The Topic

According to the introduction of textbook series (People's Educational Press, 2004), the content area of "statistics and probability" is separated from "number and algebra" and "figure and geometry", including three Chapters. First chapter focuses on the basic process of statistical data collection, classification and description at Grade 7. The second chapter deals with the analysis of data at grad 8, and third Chapter on probability is introduced at grade 9. The topic examined in this case is the second lesson of the first Chapter at grade 7. This Chapter includes three units: Statistical survey, Histogram, and Project learning. "Statistical survey" focuses on two aspects: (1) collecting and classifying data from census survey and sample survey; and (2) exposing basic process of collecting, clarifying, describing and analyzing data. Sample survey is the important content in this first unit, which is suggested to have three lessons. Lesson 1 focuses on "collecting, classifying, describing and analyzing data and making conclusion by using census survey", Lesson 2 covers "collecting, classifying, describing and analyzing data and making conclusion by using sample survey", while lesson 3 focuses on "stratified sampling and evaluating the population by investigating samples."

The Process of Developing a Lesson Design

All six teachers at the seventh grade in the school under study were required to prepare a design of teaching sampling as a school-based teaching research activity (as a required part of the nationwide project). Two weeks later, all the teachers explained their lesson designs at a meeting of the lesson preparation group. Based on the characteristics of lesson designs, and extensive discussions, the lesson preparation group decided that Miss Wang would take the principal responsibility for developing the lesson design with the support of this lesson preparation group.

Based on the discussion and suggestions, she developed an initial lesson design, and sent it to Mrs. Zhang for comments. After getting Mrs. Zhang's feedback, she had some conversations with the head of the lesson preparation group through telephone and e-mail, and completed a revised lesson design, and then had a trial lesson teaching. The members of the lesson preparation group and Mrs. Zhang observed the trial teaching, and then had a post-lesson meeting to discuss the lesson design. Based on the trial teaching and discussions, Miss Wang further revised lesson design and sent it to Mrs. Zhang for comments again. According to Mrs. Zhang's further suggestions, the teacher revised the lesson design further and gave another open lesson in another school. On the basis of that teaching and post-lesson discussion, a final version of the lesson design was formed that is used for sharing at a national workshop several months later in Beijing. According to a unified lesson design format suggested by the project, the lesson design includes five components: content and its interpretations; aims and its interpretation; diagnosing teaching and learning difficulties; selecting teaching methods and measures; and instructional process.

Data Resources and Analysis

Data resources. The data collected for this case study include initial and final lesson designs, and two tracked comments on lesson plans in progress. Two interviews with each participant were conducted. The first interview includes five types of questions, including personal information, the forms and requirement of coaching, the roles of coaching for professional development, the learning from developing the open lesson, and learning from the whole process of participating in the activity. The interview questions were required to answer in written format. Based on the analysis of lesson design documents, and written interviews, the researchers identified the substantial changes in the lesson design, and teacher's possible gains and associated factors. The researchers conducted a follow-up interview with the teacher through Skype probing further explanations and interpretations of the teacher's changes and gains. The follow-up interview lasted 45 minutes and was recorded. And the record was transcribed into Chinese literally. Similarly, a follow-up interview with the coach was made to get further explanations of her roles in helping the teacher's developing this open lesson and her own gains through this coaching process (Hereafter TI1~2 refer to interviews 1 and 2 with the teacher, while CI1~2 refer to interviews 1 and 2 with the coach).

Data analysis. The data analysis focused on two main themes: what changes made with regard to the lesson design and teacher's learning, and what factors associated with these changes. Through comparing the initial design and final design, the changes in lesson design were identified. Through analyzing the track comments and interviews, we identified how the teacher achieved these changes, what she might learn from developing the lesson design. Meanwhile, we also paid attention to identifying possible contributions made by the coach. All data analysis was based on Chinese documents; only quoted parts were translated into English as needed.

Results

The findings are presented in three parts. First, main changes in lesson design and factors associated with these changes are presented and discussed. Then, the teacher's gains through participating in the open lesson development are described. Finally, the section ends with a summary.

The Substantial Changes in Lesson Design and Factors Associated with the Changes

Changes in instructional objectives and important content points. With regard to the changes in instructional objectives and important content points, the following points were identified: (1) developing a deep understanding of teaching materials and setting reasonable instructional objectives; (2) getting a proper understanding of students' learning readiness and difficulties.

Developing a deep understanding of teaching materials and setting reasonable instructional objectives. One essential improvement was to get a deep understanding of the teaching materials and to identify appropriate instructional objectives. In the initial lesson plan, the teacher identified the instructional objectives as to help students learn and differentiate relevant basic concepts, including sample survey, population, element, sample, size of sample, simple random sampling.

In the traced comments on the initial lesson plan, Mrs. Zhang made several constructive suggestions about students' readiness, content objectives, and process and method objectives. With regard to the students' preparation of learning, Miss Wang only mentioned what knowledge student had learned. Mrs. Zhang specified students' readiness as they should already accustom to innovative teaching approach that the new curriculum suggests rather than what they used to learn using traditional textbooks.

> These students at grade 7 are the last graduates from elementary school where the traditional textbooks were used. Thus, the statistic knowledge they have learned are relatively weak. However, since the new curriculum has been implemented for quite a number of years, these students have familiarized with innovative learning approaches such as collaborative learning, group activities although they used the traditional textbooks.

Regarding to the content objectives, the teacher originally tried to help students understand census survey and sample survey, and master four basic concepts through contextual examples, and master four commonly used sampling methods, including simple random sampling, stratified sampling, and same distance sampling and group

251

sampling. However, the coach suggested that mastering four sampling methods may be not feasible, and lesson design should focus on "the understanding of the necessity of sample survey, appropriateness and representativeness of sample, and relevant concepts through contextual situations and case studies. The emphases would be placed on preliminarily mastering simple sampling method, understanding the statistical thinking of population estimation based on sample." These suggestions were fully taken by the teacher in her final lesson plan.

Regarding the process and method, the coach suggested the following points: (1) experience the universality and necessity of sample survey through daily life examples; (2) experience the statistical thinking of estimating population based on sample; and (3) experience the methods and process of sample survey through experiment and exploration.

The coach also indicated that "if she is not able to reflect my suggestions, she just adopts her own objectives." Evidently, the teacher took in all these points in the further design.

Finally, the teacher developed more essential and concrete understandings of designing this lesson: (1) population estimation based on sample is not an inductive thinking method; (2) two basic conditions of sampling: appropriate size of a sample and representativeness of a sample; (3) statistical thinking is the core and essence; (4) selecting more daily life and social cases; and (5) letting students develop collaborative learning and enjoy the process.

Getting a holistic understanding of students' learning readiness and difficulties. In the teacher's initial lesson design, she only paid attention to the knowledge that students learned at elementary school. But in the final lesson plan, the teacher not only recognized what students learned at elementary school but also what students learned and reviewed in the first lesson.

With regard to learning difficulty, the teacher described in the initial lesson design that the difficulty is "how to conduct sample survey" in general terms. In the final lesson plan, she identified more concrete difficulties, including "how to select a representative sample, what is an appropriate sample size" and "how students may master relevant methods progressively through concrete activities."

Changes in instructional procedures. "Since the instructional objectives and emphases were changed, I had to redesign the lesson again with substantial changes. In fact, I totally changed my design except keeping several examples" (TI1). Based on the tracked coach's comments on the teacher' lesson plan, we identified major changes in the following aspects: (1) review of census survey; (2) highlighting the differences between census survey and sample survey; (3) introducing new concepts through problem solving; (4) carefully developing relevant concepts of random sampling; (5) developing the simulation experiment from special case to general situation; and (6) extending learning to next lesson. The details were displayed in Figure 1.

IMPROVING TEACHERS' EXPERTISE

Category	Initial	Final	Change
Introduction	Situational problem 1: if you want to know the taste of a pot of soup, what will you do? Discussing situational problem 1; Introducing sampling concepts; Student discussion of sampling	Review of census survey; Discussing situation 1; Introducing the concept of sampling method; Students provide examples	Learning new concepts through reviewing relevant concepts
Census survey and sample survey	Example 1: introducing simple random sampling survey and census survey Exercise 1: differentiating these two survey methods	Example 1: Comparing the features of simple random sampling survey and census survey;	Contrasting census survey and sample survey;
Sampling related concepts	Introducing four different sampling methods; Introducing relevant concepts of random sampling directly; Exercise 2: Exercise for clarifying relevant concepts.	Example 2: By using the problems in Example 1, introducing relevant concepts of random sampling	Introducing new concepts through problem solving
Application of simple random sampling	Example 2: Design survey: students' eye vision in the city	Example 3A: survey students' preferences of TV programs in the school. Considering a special case: the class first	Adapting and modifying Example 3 from textbook
Consolidation exercise	No	Example 3B: survey student's preferences in TV programs in the school.	Consolidating and applying learned concepts and methods
Summary	Simple random sampling Comparing the features of census and simple random sampling	Relevant concepts; Simple random sampling	Emphasizing relevant concepts and process of sampling
Paving for further learning	No	Exploration: Design survey of students' eye vision in the city, according to the detailed instruction	Extending learning for further study

Figure 1. Substantial changes in instructional procedures

Review of census survey. The situational problem 1 is the only one kept from initial design in the final lesson plan. The slight improvement was to have a brief review through asking what students learned about census survey before introducing the problem. It makes the lesson connected to the previous lesson.

253

Highlighting the differences between census survey and sample survey. In the initial lesson design, one example including four sub-questions and three exercises were used for helping students identify the features of sample survey and census survey. However, in the final design, the teacher used Example 1, including five sub-questions, to compare the features of two survey methods and summarize key points.

Developing relevant concepts of sampling through problem solving. In the initial lesson plan, after introducing four different sampling methods, the teacher directly introduced four relevant concepts and then provided two multiple choice exercises for students to clarify differences among these concepts. However, in the final design, the teacher asked students to discuss the features of three simple random sampling problems that are modified from Example 1. Through tabulating and comparing the population, elements, sample, and sample size with regard to the three questions, the four concepts were introduced contextually.

Developing a simulation project through deliberative modifications of problems from textbook. In the initial lesson plan, the teacher gave a survey problem as follows: in order to know the eye vision of students at elementary and secondary schools in the city, please design a survey. Step 1: determine which survey methods will be used (census survey or sample survey); Step 2: determine the population and elements; Step 3: discuss the sampling method, determining the sample and its size. According to suggestions from the coach, the teacher adopted a problem from the textbook (example 3) to survey students' preferences of TV programs. The problem was deliberately refined. She realized that the refined problems are "more meaningful and interesting than the previously selected one" (TI1). According to the textbook, in the first lesson, students did a census survey on their own preferences of TV programs. In the second lesson, students would be asked to design a survey on all students' (2447 students in school) preferences of TV programs. However, the teacher added one sub-problem: if the class is the population and we use different sizes of samples (e.g., 5 students vs. 20 students) to estimate the whole class's preferences with a simple random sampling method (e.g., numbering students, and then drawing numbers randomly). The simulation activity aimed to help students experience the process of data collection, classification and analysis. It also tended to draw students' attention to two key issues: representativeness of sample and sample size. After that, the teacher organized another activity in groups to design a similar survey of the whole school as population. Pedagogically, this heuristic approach helps students understand and solve the problem with increasing complexity.

Preparation for further learning. The survey on students' eye vision presented in the initial plan was used as an exploratory problem for further learning in the final plan. When solving the problem by a simple random sampling, students may face a new challenge: sometimes the sample selected through a random sampling is not representative, how to deal with the problem becomes a pressing issue which actually is the topic of the next lesson.

In summary, through these major improvements, the lesson plan showed the following features: (1) connections in content: the new content is linked to the previous knowledge and lays the foundation for further learning in next lesson; (2) coherence and development: knowledge development and application are based on contextual, interconnected, and varying mathematical task explorations; (3) students' participation in the process of developing knowledge and statistical thinking methods through collaborative learning and group activities.

The Teacher's Gains through Participating in the Public Lesson Development

During the teacher's interview, the teacher further explained her major gains from developing this lesson with Mrs. Zhang's guidance. In particular, she made substantial progresses in the following aspects: improving her teaching research ability; developing her conception of quality teaching, deepening her understanding of the content and enhancing her teaching skills.

Improving the teacher's teaching research ability. Teaching research ability basically includes two aspects. One refers to the process of developing and reflecting on lessons, and writing reflection reports or papers. Another means the study of teaching materials. In this particular case, Mrs. Wang needed not only to complete several cycles of lesson design, delivery and revision, but also to deliver public lessons at the city and national levels and to write reflection report for sharing or publication. In our follow-up interview, she further explained the benefits from participating in this activity as follows:

> If I had not participated in this activity, I would not have gotten such a deep understanding of statistical content and the use of examples in the textbook. After participating in this activity, when I teach other contents, I will pay particular attention to these questions: if I use this approach, what may be its impact on students' learning? If I adopt another approach, what may be the differences regarding students' learning? How to understand and implement the guidance stated in teaching reference books? If I had not attended this project, I would not have considered these aspects so deeply. Nor would I have known how to consider and solve these problems. Sometimes, I even need to go further beyond the textbook in order to achieve my instructional objectives. For example, how to use introductory situations? How to teach mathematics with a focus on students' understanding of concepts? Moreover, how to improve students' learning? I have learned a lot through developing the open lesson and exchanging with experts. (TI2)

In above explanations, the teacher emphasized that she did get a profound understanding of the particular content knowledge and relevant pedagogical knowledge. More importantly, she developed an awareness of the importance to study teaching materials, consider relevant problems (e.g., instructional objectives), and master methods to deal with relevant issues.

With regard to reflecting on practice and writing research reports and papers, the teacher explained how she had learned from Mrs. Zhang's face-to-face guidance, and written comments on lesson plans at different stages in particular.

> Mrs. Zhang gave me a lot of guidance. For example, I can learn from her comments on our instructional objectives and teaching material analysis. Previously, I am not able to express what I thought. Now I can write what I thought so as to let others understand and make sense. I feel that I have made some progress. But I need to put in more efforts to learn from Mrs. Zhang for a long time. (TI2)

In addition, the teacher also explained that Mrs. Zhang's attitude toward research has a great impact on her thoughts and behavior as follows:

> In particular, when Mrs. Zhang led us to Beijing to attend the fourth symposium of the research project, I observed how she raised questions, how she commented on others from very unique and critical perspectives. She just focused on important issues, and expressed her own opinions on the questions. She did not care about the status of the person who has the problems, even if the person is an expert. If there are any problems, she always expressed her critical opinion to the problems but not to the persons. I do learn a lot from her.

In her reflection report on lesson plans, the teacher summarized the strengths and weaknesses of the lesson plan. With regards to the strengths of the lesson, she pointed out four features as follows: (1) deliberated design and appropriate guidance; (2) holistic consideration and optimizing instruction process; (3) illustrating basic concepts through concrete examples; (4) students' high engagement and active collaboration in groups. Meanwhile, she also realized several weaknesses: (1) the teacher explained too much and students' discussion was not enough at some stages; (2) some particular treatments of contents need to be improved. In the interview, the teacher further explained that the lesson could be improved further in the following aspects: (1) refining the "random sampling survey and census survey", (2) when conducting simulation experiment, more students should be asked to participate in the process of survey; and (3) when introducing the four concepts, students do not need to write their answers in the table as suggested in the final design, rather they could express their answers orally.

In a reflection on the whole process of developing the lesson, she highlighted three important aspects: (1) how to select situational problems; (2) how to teach mathematical core concepts and thinking methods; and (3) how to deal with relevant statistical concepts.

Developing the teacher's conception of quality teaching and improving her ability in instructional design. Through developing and revising this lesson plan, the teacher realized that "teachers should arrange instructional content and teaching strategies, and create appropriate instruction situations according to students' cognitive level and cognitive development progression. Then, they can get students engaged in classroom activities and achieve an effective instruction" (TI1). The teacher further provided one example. In one trial teaching, she introduced "population, individual, sample, and size of sample" after conducting the simulation experiment (survey on the students' references of TV programs). After the trial teaching, Mrs. Zhang (coach) pointed out that these four concepts should be introduced using concrete examples before the simulation experiment. This revision in the final plan could help students get a deep understanding of these concepts due to the immediate application of these concepts in the simulation experiment.

Through the open lesson development, the teacher realized that it is very important to have clear, reasonable and feasible instructional objectives first, then she can develop the instruction procedures based on the overarching goals as follows:

> We first analyzed instructional objectives, identified important and difficult teaching points, and analyzed teaching materials and its treatments. Then, we designed instructional procedure focusing on the implementation of these instructional objectives. Through this activity, for each lesson, I had to think what overarching goals are, what expectations to students are in terms of knowledge and ability. Then I thought of methods how I can achieve these goals. I made a continuous effort to find appropriate methods to help student's master skills and develop ability. (TI2)

Deepening the teacher's understanding of the content. The teacher indicated that the biggest gains through the whole process of developing the lesson include: (1) getting a profound understanding of statistical content. Statistics is different from other mathematics contents. We should help students make sense of statistical activities and reasoning; and (2) the concepts of statistics are not the difficult content points and the emphasis should be placed on statistical thinking methods.

Enhancing the teacher's teaching skills. The teacher realized that it is important to arrange classroom exercises scientifically and reasonably, to design introductory activities or situational problems that meet students' cognitive levels, and to have formal instructional language. For example, selection, presentation, explanation and evaluation of classroom exercises are important. The exercises should be presented progressively in terms of their difficulty and with a proper amount of different exercises. Students' answers should be used for eliciting and assessing their thinking. Questioning should be directed to students as many as possible and cater for individual differences. In particular, the teacher learned from the coach with regard to instructional skills as follows: (1) how to explain and summarize classroom exercises; (2) how to use instructional language more scientifically, for example, 'which survey method should be adopted" should be changed into "which survey method is suitable for"; and (3) how to create or adapt a situational problem. For example, how to adapt the survey on students' preferences of TV programs from the textbook (TI1).

DISCUSSION AND CONCLUSION

We described basic features of apprenticeship practices in Mainland China and illustrated teachers' learning and changes through these practices by case studies. The two case studies revealed that: (1) beginning teachers can learn from their mentors about effective teaching strategies, understanding of students' learning, and effective classroom management; (2) beginning and mid-career teachers can develop the conception of culturally-valued quality teaching, learn how to analyze teaching materials, identify proper instructional objectives, deal with difficult and important content points, and design instructional procedures to achieve instructional objectives; and (3) the apprenticeship practices can help establish a harmonious and collaborative subject-based teachers' learning community. In addition, for the mid-career teacher, she is able to apply the learned principles and methods of studying textbooks, identifying instructional objectives and achieving the objectives in daily teaching. She also made essential progress in reflecting on practice and writing teaching research report or papers, which is crucial for getting a promotion to the senior rank.

Comparing these two cases, it was found that both practices focus on studying teaching materials, developing teachers' capacity in identifying appropriate instructive objects, adopting appropriate instruction strategies based on a deep understanding of content and students learning readiness, implementing lesson effectively. Yet, the content-focused coaching practice may benefit teachers to develop teaching research ability and change their views about mathematics teaching and learning, while the school-based mentoring practice may benefit teachers more to building a productive learning community in school.

A Coherent Teacher Education System as a Foundation for Teachers' Professional Development

Many studies found that Mainland China has a coherent and cooperative system to support teachers' professional development through studying teaching materials, mastering important and difficult points of teaching (Ma, 1999; Wang & Paine, 2003), and extensively observing experienced teachers' instruction and being observed by others (Steward, 2006). Teachers' ranking and promotion system applies to every teacher in Mainland China (Huang et al., 2016; Li et al., 2009). The expectations for teachers at different ranks specify what teachers need to pursue for promotion. In particular, the promotion from primary to intermediate level contains much emphasis on the teacher's own professional development in terms of his/her teaching competence and instructional quality. The expectation matches very well with what novice teachers would need to learn about the teaching routine, develop basic teaching skills and lesson planning, deepen their understanding of teaching materials, and become competent teachers (Huang & Li, 2008). Through participating in the apprenticeship practices and other supportive school-based teaching research activities, those qualified teachers can be promoted to the intermediate rank and further to the senior rank, namely, master teachers (Huang et al., 2016).

Moreover, the apprenticeship practices, together with other professional development approaches and practices such as teaching research activities at different levels (from school-based to municipal wide) (Yang, 2009) and Chinese lesson study (exemplary lesson development, and teaching contest) (Huang & Han, 2015; Huang & Li, 2009; Li & Li, 2009), provide supportive and competitive environments for teachers to develop their expertise.

Apprentice Practice as a Catalyst of Professional Development System

As discussed previously, the ranking and promotion system provides a guideline that drives teachers' professional development in Mainland China. At the same time, this ranking and promoting system also places expectations for the teachers with a senior rank to mentor novice and junior teachers. The merit of mentoring junior teachers becomes important when senior teachers are to be considered for promotion to the exceptional class (Huang et al., 2016; Li et al., 2009; Ministry of Education, 1995).

Thus, for senior teachers, it is their honor and responsibility to serve as mentors to help novice and/or mid-career teachers develop their professional knowledge and skills. For novice and mid-career teachers, the apprenticeship practice provides them with feasible and pertinent learning opportunities.

In summary, both the school-based and content-focused across-school apprenticeship practices provide teachers with professional development opportunities that are pertinent to their daily teaching. The apprenticeship practice is not only adopted by the school locally, but also supported by the professional promotion system in Mainland China. As a public policy, the ranking and promotion system supports teachers' collaborative learning of their classroom instruction as a public professional practice. The specifications of teachers' expertise at different stages also support the mentoring practice for both master teachers and novice teachers.

What Can We Learn from the Apprenticeship Practice in Mainland China?

In the United States, a mentoring and coaching approach has been proposed and used in some schools for a number of years (Russo, 2004; West & Staub, 2003). The approach aims to embed professional development in teachers' classroom work with children, academic content and research-based approaches, and to create more collaborations and a sense of community among teachers in a school (Russo, 2004). This approach reflects a situated learning perspective for teachers (Peressini, Borko, Romagnano, Knuth, &Willis, 2005), which emphasizes the development of teachers' knowledge for teaching as embedded in their learning community.

In the case of Mainland China, the development and use of the professional development system and various strategies focus on improving classroom instruction, dealing with daily teaching issues and implementing new curriculum in classrooms. The essential features include content-focused, teachers-oriented, aiming at developing high-quality classroom instruction, and ultimately the improvement of students' learning. The professional development system provides a context for teachers to develop their knowledge and skills needed for teaching within a multilevel learning community (from school-based to district to city based) (Huang et al., 2016). Thus, the apprenticeship practices in Mainland China demonstrate certain features of teachers' learning which are in line with the approach proposed and used in the U.S.

As the issue of developing teachers' expertise through professional development has been a perplexing one in many education systems, an explicit examination of relevant practices in Mainland China can hopefully advance our understanding of the issue. Investigating the particular apprenticeship practice in Mainland China in detail should provide an important case for mathematics education researchers to reflect on their own practices and carry out critical examinations of what can be learned from other education systems.

NOTE

[1] Master teachers refer to those with the senior and full senior rank in this chapter.

REFERENCES

An, S., Kulm, G., & Wu, Z. (2004). The pedagogical content knowledge of middle school mathematics teachers in China and the U.S. *Journal of Mathematics Teacher Education, 7*, 145–172.

Beishuizen, J. J., Hof, E., van Putten, C. M., Bouwmeester, S., & Asscher, J. J. (2001). Students' and teachers' cognition about good teachers. *British Journal of Educational Psychology, 71*, 185–201.

Guskey, T. R. (2003). What makes professional development effective? *Phi Delta Kappan, 84*, 748–750.

Huang, R. (2014). *Prospective mathematics teachers' knowledge of algebra: A comparative study in China and the United States of America*. Wiesbaden: Springer Spektrum.

Huang, R., & Bao, J. (2006). Towards a model for teacher professional development in China: Introducing keli. *Journal of Mathematics Teacher Education, 9*, 279–298.

Huang, R., & Han, X. (2015). Developing mathematics teachers' competence through parallel lesson study. *International Journal for Lesson and Learning Studies, 4*(2), 100–117.

Huang, R., & Li, Y. (2008). Challenges and opportunities for in-service mathematics teacher professional development in China. *Journal of Mathematics Education, 17*(3), 32–38.

Huang, R., & Li, Y. (2009). Pursuing excellence in mathematics classroom instruction through exemplary lesson development in China: A case study. *ZDM: The International Journal on Mathematics Education, 41*, 297–309.

Huang, R., Peng, S., Wang, L., & Li, Y. (2010). Secondary mathematics teacher professional development in China. In F. K. S. Leung & Y. Li (Eds.), *Reforms and issues in school mathematics in East Asia* (pp. 129–152). Rotterdam, The Netherlands: Sense Publishers.

Huang, R., Ye, L., & Prince, K. (2016). Professional development system and practices of mathematics teachers in Mainland China. In B. Kaur, K. O. Nam, & Y. H. Leong (Eds.), *Professional development of mathematics teachers: An Asian perspective* (pp. 17–32). New York, NY: Springer.

Jin, L., & Cortazzi, M. (1998). Dimensions of dialogue: Large classes in China. *International Journal of Educational Research, 29*, 739–761.

Lee, J. C., & Feng, S. (2007). Mentoring support and the professional development of beginning teachers: A Chinese perspective. *Mentoring and Tutoring: Partnership in Learning, 15*, 243–263.

Li, Y., & Huang, R. (2008). Chinese elementary mathematics teachers' knowledge in mathematics and pedagogy for teaching: The case of fraction division. *ZDM: The International Journal on Mathematics Education, 40*, 845–859.

Li, Y., Huang, R., Bao, J., & Fan, Y. (2009). Facilitating mathematics teachers' professional development through ranking and promotion practices in the Mainland China. In N. Bednarz, D. Fiorentini, & R. Huang (Eds.), *The professional development of mathematics teachers: Experiences and approaches developed in different countries* (pp. 72–87). Ottawa: Ottawa University Press.

Li, Y., Kulm, G., & Smith, D. (2006, October). *Facilitating mathematics teachers' professional development in knowledge and skills for teaching in China and the United States*. Invited presentation given at the Second International Forum on Teacher Education, Shanghai.

Li, Y., & Li, J. (2009). Mathematics classroom instruction excellence through the platform of teaching contests. *ZDM: The International Journal on Mathematics Education, 41*, 263–277.

Li, Y., Ma, Y., & Pang, J. (2008). Mathematical preparation of prospective elementary teachers. In P. Sullivan & T. Wood (Eds.), *International handbook of mathematics teacher education: Knowledge and beliefs in mathematics teaching and teaching development* (Vol. 1, pp. 37–62). Rotterdam, The Netherlands: Sense Publishers.

Li, Y., & Shimizu, Y. (2009). Exemplary mathematics instruction and its development in East Asia. *ZDM: The International Journal on Mathematics Education, 41*, 257–262.

Ma, L. (1999). *Knowing and teaching elementary mathematics: Teachers' understanding of fundamental mathematics in China and the United States*. Mahwah, NJ: Lawrence Erlbaum Associates.

Ministry of Education, P. R. China. (1993). *Regulations for promoting teachers to exceptional class* [in Chinese]. Retrieved February 21, 2008, from http://www.moe.edu.cn/edoas/website18/info5947.htm

Ministry of Education, P. R. China. (1994). *The teachers act in P. R. China* [in Chinese]. Retrieved October 16, 2007, from http://www.moe.edu.cn/edoas/website18/info1428.htm

Ministry of Education, P. R. China. (1995). *Regulation of teacher's qualification* [in Chinese]. Retrieved October 16, 2007, from http://www.moe.edu.cn/edoas/website18/info5919.htm

Ministry of Education, P. R. China. (1999). *Education revival agenda for 21st century* [in Chinese]. Retrieved September 22, 2008, from http://www.202.195.144.106/jxzlhb/zlhb2_1.htm

Ministry of Human Recourses and Social Security and Ministry of Education of China. (2015). *Guidance for deepening the reform of secondary and primary school teachers' promotion system* [in Chinese]. Retrieved December 12, 2015, from http://www.mohrss.gov.cn/SYrlzyhshbzb/ldbk/rencaiduiwujianshe/zhuanyejishurenyuan/201509/t20150902_219575.htm

People Educational Press. (2004). *Compulsory education standards-based experiment textbook, Mathematics (7B)* [in Chinese]. Beijing: Author. Retrieved June 29, 2009, from http://www.pep.com.cn/czsx/jszx/qnxc/dzkb/

Peressini, D., Borko, H., Romagnano, L., Knuth, E., & Willis, C. (2005). A conceptual framework for learning to teach secondary mathematics: A situative perspective. *Educational Studies in Mathematics, 56*, 67–96.

Russo, A. (2004). School-based coaching: A revolution in professional development or just a fad? *Harvard Education Letter, 20*(4), 1–3.

Stewart, V. (2006). China's modernization plan: What can US learn from China? *Education Week, 25*(28), 48–49.

Wang, J. (2009). *China mathematics education: Tradition and reality*. Jiangsu: Jiangsu Educational Press.

Wang, J., & Paine, L. W. (2001). Mentoring as assisted performance: A pair of Chinese teachers working together. *The Elementary School Journal, 102,* 157–181.
Wang, J., & Paine, L. W. (2003). Learning to teach with mandated curriculum and public examination of teaching as contexts. *Teaching and Teacher Education, 19,* 75–94.
West, L., & Staub, F. C. (2003). *Content-focused coaching: Transforming mathematics lessons.* Portsmouth, NH: Heinemann.
Yang, Y. (2009). How a Chinese teacher improved classroom teaching in teaching research group: A case study on Pythagoras theorem teaching in Shanghai. *ZDM: The International Journal on Mathematics Education, 41,* 279–296.

Yeping Li
College of Education and Human Development
Texas A&M University
Texas, USA
and
Shanghai Normal University
Shanghai, China

Rongjin Huang
Department of Mathematics Sciences
Middle Tennessee State University
Tennessee, USA

GLORIA ANN STILLMAN

14. LEARNING AND IMPROVING MKT THROUGH TEACHING AND PROFESSIONAL DEVELOPMENT MECHANISMS

INTRODUCTION

Mathematics teachers' knowledge of mathematics and their knowledge of how to teach it has been the object of increasing research attention in recent decades (e.g., Barwell, 2013; Burgess, 2009; Rowland & Ruthven, 2011; Sfard, 2005; Speer, King, & Howell, 2015). Ball and colleagues (Ball & Bass, 2003; Hill & Ball, 2004; Ball, Hill, & Bass, 2005) are attributed with coming up with the term *Mathematics Knowledge for Teaching* (MTK) which they tied to "the work of teaching mathematics" (Hill, Rowan, & Ball, 2005, p. 373) as their purpose was to develop a practice-based theory of the nature of this knowledge. Ball, Bass, Hill, and Thames (2006) initially specified only four components but these were expanded by others (e.g., Chick, Baker, Pham, & Cheng, 2006; Li & Kulm, 2008) and in Ball's own research group's work. The practice-based framework of teachers' mathematical knowledge for teaching of Ball, Thames, and Phelps (2008) has six domains, three are sub-domains of what Shulman (1986) called subject matter knowledge, namely, *common content knowledge*—CCK, *specialised content knowledge*—SCK, and *horizon content knowledge*—HCK and there are three sub-domains of pedagogical content knowledge, namely, *knowledge of content and students*—KCS, *knowledge of content and teaching*—KCT, and *knowledge of content and curriculum*—KCC. Despite these domains being derived mainly from analyses of elementary school teachers and their practices, "researchers have not explicitly examined the potential impact" this lineage has had on the theorising of MKT (Speer et al., 2015, p. 110) in a concerted and coordinated effort and its use at the secondary level. Speer et al. (2015) question the validity of "potentially overgeneralising the current theoretical framing from elementary to secondary settings" (p. 120) and use an example vignette from a research project in a secondary classroom to show the difficulty of distinguishing between CCK, SCK and PCK. In the analyses that follow of the outcomes of the studies used to demonstrate the utility of different professional development mechanisms in Part III of this volume, Ball et al. still proved useful even for studies that involved teachers at secondary level as it was possible to discriminate between domains of MKT to the extent that evidence was provided by the chapter authors.

Shulman (2004) promoted the idea of career-long learning "communities" for teachers "where they can actively and passionately investigate their own teaching" (p. 498) in order to build a knowledge base for teaching that goes beyond what is possible within a subject discipline department in one school. This focus on the generation of distributed knowledge for teaching amongst "a community of teachers as a collective subject" is taken up by Williams (2011, p. 176) within the context of auditing and evaluation of teacher knowledge in the UK. In many countries there has been a recent focus on facilitation of teachers' learning throughout their professional careers. From Singapore, for example, Chua reports on the roles of Learning Circles and Centres of Excellence for Mathematics in facilitating the meeting of practising mathematics teachers' learning needs.

The chapters in Part III of this volume add to the literature on professional learning and development of practising mathematics teachers through the presentation of qualitative case studies from mainland China evaluating the development of MKT appropriately in situ (Williams, 2011). According to Liang (this volume), engagement in professional development is mandatory in China throughout a teacher's career. The knowledge teachers acquire and continually seek to improve through professional development has for them "both use-value (knowledge needed for them to be able 'to teach') and exchange value" (Williams, 2011, p. 168) (as a means to progress professional status through the promotion system to different levels). The particular professional development mechanisms foregrounded, except for the master teacher workstation program, have been used for many years in mainland China (Huang, Ye, & Prince, 2016) but are evolving and increasing in reach (or have the potential to do so) as digital tools allow communication between widely geographically located individuals and learning collectives to be bridged (see Huang & Huang, this volume, for an example). What is distinctive about these Chinese examples is a deliberate focus on nourishing practical knowledge – that is, the wisdom of practice cultivated in teachers' own classrooms as sites of inquiry into improving their own practice (Gu & Gu, 2016).

The particular professional development mechanisms for growth of MKT once teachers have entered the teaching workforce, that are featured here, are overviewed in Table 1. These are: (a) in-depth study of textbooks for teaching; (b) the master teacher workstation (MTW) program; (c) teaching research group (TRG) activities; (d) public lessons (or Open Class as they are also called); and (e) apprenticeship practice. All but one chapter incorporates the reporting of one or two in-depth qualitative case-studies undertaken by the chapter authors. The chapter by Liang differs in that it is more descriptive of different forms of the professional development mechanism used. These are then illustrated by excerpts of findings from several different source studies. Thus, even though this chapter is included as a row in Table 1, some cells are not applicable (N/A) or unable to be specified as explicitly as in the other chapters. For the other chapters, details, if given by the authors, of studies, the "research community" in the sense of the community researched as part of the case-study which often focussed on a subset of participants for reporting, the focus

Table 1. Overview of mechanisms for MKT development in Part III

Authors	Mechanism for MKT development	Nature of studies	Research community (RC)	Participants	MKT aspects developed (Ball et al.)
Pu, Sun & Li	In-depth textbook study: planning in school-based Teaching Research Group (TRG)	Qualitative case-study (2 weeks)	Organised TRG: 10 teachers in a primary school, 3 master teachers, 1 university teacher, 2 teaching research staff	3 primary (Grade 6) maths teachers (with approx. 3, 6, 13 years experience)	SCK, KCS, KCT, KCC
Huang, X. & Huang, R.	Master teacher workstation (MTW) program`	Intensive qualitative case-study	MTW (6 + Mr Wan) + 100 social media group members led by Mr Zhang	1 experienced (>15 years) senior secondary maths teacher (Mr Wan) in a rural school	KCT, KCC
Yang & Zhang	Teaching Research Groups (TRG)	Qualitative case study of TRG	Hangzhou TRG: 10 primary maths teachers – 3 <5years experience; 3 approx. 10 years experience; 3 15–20 years experience; 1 master teacher	1 inexperienced. primary teacher taught 1st & 2nd iteration of lesson to her two Grade 6 classes; other TRG members observed & participated in post-class discussion	SCK, HCK, KCS, KCT, KCC
Liang	Public lessons	N/A	various	various	CCT/SCK, KCT, KCS, KCC
Li, & Huang, R	School-based Apprenticeship practice	Qualitative case-study	N/A	2 mentee-mentor pairs in secondary school (one pair where mentee new graduate other mentee experienced middle school teacher)	SCK of experienced mentee, KCS, KCT, KCC
Li, & Huang, R	Content-focussed coaching practice	Qualitative case-study	Lesson Preparation Group (6 teachers) + coach	2 secondary maths teachers – 1 mid-career teacher (Grade 7) and 1 Master teacher as coach	SCK, KCS, KCT, KCC

265

participants including schooling level, teaching grade and teaching experience, and the aspects of MKT developed according to the broad domains of Ball et al. (2008) are brought together for easy reference and comparison. I now briefly examine how the authors frame each method to show how it improves knowledge of teaching and teacher knowledge.

MKT DEVELOPMENT THROUGH STUDYING TEXTBOOKS

There is relatively little research internationally on in-service teacher development by studying of the textbook(s) used for teaching. Most textbook studies are textbook analyses or comparisons (Fan, Zhu, & Miao, 2013) of content for particular purposes such as level of challenge of tasks (Brändström, 2005). suitability for teaching mathematical literacy (Gatabi, Stacey, & Gooya, 2012), or development of a socio-critical perspective on mathematical modelling (Stillman, Brown, Faragher, Geiger, & Galbraith, 2013). Clearly, results of this research are intended to inform teaching but often these point to inadequcies rather than enabling practices. For instance, Swedish textbook tasks are meant to be grouped into strands according to difficulty in order to assist teachers to choose tasks for differentiation of learning; however, Brändström (2005) found that the level of challenge was low in Year 7 textbooks in all strands even when meant to be higher.

In Western professional development teaching tasks are often the focus as they are considered the mainstay of the lesson whether it be for conceptual development, development of techniques, practice, consolidation, problem solving, investigation or mathematical modelling. Rarely is it considered in textbook research and scholarly literature that in exploring textbook tasks in planning lessons for the next day the teacher might be using this to increase their MKT of their own accord (Fan et al., 2013). This is despite evidence (Pepin, Gueudet, & Trouche, 2013) of the role of textbooks as an interface between cultural traditions, educational policy and practice and a primary resource for preparation in teaching in several countries. For instance, Pepin, Gueudet, and Trouche (2013, pp. 695–696) found in their study of textbooks in France and Norway that textbooks are "a vital mediating object" between policy documents such as national curricula and what occurs in the classroom which is the site of teachers' work.

The perceived important role of textbooks in the professional growth of Chinese elementary school teachers was brought to the attention of many in Western countries with the publication of Ma's (1999) comparative study of the mathematical understandings of Chinese and USA elementary teachers. The combination of Chinese elementary school teachers' use and intensive study of textbooks was considered to be the most important contributing factor to their professional knowledge growth over their career. Pu, Sun and Li in their chapter attempt to add to the sparse research results on knowledge improvement of in-service teachers through textbook study. They report on the results of a two-week intensive study focussing on teaching the topic of Knowing and Understanding Circles in Grade 6. The primary school with its 10 teachers partnered with three master teachers, a university teacher and two teaching research staff to form a research community to support the mathematics teachers in the school master textbook analysis whilst improving their knowledge and teaching skills. The three teachers of Grade 6 participated fully in the textbook analysis process and carried out the classroom teaching with the support of the organised research community, a school-based teaching research group (see Table 1). From a MKT perspective, there was growth in both subject matter knowledge and pedagogical content knowledge. All three teachers knew the basic definitions of circles before the study and there was no change in depth of this knowledge during the study; however there was growth in SCK as the least experienced teacher came to understand the relationships between the components of circles and all three teachers developed knowledge of the history of circles and how this could be used in teaching. With respect to pedagogical content knowledge, all teachers increased their curricular knowledge (KCC) and knowledge of instructional strategies (KCT) whereas the two less experienced teachers also increased their knowledge of student response and potential difficulties with approaches to teaching Knowing and Understanding Circles (KCS).

MKT DEVELOPMENT THROUGH MASTER TEACHER WORKSTATION PROGRAM

One of the most innovative ideas in this collection of professional learning mechanisms is the notion of a master teacher workstation (MTW) (or studio) program to help up-and-coming experienced teachers become experts (see Zhang & Leung, 2013, for conceptions of expert teachers in China) and so facilitate the growth of their professional knowledge towards that of master teachers. There seems to be little existing research (Chen & Wu, 2016; Li, Tang, & Gong, 2011) on MTW published in English to date. Li, Tang, and Gong (2011) have provided evidence through case studies of two experienced senior-rank, elementary teachers improving lessons through participation in a MTW led by a nationally recognised elementary school teacher but it is still unknown what, and how, these experienced teachers learn and whether or not what they do transfers to other lessons. The chapter by Huang and Huang is thus a unique contribution in the sense that it provides an exemplar from practice that serves as a paradigmatic case (Freudenthal, 1981) that answers these questions.

Through a detailed analysis of the case of an experienced secondary school teacher, Mr Wan, as a member of master teacher Mr Zhang's MTW, they show the development of Mr Wan's knowledge of content and teaching (KCT) by refining his knowledge of instructional design and knowledge of manipulative materials and their effective use (KCC). This occurs through the process of *participation* in the MTW; *inspiration* from the master teacher's reflective comments and pedagogical approach; *development* of a problem-posing content review lesson (*Fuxie Ke*) with the support of the MTW members and leader through planning, implementation, reflection and refinement; *experimentation* in an attempt to perfect the teaching approach in the *Fuxie Ke* context; *success* in discovering the most effective strategies to use student problem posing for content review; *extension* of the problem posing approach to another lesson type, namely an introductory lesson; and finally *permeation* throughout his teaching repertoire as he naturally and flexibly finds ways to incorporate student problem-posing into other lesson types. Mr Wan had *dabbled* with the idea of using problem-posing several years previously when he used it in development of a public lesson but it was only when he made a *commitment* to changing his practice as an experienced teacher with a well-developed knowledge of content and students (KCS) that he was able to bring his change in practice to fruition with the support of the MTW.

MKT DEVELOPMENT THROUGH TEACHING RESEARCH GROUP ACTIVITIES

Teaching research group (TRG) activities have played a role in Chinese teachers' professional development since they were set up in the 1950s as a national system through government regulation. They are a form of lesson study that differs from that practised in Japan. Chinese lesson study is receiving increasing attention outside China as more Chinese authors (e.g., Chen & Yang, 2013; Gu & Gu, 2016; Huang, Gong, & Han, 2016; Pang, Marton, Bao, & Ki, 2016; Yang, 2009; Yang & Ricks, 2012, 2013) publish in English in high profile journals, both generalist and dedicated to lesson study, and in books. Although TRG activities can be used for other purposes such as studying the development of exemplary lessons (called *Keli*) or public lessons, Yang and Zhang chose to report on the development of MKT during the study of lesson development by an inexperienced teacher within the support system of a mathematics, school-based TRG in what they categorise as an average primary school in Hangzhou.

The focus of the research is on the preparation, implementation, and refinement of a lesson on Position for Grade 6. A novice teacher who had not taught the topic previously worked with a mentor teacher to prepare, implement and refine the lesson observed and discussed by her TRG (see Table 1). The culturally derived *Three Points Framework* (Yang & Ricks, 2013) was used as a heuristic tool (lesson's key point, difficult point, critical point) to frame their discussion during the two post-lesson discussions that followed the two implementations of the lesson during the lesson-improving process. This tool not only was useful to the teachers but also it allowed the chapter authors insight into the nature of the knowledge acquired during this process. Yang and Zhang relate this knowledge to Grossman's (1990) take on PCK. They link the two as follows:

< knowledge & beliefs about purpose; curricular knowledge > ←→ < key point >
< knowledge of student conceptions > ←→ < difficult point>
< knowledge of instructional strategies > ←→ < critical point>.

Using Ball et al.'s Framework (2008), the knowledge domains of MKT that are highlighted, and thus potentially developed, in both the lesson developer's reflection on the lessons and the lesson preparation and the post-lesson reflections of the others in the research community reveals that, in this instance. articulating the *key point* of the lesson surfaced a deeper understanding of the mathematical nature of locating and representing position (SCK), an awareness of how the mathematical focus of position is related and differs over the primary years becoming a foundation for co-ordinates in secondary school (HCK), as well as an understanding of how the big mathematical idea for this level could be effectively presented using concrete resources (KCC). Discussion of the *difficult point* was a clear learning experience for the novice lesson developer as the more experienced members of the TRG drew on their case knowledge and shared their understanding of how students would think about position in the real world and be confused, relating it to the mathematical representation of position as a point (KCS). Consideration of how the teacher would help the class reach the objective of the lesson (i.e., the *critical point*) relied on knowledge of global and specific effective teaching strategies (KCT). Again, at this point, members of the TRG drew on case knowledge of teaching strategies which to some extent the lesson developer was able to transform into strategic knowledge in the second lesson implementation. Whichever framework is used, it is clear that lesson improvement in a TRG environment is a potential site for more nuanced investigation of teaching and teacher's knowledge growth.

MKT DEVELOPMENT THROUGH PUBLIC LESSONS

Liang's chapter differs from the others in this Part of this volume. It is predicated on the assumption that experience of public lessons whether that be through participation in preparing and teaching them or merely observing and reflecting on them allows a teacher, at whatever level of experience, to acquire mathematics knowledge for teaching and increased such experience leads to an increasing accumulation of such knowledge. Public lessons started in pre-service teacher education as demonstration lessons by experts but their transformative nature was recognised as a means to facilitate the professional growth of all teachers. Different forms of public lessons, their purposes and roles in the professional learning of in-service teachers are described and supported by anecdotal transcripts from studies conducted by others or the chapter author. The forms of public lesson dealt with are: evaluative, visiting-exchange, research, exemplary (elsewhere referred to as *Keli*) and competitive lessons. Arguments are put forth that evaluative and competitive lessons are particularly beneficial to the growth of novice teachers although competitive lessons appear to have been foundational experiences on the pathway of many master/expert teachers.

In a paragraph in the conclusion, the author tells the reader that from her analysis of interview data, and teachers' reflections in research papers in Chinese and online documentation, MKT derived through these activities takes nine different forms which I was able to classify as CCK/SCK (more detail would be needed to definitively classify as CCK but it is potentially present), KCT, KCS, and KCC. In reality there is no way of confirming this list other than from the brief anecdotal transcripts included in the chapter which do not cover all the areas of knowledge listed by the author.

MKT DEVELOPMENT THROUGH APPRENTICESHIP PRACTICE

Practice is another form of professional development in mainland China with a long history in the form of school-based coaching for the induction of novice teachers or teachers changing career pathways (e.g., from middle school to high school). More recently, however, as in several other countries (e.g., Larkin, Grootenboer, & Lack, 2016; Hunter, Hunter, Bills, & Thompson, 2016; Gibbons & Cobb, 2016; Poly, 2012) another form of apprenticeship practice has appeared as subject content-focussed coaching of mid-career teachers usually from several schools. The former program has as main goal knowledge needed for them to be able 'to teach' so use-value (Williams, 2011) whereas the latter program seems to have as its goal, in China at least, acquiring of knowledge for both its "use-value" and "exchange value" as a means to progress professional status through the promotion system.

In their chapter, Li and Huang provide two case studies of apprenticeship practice (see Table 1). In the first of these the focus is on school-based coaching for induction of a beginning, recently graduated, teacher and a career changer from middle school, both being inducted into high school in mentee-mentor pairs. As well as enculturation into mathematics teaching practices in the high school setting, these coaching/mentoring pair activities provide the extra benefit of fostering a mathematics education community within the school that is a respectful learning community. In terms of MKT, the core practices employed of collaborative lesson planning, mutual lesson observation and informal post-lesson exchange serve to develop both the novice and career changer's mastery of instructional materials for the topics at this particular schooling level (KCC), appropriate teaching strategies for topics at this level (KCT) and students' typical responses and difficulties in the topics (KCS). Again, *The Three Points* framework (Yang & Ricks, 2013) works as a heuristic tool to effect this change. In addition, the mathematical knowledge and skill associated with teaching such mathematical content (SCK) is deepened for the career changer. Although the terms, mentor and mentee and coaching are used, it does not seem appropriate to make the coaching versus mentoring distinction as others do (e.g., Fletcher, 2012) but rather acknowledge that coaching in this quite practice-based context, as in others (e.g., Hunter et al., 2016), leads to reflection.

In the second case study the mentee, a mid-career teacher was school-based whilst the mentor was a master teacher working off-site in the teaching research department of the city education bureau pursuing doctoral studies part-time.

The mentor was also the leader of a research project at the time focussing on a unified open lesson design. The teacher was supported in her school by her lesson preparation group for Grade 7 (see Table 1). The mechanism for the coaching was preparation and teaching of an open lesson on Sampling for Grade 7. Although the mentee had a well-developed mathematical knowledge to draw on from her bachelor of mathematics, participation in the coaching experience allowed her to develop a much deeper understanding of SCK by making its meaning and connection within the topic, and to other mathematical topics in Grade 7, more concrete. In the process, she also deepened her understanding of teaching materials particularly textbook examples (KCC), students' thinking and possible difficulties in statistical thinking if she proceeded with her initial approach to their knowledge preparation (KCS), and a much deeper combination of her teaching and mathematical knowledge in instructional design through a deeper understanding of what particular teaching strategies would achieve in the Sampling context (KCT). A further spin-off which was a requirement of her promotion was a deeper means of reflecting on her lesson preparation and implementation so that she could express herself more articulately in writing reflection reports or research papers.

POSSIBLE FUTURE IMPACTS ON PROFESSIONAL DEVELOPMENT FOR MKT DEVELOPMENT

As indicated in the last column of Table 1, nearly all of the Ball et al. domains of MKT have been developed by one or other of the professional development mechanisms featured in Part III. Which domains are developed depends on both the purpose of the activity and the teaching experience, intentions and educational background of the teachers involved. Unlike many other countries, the expectations and standards of the ranking and promotional system in teaching in mainland China are drivers of teachers' professional development throughout their teaching careers through promulgated official guidelines (MHRSS & MOE, 2015). Practical accomplishment and practical experience are emphasized but writing academic articles and possessing of academic degrees are de-emphasized. These drivers of rank standards and promotion are likely to lead to the strengthening of the newer professional learning mechanisms mentioned in these chapters and confirm the place of others in the professional learning of the education community in mainland China; however, personal focuses by individual teachers on work-life balance might slow the pace of their trajectory of professional development (Chen & Wu, 2016). The winds of change now blow from another corner – digital technologies as the means of delivery or as cognitive collaborative communicative tools. Thus collaboration, mentoring and coaching can expand across school and geographical boundaries and overcome lack of provision locally (Borba & Gadanidis, 2008). Research into such electronic environments for teacher collaboration and the extent to which they can successfully lead to development of teachers' MKT in mainland China (e.g., Li & Qi, 2011), as elsewhere, is in its infancy.

REFERENCES

Ball, D. L., & Bass, H. (2003). Towards a practice-based theory of mathematical knowledge for teaching. In B. Davis & E. Simmt (Eds.), *Proceedings of the 2002 annual meeting of the Canadian Mathematics Education Study Group* (pp. 3–14). Edmonton: CMESG/GDEDM.

Ball, D. L., Bass, H., Hill, H. C., & Thames, M. (2006, May). *What is special about knowing mathematics for teaching and how can it be developed?* Presentation at Teachers' Program and Policy Council, American Federation of Teachers, Washington, DC.

Ball, D. L., Hill, H. C., & Bass, H. (2005). Knowing mathematics for teaching: Who knows mathematics well enough to teach third grade, and how can we decide? *American Educator, 29*(1), 14–17, 20–22, 43–46.

Ball, D. L., Thames, M. H., & Phelps, G. (2008). Content knowledge for teaching. *Journal of Teacher Education, 59*, 389–407.

Barwell, R. (2013). Discursive psychology as an alternative perspective on mathematics teacher knowledge. *ZDM Mathematics Education, 45*(4), 595–606.

Borba, M., & Gadanidis, G. (2008). Virtual communities and networks of practicing mathematics teachers: The role of technology in collaboration. In K. Krainer & T. Wood (Eds.), *Participants in mathematics teacher education* (pp. 181–206). Rotterdam, The Netherlands: Sense Publishers.

Brändström, A. (2005). *Differentiated tasks in mathematics textbooks: An analysis of the levels of difficulty* (Licentiate thesis). Luleå University of Technology, Luleå.

Burgess, T. (2009). Statistical knowledge for teaching: Exploring it in the classroom. *For the Learning of Mathematics, 29*(3), 20–29.

Chen, X., & Wu, L.-Y. (2016). The affordances of teacher professional learning communities: A case study of a Chinese secondary school. *Teaching and Teacher Education, 58*, 54–67.

Chen, X., & Yang, F. (2013). Chinese teachers' reconstruction of the curriculum reform through lesson study. *International Journal for Lesson and Learning Studies, 2*(3), 218–236.

Chick, H., Baker, M., Pham, T., & Cheng, H. (2006). Aspects of teachers, pedagogical content knowledge for decimals. In J. Novotná, H. Moraová, M. Kráták, & N. Stehlíková (Eds.), *Proceedings of 30th conference of the international group for the psychology of mathematics education* (pp. 297–304). Prague: Program Committee.

Chua, P. H. (2009). Learning communities: Roles of teachers' network and zone activities. In K. Y. Wong, P. Y. Lee, B. Kaur, P. Y. Foong, & S. F. Ng (Eds.), *Mathematics education: The Singapore journey* (pp. 85–103). Singapore: World Scientific.

Fan, L., Zhu, Y., & Miao, Z. (2013). Textbook research in mathematics education: Development status and directions. *ZDM Mathematics Education, 45*(5), 633–646.

Fletcher, S. (2012). Editorial of the inaugural issue of the. *International Journal of Mentoring and Coaching in Education, 1*(1), 4–11.

Freudenthal, H. (1981). Major problems of mathematics education. *Educational Studies of Mathematics, 12*(2), 133–150.

Gatabi, A. R., Stacey, K., & Gooya, K. (2012). Investigating grade nine textbook problems for charactersitics related to mathematical literacy. *Mathematics Education Research Journal, 24*(4), 403–421.

Gathumbi, A. W., Mungai, N. J., & Hintze, D. L. (2013). Towards comprehensive professional development of teachers: The case of Kenya. *International Journal of Process Education, 5*(1), 3–14.

Gibbons, L. K., & Cobb, P. A. (2016). Content-focused coaching. *The Elementary School Journal, 117*(2), 237–260.

Grossman, P. L. (1990). *The making of a teacher: Teacher knowledge and teacher education.* New York, NY: Teachers College Press.

Gu, F., & Gu, L. (2016). Characterizing mathematics teaching research specialists' mentoring in the context of Chinese lesson study. *ZDM Mathematics Education, 48*(4), 441–454.

Hill, H. C., & Ball, D. L. (2004). Learning mathematics for teaching: Results from California's mathematics professional development institutes. *Journal for Research in Mathematics Education, 35(5)*, 330–351.

Hill, H. C., Rowan, B., & Ball, D. L. (2005). Effects of teachers' mathematical knowledge for teaching on student achievement. *American Educational Research Journal, 42*, 371–406.

Huang, R., Gong, Z., & Han, X. (2016). Implementing mathematics teaching that promotes students' understanding through theory-driven lesson study. *ZDM Mathematics Education, 48*(4), 425–439.

Huang, R., Ye, L., & Prince, K. (2016). Professional development system and practices of mathematics teachers in Mainland China. In B. Kaur, K. O. Nam, & Y. H. Leong (Eds.), *Professional development of mathematics teachers: An Asian perspective* (pp. 17–32). New York, NY: Springer.

Hunter, R., Hunter, J., Bills, T., & Thompson, Z. (2016). Learning by leading: Dynamic mentoring to support culturally responsive mathematical inquiry communities. In B. White, M. Chinnappan, & S. Trenholm (Eds.), *Opening up mathematics education research* (Proceedings of the 39th annual conference of the Mathematics Education Research Group of Australasia) (pp. 59–73). Adelaide: MERGA.

Larkin, K., Grootenboer, P., & Lack, P. (2016). Staff development: The missing ingredient in teaching geometry to year 3 students. In B. White, M. Chinnappan, & S. Trenholm (Eds.), *Opening up mathematics education research* (Proceedings of the 39th annual conference of the Mathematics Education Research Group of Australasia) (pp. 381–388). Adelaide: MERGA.

Li, Y., & Kulm, G. (2008). Knowledge and confidence of pre-service mathematics teachers: The case of fraction division. *ZDM Mathematics Education, 40*(5), 833–843.

Li, Y., & Qi, C. (2011). Online study collaboration to improve teachers' expertise in instructional design in mathematics. *ZDM Mathematics Education, 43*, 833–845.

Li, Y., Tang, C., & Gong, Z. (2011). Improving teacher expertise through master teacher workstations: A case study. *ZDM Mathematics Education, 43*(6–7), 763–776.

Ma, L. (1999). *Knowing and teaching elementary mathematics: Teachers' understanding of fundamental mathematics in China and the United States.* Mahwah, NJ: Lawrence Erlbaum Associates.

Ministry of Human Resources and Social Security & Ministry of Education (MHRSS & MOE). (2015). *Guidance for deepening the reform of secondary and primary school teachers' promotion system.* Retrieved from: http://www.mohrss.gov.cn/SYrlzyhshbzb/ldbk/rencaiduiwujianshe/zhuanyejishurenyuan/201509/t20150902_219575.htm [in Chinese]

Pang, M. F., Marton, F., Bao, J., & Ki, W. W. (2016). Teaching to add three-digit numbers in Hong Kong and Shangahai: Illustration of differences in the systematic use of variation and invariance. *ZDM Mathematics Education, 48*(4), 455–470.

Pepin, B., Gueudet, G., & Trouche, L. (2013). Investigating textbooks as crucial interfaces between culture, policy and teacher curricular practice: Two contrasted case studies in France and Norway. *ZDM Mathematics Education, 45*(5), 685–698.

Poly, D. (2012). Supporting mathematics instruction with an expert coaching model. *Mathematics Teacher Education and Development, 14*(1), 78–93.

Rowland, T., & Ruthven, K. (2011). *Mathematical knowledge in teaching.* Dordrecht: Springer.

Sfard, A. (2005). What could be more practical than good research? On mutual relations between research and practice of mathematics education. *Educational Studies in Mathematics, 58*(3), 393–413.

Shulman, L. (1986). Those who understand: Knowledge growth in teaching. *Educational Researcher, 15*(2), 4–14.

Shulman, L. S. (1984). Communities of learners and communities of teachers. In L. S. Shulman & S. M. Wilson (Eds.), *The wisdom of practice: Essays on teaching, learning, and learning to teach* (pp. 485–500). San Francisco, CA: Jossey-Bass.

Speer, N. M., King, K. D., & Howell, H. (2015). Definitions of mathematical knowledge for teaching: Using these constructs in research on secondary and college mathematics teachers. *Journal of Mathematics Teacher Education, 18*, 105–122.

Stillman, G., Brown, J., Faragher, R., Geiger, V., & Galbraith, P. (2013). The role of textbooks in developing a socio-critical perspective on mathematical modelling in secondary classrooms. In G. A. Stillman, G. Kaiser, W. Blum, & J. P. Brown (Eds.), *Mathematical modelling: Connecting to research and practice* (pp. 361–371). Dordrecht: Springer.

Williams, J. (2011). Title. In T. Rowland & K. Ruthven (Eds.), *Mathematical knowledge in teaching* (pp. 195–212). Dordrecht: Springer.

Yang, X., & Leung, F. K. S. (2013). Conceptions of expert mathematics teachers: A comparative study between Hong Kong and Chongqing. *ZDM Mathematics Education, 45*(1), 121–132.

Yang, Y. (2009). How a Chinese teacher improved classroom teaching in Teaching Research Group: A case study on Pythagoras theorem teaching in Shanghai. *ZDM Mathematics Education, 41*, 279–296.

Yang, Y., & Ricks, T. (2012). How crucial incidents analysis support Chinese lesson study. *International Journal for Lesson and Learning Studies, 1*(1), 41–48.

Yang, Y., & Ricks, T. (2013). Chinese lesson study: Developing classroom instruction through collaborations in school-based Teaching Research Group activities. In Y. Li & R. Huang (Eds.), *How Chinese teach mathematics and improve teaching* (pp. 51–65). New York, NY: Routledge.

Gloria Ann Stillman
Learning Sciences Institute of Australia
Australian Catholic University (Ballarat)
Victoria, Australia

PART IV

REFLECTION AND CONCLUSION

TIM ROWLAND

15. MATHEMATICS KNOWLEDGE FOR TEACHING

What Have We Learned?

INTRODUCTION

The authors of the chapters in this book have given a fascinating and informative account of mathematics teaching development in China, from pre-service teacher preparation in universities through in-service, lifeline professional development. Meanwhile Chapter 4 provides a valuable international perspective on what is meant by 'mathematics knowledge for teaching' to which I shall refer later in this chapter. As a researcher living and working in Europe, my default perspective on mathematics teacher knowledge and teaching development is inevitably Western. However, as I shall explain, a number of factors have caused researchers and especially policy-makers in the UK to "look to the East." Unfortunately, what we 'learn' from selective reading of aspects of Chinese mathematics teaching practice is limited and often controversial. This book makes an important contribution to the goal of achieving a more balanced and comprehensive Western understanding of how Chinese mathematics teachers acquire and improve knowledge to carry out their professional role. In this chapter I shall briefly outline the influence in the UK of mathematics in high-performing Eastern cultures, discuss what more we could learn from practices described in this book, and reflect on those practices from my own theoretical and practical perspectives.

KEEPING UP?

For many years, successive UK governments[1] have been dissatisfied with mathematics attainment in schools, especially since recent international surveys show the UK with a low ranking in international mathematics achievement league tables. In the 2012 OECD-PISA (Program for International Student Assessment) survey of 15-year-olds (OECD, 2013) the UK was listed 26th. The mathematics table was headed by Shanghai, followed by Singapore, Hong Kong and Taiwan. The response[2] of the UK government Minister of State of Education was predictable, if somewhat selective:

> I would like to see all schools, both primary and secondary, using high quality textbooks in all subjects, bringing us closer to the norm in high performing countries ... In this country, textbooks simply do not match up to the best in

Y. Li & R. Huang (Eds.), How Chinese Acquire and Improve Mathematics Knowledge for Teaching, 277–288.
© 2018 Koninklijke Brill NV. All rights reserved.

the world, resulting in poorly designed resources, damaging and undermining good teaching.

The reaction of the respected British educationist and scholar Robin Alexander represented that of many others in schools and universities.

> There's a lot to unpack and unpick here: the assumption that among PISA top performers it's textbooks that make the difference …; that what appears to hold for secondary students tested by PISA must therefore hold for their primary school peers; that what works for maths works for all other subjects. (Alexander, 2015)

But it didn't end there. In 2014 the government set up and funded 35 'Maths Hubs' across England. Based in local schools deemed to be successful, these mathematics professional centres coordinate individual and organisational expertise in their region to provide a programme of events and activities for mathematics teachers in all phases of schooling. In the first two years of the programme, teachers in some schools in every Maths Hub area used adapted versions of Singapore textbooks. Informal feedback from these teachers has been positive, but no formal evaluation has yet taken place.

TEXTBOOKS

Notwithstanding the irritation of Professor Alexander, the Minister of State may have had a point. As the authors of Chapter 9 point out, textbooks are highly-regarded resources in the Chinese education system, in contrast to the situation in the UK. They refer to the primary mathematics textbook series published by People's Education Press (PEP), a press under the direct leadership of the Ministry of Education of the People's Republic of China, employing many well-educated and expert authors. By contrast, the writing and marketing of textbooks in the UK is entirely a commercial free-enterprise, with no quality control from the centre. This is not to say that there are no good textbooks, but the connection between textbook writing and research is not strong. Unfortunately, in the UK tradition of the evaluation of scholarly activity in universities, textbook-writing is not highly rated compared with the writing of journal papers and books for research colleagues and graduate students. It was not always so, and two widely-acclaimed mathematics textbook series in the 1960s had their roots in universities – the secondary School Mathematics Project was the child of Brian Thwaites, Professor of Theoretical Mechanics at Southampton University; and Nuffield Primary Mathematics was overseen by Geoffrey Matthews, who became the first British professor of mathematics education at Chelsea College (now King's College), London. These textbook series were adopted in several countries outside the UK. Both series published support materials aimed at teachers coming to learn and to teach 'new' topics such as graph theory and statistics. 15 years later other commercially-driven publishers had books on offer that encouraged children to learn in isolation, and seemed to make fewer demands on teachers.

PRE-SERVICE MATHEMATICS TEACHER KNOWLEDGE

Chapters 5 and 6 on elementary and secondary mathematics teacher preparation in China are very informative. Given the success of Chinese school students in international comparisons of attainment, it comes as a surprise to read (Chapter 5) that a questionnaire survey study concluded that primary pre-service teachers' mathematics-related knowledge for teaching is in need of improvement. The design of the questionnaire took into account Lee Shulman's (1986) celebrated categories of teacher knowledge, in particular the two categories 'subject matter knowledge' (SMK) and 'pedagogical content knowledge' (PCK), and the 'mathematical knowledge for teaching' (MKT) elaboration of these categories by Ball and her colleagues (2008). It is sometimes problematic to separate SMK and PCK, and to devise test items that distinguish them. One PCK item from the questionnaire instrument included in Chapter 5 is:

A student computed 26×53, the vertical form on the right was used. How do you explain his/her error? Write down what you might say to this student.

$$\begin{array}{r} 26 \\ \times\ 53 \\ \hline 78 \\ 130 \\ \hline 208 \end{array}$$

Figure 1. PCK item from the mathematics teacher knowledge questionnaire (Chapter 5)

The need to identify a student error is particular to teaching, but arguably no pedagogical knowledge is needed to see that there is a place-value error in the positioning of the 130. So (to me) it looks like a 'Common Content Knowledge' item, an MKT-SMK domain (Ball et al., 2008; see Chapter 4 for details). Knowledge in the MKT-PCK domain 'knowledge of content and students' would be needed to know that this is a common student error (e.g. Hansen, 2005), but that's not what this item is evaluating. Indeed, Hill et al. (2008) include the following to exemplify a Specialised Content Knowledge item (still in MKT-SMK):

---

3. Imagine that you are working with your class on multiplying large numbers. Among your students' papers, you notice that some have displayed their work in the following ways:

Student A	Student B	Student C
35	35	35
x 25	x 25	x 25
125	175	25
+75	+700	150
875	875	100
		+600
		875

Which of these students would you judge to be using a method that could be used to multiply any two whole numbers?

Figure 2. SCK item from the Michigan MKT assessment (Hill et al., 2008)

Hill et al. (2008, p. 439) comment

> Here teachers inspect three different approaches to solving a multi-digit multiplication problem – 35 × 25 – and assess whether those approaches would work with any two whole numbers. To respond to this situation, teachers must draw on mathematical knowledge, including making sense of the steps shown in each example, then gauging whether the steps might make sense and work for all whole numbers. Appraising nonstandard solution methods is not a common task for adults who do not teach. Yet this task is *entirely mathematical, not pedagogical*; in order to make sound instructional decisions, teachers must be able to size up and evaluate the mathematics of these alternatives—often swiftly, on the spot. (emphasis added)

I could debate whether the item in Figure 2 is assessing CCK or SCK. I could even find out by asking some adults not working in education. But both of these MKT domains fall within SMK, not PCK. Hill's argument (above) would apply equally well to the item in Figure 1.

Pre-Service Mathematics Teacher Knowledge in England

A questionnaire study of elementary PSTs mathematics content knowledge – mainly SMK – in England (Rowland et al., 1998) came about as a response to a government paper first issued in 1997, which required teacher education programmes to "audit trainees' knowledge and understanding of the mathematics contained in the National Curriculum", and where 'gaps' are identified, to "make arrangements to ensure that trainees gain that knowledge" (Department for Education and Employment (DfEE), 1998, p. 48). This process of audit and remediation of subject knowledge within primary 'initial teacher training' (ITT) became a high profile issue following the introduction of these and subsequent government requirements. This regime provoked a body of research in the UK on prospective elementary teachers' mathematics subject-matter knowledge (e.g., Goulding et al., 2002). The proceedings of a symposium held in 2003 usefully drew together some of the threads of this research (BSRLM, 2003). A London-based project (Subject Knowledge in Mathematics: SKIMA) with 170 London-based graduate pre-service primary teachers (Rowland, Martyn, Barber, & Heal 2000) used a 16-item instrument to assess their content knowledge. In interpreting and making sense of the findings of this study, a distinction due to Joseph Schwab (1978) but also explicit in Shulman's mental map (Shulman & Grossman, 1988) becomes useful – that between *substantive* and *syntactic* subject-matter knowledge. Substantive knowledge encompasses the key facts, concepts, principles, structures and explanatory frameworks in a discipline, whereas syntactic knowledge concerns the rules of evidence and warrants of truth within that discipline, the nature of enquiry in the field, and how new knowledge is introduced and accepted in that community – in short, how to find out. This distinction comes close to that between content (substantive) and process (syntactic) knowledge, although syntactic knowledge seems to entail greater epistemological awareness than process knowledge. Ball (1990) has made a similar, perhaps identical, distinction between knowledge *of* mathematics (meanings and underlying procedures) and knowledge *about* mathematics (what makes something true or reasonable in mathematics).

A response to an audit item was coded 'secure' if it was correct, and included (where appropriate) a valid justification for the answer given. In general, the PST participants were more secure on the substantive items than on the syntactic. Responses of almost all the PSTs to the item (deemed substantive) shown in Figure 3 were secure.

I think of a number, add 15 to it, and divide the answer by 9. The answer is 10. What number did I start with? Briefly describe your method.

Figure 3. Item #10 from the SKIMA SMK assessment (Rowland et al., 2000)

By contrast, only one-third of the PST students made a secure response to the whole of the syntactic item shown in Figure 4, and 30% gave insecure (or blank) answers to all three parts.

A rectangle is made by fitting together 120 square tiles, each 1 cm^2. For example, it could be 10cm by 12 cm. State whether each of the following three statements is true or false for every such rectangle. Justify each of your claims in an appropriate way:
(a) The perimeter (in cm) of the rectangle is an even number.
(b) The perimeter (in cm) of the rectangle is a multiple of 4.
(c) The rectangle is not a square.

Figure 4. Item #16 from the SKIMA SMK assessment (Rowland et al., 2000)

More than one mode of justification is possible for each part, and a proof by exhaustion (listing the 8 possible rectangles) would meet the requirements of all three. We anticipated some deductive arguments for (a), counterexamples for (b) and perhaps contradiction ($\sqrt{120}$ is not an integer) for (c).

Comparative Studies of Pre-Service Mathematics Teacher Knowledge

It is relevant here to report that the same audit instrument was administered to a sample of 41 primary PSTs in Singapore, in 1999. The profile of easy/hard items was more or less the same for the English and the Singapore trainees (See Fong Ng, personal communication). The only differences worthy of note are on two items. The first concerns inverse operations, a topic which the Singapore PSTs found more problematic than the English sample. The second was the Figure 4 item, on which the Singapore students demonstrated superior syntactic subject matter knowledge. This may well be because the Singaporean NIE elementary teacher programme allocates more time than the English one to the development of mathematics subject matter knowledge.

The comment in Chapter 5 that Chinese primary pre-service teachers' mathematics-related knowledge for teaching is in need of improvement is illuminating in relation to the findings from a comparative study between England, the Chinese mainland and Hong Kong (Wong et al., 2009). Ten items relevant to the mathematics curricula in the Chinese mainland and Hong Kong were chosen from the SKIMA subject matter audit instrument (Rowland et al., 2000). These items (listed in Wong et al., 2009) assess aspects of arithmetic, geometry and mathematical enquiry. Both pre-service and in-service teachers from the Chinese mainland and Hong Kong participated in the study. In the Chinese mainland, data were collected from two cities, Changchun and Guangzhou. There were 158 Hong Kong participants (88 PSTs) and 198 from the Chinese mainland (79 PSTs). Participants' written responses were coded on a 5-point scale from 'very insecure' to 'very secure.' The researchers found that

> As a whole, in some items, participants from England had a higher percentage of secured mastery, and those from Hong Kong had higher ones in the others. Those from the Chinese mainland had the lowest percentage in general, though we are fully aware that they cannot be viewed as a representative sample from the Chinese mainland. (p. 195)

It then transpires that the performance of the in-service teachers in both countries, especially the Chinese mainland, was relatively weak: on the whole, the performances of the PSTs across the three countries was broadly similar. In all three, participants were stronger on the arithmetic items than those in the other two categories. In any case, this finding ought to be reassuring to policy-makers in England.

MATHEMATICS TEACHERS' LEARNING-IN-COMMUNITY

Throughout their careers, teachers must also be learners: both as an example to their students, and because the process of learning to teach never ends. From a Western perspective, it was very interesting – uplifting, even – to read the chapters in Part III of the book. The Chinese emphasis on professional development at all levels is striking. I was especially inspired by Chapter 10 (Experienced teacher's learning through a master teacher workstation program); Chapter 11 (In-service mathematics teachers' professional learning in teaching research group); and Chapter 12 (Learning from developing and observing public lessons). I suggest that these ways of developing mathematics teaching and teacher knowledge are indicative of learning communities of mathematics teachers, and have reflected on what this might mean, and how it compares with practices in the West.

This notion of a teachers' learning community is strongly exemplified in the teaching research groups (TRGs) which are the focus of Chapter 11. These TRGs have existed in schools in China since 1952, and provide an excellent vehicle for teaching development. The 'Three Points' framework for lesson analysis, the work of one of the authors of the chapter (Yang, 2009), was new to me, and I see it as an excellent structure for reflection on mathematics teaching. It will be important to understand and be clear about the characteristics of the three lesson components: the *key* point; the *difficult* point; and the *critical* point. According to Yang and Zhang (Chapter 11, this volume, pp. 220–221):

> The lesson's key point refers to the central objective of the lesson for which the lesson is constructed. The success of the lesson depends on how well students learn this core mathematical content topic. [...] The difficult point is the cognitive difficulty that the students might encounter as they try to learn the mathematical key point. [...] the metaphorical mental stumbling block. Being able to clearly state and anticipate this difficult point helps Chinese teachers to plan lessons that go beyond just attempting to transmit content knowledge to students [...] The critical point is the heart of the lesson that shapes the teaching methods. If the key point is what the lesson is about, the critical point emphasizes how to reach that objective.

In a helpful explication of the use of the Three Points framework in lesson study analysis of two primary school lessons on fractions in Singapore, Choy summarises:

> According to Yang and Ricks (2012), the *key point* refers to key mathematical ideas of the lesson; the *difficult point* refers to cognitive obstacles encountered by students when they attempt to learn the key point; and the *critical point* refers to the approach that teachers take to help students overcome the difficult point. (Choy, 2014, p. 144)

Of course these three points should be kept in mind from the beginning in planning the lesson; being able to identify the *difficult point* (or points, perhaps) is a significant – perhaps the most significant – indicator of a teacher's pedagogical content knowledge; since the person who already knows what the students are to learn (on account of common content knowledge) might be completely unaware of why students might experience difficulty in coming to know it themselves, and the obstacles they are likely to face in trying to do so (e.g. Hansen, 2005; Cockburn & Littler, 2008). Learning the critical points in mathematics instruction is surely one of the key objectives of pre-service (and then in-service) mathematics teacher education.

Indeed, what I find missing from the account (Chapter 11) of the TRG practice is reference to any authority outside the TRG itself. Identifying the difficult points in particular mathematical topics has been the subject of empirical research studies over many years, and the findings of such studies are helpfully summarised and collated in books such as Hansen (2005) and Cockburn and Littler (2008), both targeted at prospective and serving elementary school teachers; or Lester (2007) and Gutierrez et al. (2016), compiled with researchers in mind. The same point is made by Stein et al. (2008) in a research paper proposing five key practices teachers can learn in order to improve their management of student responses to challenging tasks in class discussion.

> The first practice is for teachers to make an effort to actively envision how students might mathematically approach the instructional tasks(s) that they will be asked to work on [...] In addition to drawing on their knowledge of their particular students' mathematical skills and understandings, teachers might draw on *their knowledge of the research literature* about typical student responses to the same or similar tasks or of common student understandings of related concepts and procedures. (e.g., Fennema et al., 1996, pp. 322–323; emphasis added)

A report of a detailed study of fourth graders responses to tasks involving coordinate systems can be found at Sarama et al. (2003). The difficult point related to spaces and points in the coordinate system encountered by some of the teachers in the case reported is documented in the literature. For example, in his best-selling book (now in its 4th Edition), Haylock (2010) observes that

> An important feature of this [coordinate] system is that it is the points in the system that are labelled by the coordinates, not the spaces. This is an important teaching point, because there are a number of situations that children will encounter which use coordinate systems based on the idea of labelling the spaces – for example, a number of board games and computer games, city street maps [...] The use of coordinates *to specify the location of points in a plane, rather than spaces*, as in street maps, is a significant point to be explained to children carefully. (Haylock, 2010, p. 267; emphasis added)

In Chapter 4 of this book, three European authors review various theoretical models for the construct of 'mathematics knowledge for teaching', each one rooted in the seminal insights of Lee Shulman (1986, 1987) some 30 years ago. That the Michigan 'egg model' of mathematics knowledge for teaching', now best known from Ball et al. (2008), is cited in most of the chapters of the book, is evidence of the outreach of Deborah Ball and her group, and the value of their elaboration of that part of Shulman's work related to the discipline (mathematics, here) being taught. The other two theoretical frameworks outlined in Chapter 4 – the Knowledge Quartet (UK) and COACTIV (Germany) are absent elsewhere in the book, which could be indicative of the influence of United States thinking in China relative to that in Europe. But the two European frameworks both have something to say about reference to accumulated knowledge (including the literature) in the preparation of teaching. In its Connection dimension, the KQ (e.g. Rowland, 2014) refers to (the importance of) the Anticipation of Complexity in lesson planning – this would be something very close to identifying and predicting the 'difficult points' in the lesson before it was taught. Likewise, in the COACTIV context, Baumert and Kunter (2013) write:

Some elements of practical knowledge can also be assumed to have a propositional mental representation. This applies to the act of lesson preparation and probably also to the categorization of perceived situations and typical sequences of events. (p. 32)

This has been an interesting aspect of Chinese mathematics teachers' professional development to consider. Most of all I appreciate the notion of teachers learning-in-community, with its strong social dimension. School teaching is, by definition, a social activity, whereas mathematics, as such, can be perceived as an individual matter. Countries in the West – my own included – would do well to come to know the lessons to be learned from the teaching research groups so well-established in China, but this allocation of teachers' time would need to be funded.

Mastery and Morality

I read chapters 3, 10 and 12 with a sense of admiration and also a sense of separation, and I shall try to explain why.

The notion of Master Teacher Workshop (Chapter 10) and Public Lessons (Chapter 12) is somewhat alien from a Western perspective, which envisages many different ways to achieve the teacher's objective (student learning). The account of the public lessons (Chapter 10) engaged my interest very much as I participated in the mathematics with pencil and paper; and those lessons must have engaged the teachers who observed and discussed them.

I would want to assert the value of observing and analysing the teaching of both expert and novice teachers in the professional development of mathematics teachers. In a study of primary mathematics teacher preparation in Korea, Pang (2011) used videos of both pre-service and in-service teachers' lessons to focus prospective teachers' attention on the mathematics-specific features of what they observed. Both types of video were found to be valuable, for different reasons. The expert teaching mirrored the vision of good mathematics teaching that had been promoted in the university methods course, and (unlike much of what they saw in school-based placements) demonstrated to the trainee teachers that it was possible to realise the vision in practice. On the other hand, the not-so-good teaching of novices with whom they could immediately identify was especially beneficial in the analysis of how teachers draw on content and pedagogy in effective – and ineffective – instruction, including central and peripheral components of planning and teaching. Pang comments:

[prospective teachers] differentiated effective mathematics lessons from seemingly good, but indeed unsuccessful, ones in terms of students' understanding. They were able to recognize that effective mathematics lessons were not related to splendid instructional materials or students' fun activities, but to the degree to which key mathematical content was meaningfully explored with students' thinking. The teachers claimed that this vision came

> from a vivid discussion of multiple cases in class, including unsuccessful and thought-provoking ones. (Pang, 2011, p. 787)

The distinction of Confucius between being a *Ren Shi* – a teacher for people development – and a (mere) *Jing Shi* – a teacher for the development of academic knowledge – is inspiring and very much underplayed in the West. *Ren Shi* – somehow lies outside the considerations of the Knowledge Quartet, but could be seen to underpin Foundation. According to Hsieh et al. (see Chapter 3 in this volume):

> a teacher in general, or a mathematics teacher in particular, is supposed to be an instructor, supervisor, guardian, and model for students. Accordingly, teachers usually consider that they are obligated and have the right to require students to study hard, to suggest (or arrange) students' direction in career choice, or to help students after class and act as a guardian.

This moral dimension brings into play a contract – well-understood but perhaps implicit and unspoken – between teacher and student: and perhaps students' parents/carers too. The contract promises that the teacher will do his/her very best to bring about the learning of the student: and in return for this commitment and investment of time and goodwill, the student promises to respond with their own commitment and hard work. There is something very attractive and logical in this view of the relationship between teachers and their students. It is strengthened later, in the world of work, in a similar contract – the employer is expected to fulfil the promise of giving money and providing good working conditions in return for work; the worker responds with effort and commitment. These moral contracts have to be learned, however, since they do not exist – in the West at least – between parent and child. The parent's obligation and commitment to the child is absolute and unconditional, but the child owes the parent nothing – their response usually brings joy to the parent, but there is no contract. Much more could be written here but the parent-child 'contract' lies outside the scope of my contribution to the book!

CONCLUSION

My reading of the chapters of this book has been interesting, and has added to what I had already learned from books such as Fan et al. (2004) and Fan (2014). Although, in a superficial sense, mathematics remains 'the same' in China and in England, the cultural differences between the two countries enhance the interest of both for the 'outsider' looking in, and improve the possibility that each can learn from the other – if only that can be done without compromising precious practices and values. The same can be said more generally of 'East' and 'West', but with the explosion of travel and relocation East-West and West-East, and of face-to-face and virtual communication, the line between the two cultures has become more blurred than ever before. Arguments can be made in favour of this transition, but also in regret.

Given the fact that mathematics teachers and mathematics education researchers see, talk and correspond with each other in this brave new world, it is crucial – and hopefully inevitable – that they learn from one another. For me, the most enviable dimension of mathematics education in schools, and in teachers' professional development, is what I have called learning-in-community, realised in particular in the Teaching Research Groups, but also in Master Teacher Workshops and Public Lessons. As I have already remarked, a large investment of public money would be needed to realise these in the UK. In my judgement, teachers would welcome them. Looking from East to West, I get the impression, from the book, that European research on teaching and learning mathematics is not so well-known as that from the United States. The potential for us to learn from each other is great, and exciting, and a realistic and achievable ambition in the twenty-first century.

NOTES

[1] It should be noted that legislation on education policy in the United Kingdom, including curriculum and teacher education, is largely devolved to local assemblies in the provinces of Wales, Scotland and Northern Ireland. The Department of Education of the UK government, under the leadership of the Secretary of State for Education, is responsible for education and children's services in England (only).

[2] Nick Gibb speech to education publishers, 20 November 2014.

REFERENCES

Alexander, R. (2015). *Teaching to the text: England and Singapore.* York: Cambridge Primary Review Trust. http://www.cprtrust.org.uk/cprt-blog/teaching-to-the-text/

Ball, D. L. (1990). Prospective elementary and secondary teachers' understanding of division. *Journal for Research in Mathematics Education, 21*(2), 132–144.

Ball, D. L., Thames, M. H., & Phelps, G. (2008). Content knowledge for teaching: What makes it special? *Journal of Teacher Education, 59*, 389–407.

Baumert, J., & Kunter, M. (2013). The COACTIV model of teachers' professional competence. In M. Kunter, J. Baumert, W. Blum, U. Klusmann, S. Krauss, & M. Neubrand (Eds.), *Cognitive activation in the mathematics classrooms and professional competence of teachers* (pp. 25–48). New York, NY: Springer.

BSRLM. (2003). *Proceedings of the British Society for Research into Learning Mathematics, 23*(2).

Choy, B. H. (2014). Noticing critical incidents in a mathematics classroom. In J. Anderson, M. Cavanagh, & A. Prescott (Eds.), *Proceedings of the 37th annual conference of the Mathematics Education Research Group of Australasia* (pp. 143–150). Sydney: MERGA.

Cockburn, A. D., & Littler, G. (Eds.). (2008). *Mathematical misconceptions.* London: Sage Publications. Discussion: Productive Mathematical Discussion. Svein Arne to lead.

DfEE. (1998). *Teaching: High status, high standards: Circular 4/98.* London: HMSO.

Fan, L. (2014). *Investigating the pedagogy of mathematics: How do teachers develop their knowledge.* London: Imperial College Press.

Fan, L., Wong, N. Y., Cai, J., & Li, S. (Eds.). (2004). *How Chinese learn mathematics: Perspectives from insiders.* Singapore: World Scientific.

Fennema, E., Carpenter, T. P., Franke, M. L., Levi, L., Jacobs, V. B., & Empson, S. B. (1996). A longitudinal study of learning to use children's thinking in mathematics instruction. *Journal for Research in Mathematics Education, 27*(4), 403–434.

Goulding, M., Rowland, T., & Barber, P. (2002). Does it matter? Primary teacher trainees' subject knowledge in mathematics. *British Educational Research Journal, 28*(5), 689–704.

Gutierrez, A., Leder, G. C., & Boero, P. (2016). *The second handbook of research on the psychology of mathematics education.* Rotterdam, The Netherlands: Sense Publishers.

Hansen, A. (Ed.). (2005). *Children's errors in mathematics: Understanding common misconceptions in primary schools.* Exeter: Learning Matters.

Haylock, D. (2010). *Mathematics explained for primary teachers* (4th ed.). London: Sage Publications.

Hill, H., Blunk, M., Charalambous, C., Lewis, J., Phelps, G., Sleep, L., & Ball, D. (2008). Mathematical knowledge for teaching and the mathematical quality of instruction: An exploratory study. *Cognition and Instruction, 26*(4), 430–511.

Lester, F. (Ed.). (2007). *Second handbook of research on mathematics teaching and learning.* Charlotte, NC: Information Age.

OECD. (2013). *PISA 2012 results: What students know and can do: Student performance in mathematics, reading and science* (Vol. 1). Paris: OECD Publishing.

Pang, J. S. (2011). Case-based pedagogy for prospective teachers to learn how to teach elementary mathematics in Korea. *ZDM: The International Journal of Mathematics Education, 43,* 777–789.

Rowland, T. (2014). The Knowledge quartet: The genesis and application of a framework for analysing mathematics teaching and deepening teachers' mathematics knowledge. *SISYPHUS Journal of Education, 1*(3), 15–43.

Rowland, T., Heal, C., Barber, P., & Martyn, S. (1998). Mind the gaps: Primary trainees' mathematics subject knowledge. *Proceedings of the British Society for Research into Learning Mathematics, 18*(1–2), 91–96.

Rowland, T., Martyn, S., Barber, P., & Heal, C. (2000). Primary teacher trainees' mathematics subject knowledge and classroom performance. *Research in Mathematics Education, 2*(1), 3–18.

Sarama, J., Clements, D. H., Swaminathan, S., McMillen, S., Rosa, M., & González Gómez, M. (2003). Development of mathematical concepts of two-dimensional space in grid environments: An exploratory study. *Cognition and Instruction, 21*(3), 285–324.

Schwab, J. J. (1978). Education and the structure of the disciplines. In I. Westbury & N. J. Wilkof (Eds.), *Science, curriculum and liberal education* (pp. 229–272). Chicago, IL: University of Chicago Press.

Shulman, L. S. (1986). Those who understand: Knowledge growth in teaching. *Educational Researcher, 15*(2), 1–22.

Shulman, L. S. (1987). Knowledge and teaching: Foundations of the new reform. *Harvard Educational Research, 57,* 1–22.

Shulman, L., & Grossman, P. (1988). *Knowledge growth in teaching: A final report to the Spencer Foundation.* Stanford, CA: Stanford University.

Stein, M. K., Engle, R. A., Smith, M. S., & Hughes, E. K. (2008). Orchestrating productive mathematical discussions: Five practices for helping teachers move beyond show and tell. *Mathematical Thinking and Learning, 10*(4), 313–340.

Wong, N.-Y., Rowland, T., Chan, W.-S., Cheung, K.-L., & Han, N.-S. (2010). The mathematical knowledge of elementary school teachers: A comparative perspective. *Journal of the Korea Society of Mathematical Education Series D: Research in Mathematical Education, 14*(2), 187–207.

Yang, Y. (2009). How a Chinese teacher improved classroom teaching in Teaching Research Group. *ZDM: The International Journal on Mathematics Education, 41*(3), 279–296.

Tim Rowland
University of Cambridge
Cambridge, UK

HUNG-HSI WU

16. SOME REMARKS ON THE PREPARATION OF MATHEMATICS TEACHERS IN CHINA

A nation's mathematics education is only as good as its mathematics teachers. The professional development of teachers is therefore serious business. This is an inherently complex subject, because, into the complex human dynamics between students and their teachers, professional development must try to inject an optimal strategy for the transfer of knowledge from teachers to students—no matter how this "transfer" is defined. (The last word on the pedagogical aspect of professional development will never be written.) What further complicates matters in this context is that, mathematics being the highly technical discipline that it is, mathematical content knowledge is bound to play a dominant role in mathematics instruction. Therefore content knowledge has to play a key role in any successful mathematics professional development. Unfortunately, there has been some serious misunderstanding in the U.S. over what this "content knowledge for teaching" ought to be, apparently all through the last century and up to 2017 (cf. Wu, 2011b; also Shulman, 1986; Ball, Thames, & Phelps, 2008). The first section of Chapter 4 of this volume points to the additional difficulty that this body of knowledge also seems to vary across nations:

> These different studies point to the difficulty to reach international agreement on a definition of mathematical knowledge for teaching and how to acquire it. (Döhrmann et al., Chapter 4, this volume)

The complexity of mathematics professional development cannot be denied.

For all these reasons, it is not likely that any one nation will ever have a monopoly on excellence in the professional development of mathematics teachers. A continuous exchange of ideas between nations will always be beneficial to the health of the enterprise that we call school mathematics education. From this perspective, the present volume promises to provoke a fruitful international dialog. It seems to this author that there are some striking features in the Chinese system that the U.S. should diligently study, and perhaps emulate, for its own benefit. At the same time, other elements in the Chinese system may be ripe for a reappraisal. The following sections will amplify on these claims.

SOME NOTEWORTHY FEATURES OF THE CHINESE SYSTEM

The pre-service preparation of mathematics teachers in China puts great emphasis on the acquisition of content knowledge but gives less attention to the acquisition of experience in student teaching or *pedagogical content knowledge* (PCK, see Shulman, 1986). The rationale for the emphasis on content knowledge is straightforward and unimpeachable:

> [The consensus is] that it is impossible to develop pedagogical content knowledge without appropriate content knowledge. (Wu & Huang, Chapter 6, this volume)

For elementary *mathematics* teachers, a survey of 16 major teacher-training programs shows that, in general, the required mathematics courses include:

> Mathematical Analysis, Spatial Analytic Geometry, Projective Geometry, Non-Euclidean Geometry, Theory of Probability, The Structure of Algebra, a Brief History of Mathematics, Mathematical Modeling, Advanced Algebra, and Elementary Number Theory. (Xie et al., Chapter 5, this volume)

Also included are modified versions of Calculus and Linear Algebra.[1] As for secondary teachers, one learns that in the teaching program of a "superior" normal university, the "credit point of the mathematics curriculum is three times that of the teacher education curriculum" (Wu & Huang, Chapter 6, this volume). The required courses include:

> Ordinary Differential Equations, Classical Geometry, Complex Analysis, Probability and Statistics, Abstract Algebra I and II, Differential Geometry, Number Theory, Real Analysis, and Combinatorics and Graph Theory. (Wu & Huang, Chapter 6, Table 1, this volume)

The 2011 *Teacher Education Curriculum Standards* try to address the imbalance of content over PCK in pre-service education. It is believed that "mathematics teacher preparation in China has undergone a significant transition from solely focusing on content knowledge to balancing content and practice-based knowledge" (Wu & Huang, Chapter 6, this volume). However, how to develop high-quality courses by integrating mathematics content and practice-based mathematic pedagogical knowledge remains an ongoing challenge. That said, it is well to observe that this imbalance is offset by the system of in-service professional development in China, which has a long tradition of supporting *professional growth and lifelong education* for the acquisition of classroom management skills and PCK. There is a well-established multi-tiered system of mentorship by experienced teachers, collaborative lesson planning with peers, lesson observation and post-lesson reflection and improvement (Huang & Huang, Chapter 10, this volume), and developing and observing public lessons (Liang, Chapter 12, this volume). Experienced teachers can further hone their craft by engaging in a Master Teacher Workstation program (Huang & Huang, Chapter 10, this volume). Lifelong learning for Chinese teachers is in fact a requirement:

> Once a teacher starts teaching career, s/he has to actively and constantly engage in a variety of professional development (*PD*) activities which are mandatory, including pairing up with an experienced teacher of the same grade level for one-on-one mentoring, participating in teaching research groups in which teachers study textbooks and develop lessons together, and giving public lessons for colleagues and administrators to observe. The PD activities are continuous, practice-oriented, relevance-focused, and integrated as a whole to be an important part of a teacher's entire teaching professional life. A teacher never stops engaging in PD activities unless s/he leaves the teaching profession. (Liang, Chapter 12, this volume)

Lifelong learning is therefore an integral part of the Chinese teaching culture so that all teachers come to feel that they are not alone but are surrounded by a strong support group. They gain much of their knowledge for teaching through school-based continuous professional development activities:

> It is very common in China that teachers observe each other's class and then discuss teaching ideas and strategies on the purpose of learning knowledge for teaching from each other and growing professionally together. It could happen between a new teacher and an experienced teacher. (Liang, Chapter 12, this volume)

> In [the research group] activities, Chinese teachers usually discuss how to improve classroom instruction. The emergence of knowledge is quite similar to PCK... That is, when a teacher copes with a special topic, he/she will organize, adjust and present subject matter knowledge to do task design by considering a special group of students' interest and capacity. (Yang & Zhang, Chapter 11, this volume)

To Chinese mathematics teachers, doing lesson preparation, observation, and post-lesson reflection together is therefore a way of life. Their American counterparts may likely read about such a teaching culture with longing and envy.

SOME CONCERNS FROM AN AMERICAN PERSPECTIVE

Here are the two features in the Chinese preparation of mathematics teachers that should be of special interest to the U.S.:

a. A strong emphasis on content knowledge in pre-service PD.
b. Making professional growth and lifelong education integral parts of the teaching culture.

Although many Chinese educators consider the advocacy of (A) at the expense of PCK to be a weakness of their system, the opposite has occurred in the U.S. where there has been a longstanding over-emphasis on pedagogy over content knowledge. At the present time (2017) when the urgent need for mathematically (and scientifically) knowledgeable teachers in the U.S. is beyond dispute, the current strategy in the empowerment of mathematics teachers still tends to be of the "unleashing students' potential through innovative teaching" variety. *Not* about learning correct mathematics for their teaching. In this climate, (A) should come as a welcome correction.

Although (B) has not been central to current discussions about mathematics education in the U.S., it should be. There is a teacher dropout crisis at the moment (cf. NPR Ed, 2014). In each of the years from 1988 to 2013, between 12.4 and 16.5 percent of public school teachers moved to a different school or left the teaching profession entirely (Goldring et al., 2014, Table 1). More than 42% of new teachers leave teaching within 5 years of entry (Ingersoll, Merrill, & May, 2014, p. 5). The reasons vary, but one of them is apparently teachers' strong desire for learning from each other, a desire which often goes unmet in the teaching profession. A telling statistic is that, of teachers who left teaching in 2012–2013 for a different occupation, about 41.7 percent of *public school* teachers reported that "opportunities for learning from colleagues" were better in their new (non-teaching) positions, compared with only 15.9 percent reporting that they were worse (Goldring et al., 2014, p. 13, Table 7). Moreover, 45.7 percent reported that "opportunities for professional development" were better in their new positions.[2] In a 2014 survey of *all* teachers (not just mathematics teachers), it was found that 84 percent wanted to spend more time on lesson observation, 82 percent on coaching, and 74 percent on professional learning communities (Bill & Melinda Gates Foundation, 2014, p. 5). It was almost as if teachers were asked whether they wanted (B) in their professional lives and they overwhelmingly voted *yes*! What these statistics point to is an undeniable need for the American teaching culture (at least in the public sector) to embrace some version of (B).

However, we cannot recommend (A) and (B) without reservation as of 2017 for at least two reasons. The first is that we have to clear up the meaning of "content knowledge for teaching" before (A) and (B) can be understood in the proper context. A second reason is that (B) would be meaningless without some large-scale restructuring of the teaching profession.

We will deal with the content issue first. Let us begin with the fact that, in the U.S., the mathematical knowledge that our teachers bring to their classrooms has typically not been *mathematics* but a corrupted version of mathematics that has resided in the standard school textbooks for (at least) the past four decades. This body of knowledge is as different from *mathematics* as margarine is from butter. We propose to call it *Textbook School Mathematics* (TSM) (see Wu, 2011b, 2011c) to distinguish it from *school mathematics*, which is the mathematics of the K-12 curriculum. TSM is *unlearnable* because of its lack of what may be called *logical transparency*. In greater detail, the unlearnability of TSM is the result of its pervasive violation of the following five *Fundamental Principles of Mathematics* (FPM) (Wu, 2011b, pp. 379–380):

1. Every concept has a precise *definition*, and definitions furnish the basis for logical reasoning. (Definitions leave no doubt about what exactly students have to learn.)
2. Mathematical statements are *precise*. Precision makes possible the distinction between what is known and what is not known, and what is true and what is false. (Precision eliminates the need to *guess* in learning mathematics.)
3. Every assertion is supported by logical *reasoning*. There are no arbitrary or irrational decrees in mathematics. (Mathematics is learnable *because* it is reasonable.)
4. Mathematics is *coherent*. Mathematics is a living organism in which all the different parts are interconnected. (Mathematics is not a bag of isolated tricks for students to memorize.)
5. Mathematics is *purposeful*. Every concept and skill in the school mathematics curriculum is there for a purpose. (Students get to know where they are headed.)

For those readers who are shocked to find that textbooks that responded to the mathematics education reform of 1989 are lumped together in this analysis with those from traditional major publishers, it suffices to point out that while the reform texts teach K-12 mathematics on a different pedagogical platform, their *mathematical content*—seen through the lens of FPM—is still just TSM.

Because TSM is an overriding theme of this chapter, we will digress a bit to illustrate how TSM violates the fundamental principles of mathematics. Some examples are given in Wu, 2011b and 2011c, but we will mention three of the most obvious to make our point here. The first illustrates (among other things) the lack of precision in TSM. TSM teaches that "22 divided by 5 has quotient 4 and remainder 2" should be written as $22 \div 5 = 4\ R2$. (The idea seems to be: since it is so convenient, why not just use the equal sign to denote "here is the result of my calculation"?) But by the same token, we will also have $42 \div 10 = 4\ R2$ and therefore $22 \div 5 = 42 \div 10$ because both are equal to $4\ R2$. In terms of fractions, we now have $42 \div 10 = \frac{42}{10} = \frac{21}{5}$. TSM has therefore led us to an apparent equality $22 \div 5 = 42 \div 10$ which implies this absurdity:

$$\frac{22}{5} = \frac{21}{5}$$

A second example illustrates the damage inflicted by TSM on student learning due to its failure to provide precise definitions of fundamental concepts. Thus a fraction in TSM is "parts of a whole" or, more often, pieces of pizza. This then begs the question of how to explain the division of $\frac{3}{7}$ by $\frac{11}{5}$. There is no reasoning on earth that can possibly convince young learners how to divide "a *quantity* that consists of 3 of the parts when a pizza is divided into 7 equal parts", by another "*quantity* consisting of 11 of the parts when a pizza is divided into 5 equal parts." Such a *quantity* is neither beast nor man, nor in fact anything students know how to deal with. They are thus forced to conclude *ours is not to reason why, just invert and multiply*.

A final example on the teaching of high school geometry illustrates the frequent lack of coherence in TSM. Since students are almost never given a definition of a concept in TSM (e.g., "fraction"), they come into their high school geometry course having no experience with how a definition can be used for reasoning.[3] Since TSM does not give students any hint of the *logical* hierarchical structure of mathematics, they have no conception of what "axioms" are before they run into them in high school geometry. Since TSM hardly ever exposes students to reasoning before high school geometry, "proving a theorem" is also a completely foreign concept. Yet, in this geometry course, and *only* in this course in TSM, students are suddenly confronted with a litany of definitions, axioms, theorems and proofs. They are called upon to *prove every single assertion no matter how trivial*, and in fact, most of the early theorems are unbelievably trivial and therefore extremely difficult to "prove." Under these circumstances, geometry teaching and learning can easily degenerate into a farce (see, for example, the documentation in Schoenfeld, 1988). Being aware of this farce, some teachers and school districts have reacted by going to the opposite extreme of teaching high school geometry with *no proofs*, relying on the use of computer software to bring mathematical conviction to geometric theorems. Either way results in an incomplete and incoherent presentation of mathematics to students.

One can get an extensive documentation of how TSM abuses the mathematics of grades 6–8 by looking up "TSM" in the indices of Wu (2016a, 2016b). A little extrapolation will give a fairly accurate picture of what TSM has done to the teaching and learning of mathematics in K-12 as a whole.

Important as FPM may be to school mathematics, we caution that teachers' content knowledge must do more than respect FPM. The mathematics we teach college math majors—groups, rings, fields, Dedekind cuts, Cauchy integral theorem, Gauss-Bonnet theorem—also respects FPM, but it would be bad education to make it the required content knowledge for all teachers because college mathematics is too advanced to be usable in the K-12 classroom. The content knowledge we need our mathematics teachers to possess therefore has to meet both of the following conditions:

I. It closely parallels the K-12 mathematics curriculum.
II. It is consistent with FPM.

For brevity, we will refer to the body of mathematical knowledge that satisfies both (I) and (II) as *principle-based mathematics* (see Poon, 2014; Wu, 2018). TSM satisfies (I) but emphatically not (II), and the mathematics embodied in the usual requirements of a college math major satisfies (II) but definitely not (I). If we want the above outcomes (A) and (B) to materialize in the preparation of American mathematics teachers, then *"content knowledge" will have to be understood to be principle-based mathematics.* The relevance of principle-based mathematics to (A)—a strong emphasis on content knowledge in pre-service PD— is obvious, but its relevance to (B), regarding professional growth and lifelong education among in-service teachers, should be no less clear. If teachers only know TSM and not principle-based mathematics, then their collaborative lesson planning will produce nothing more than pedagogical embellishments of TSM. Their critiques of each other's teaching in their lesson observations will all be based on TSM, but TSM is too inherently defective to serve as an arbiter of mathematical truth. For example, one can imagine one teacher suggesting to another, "Don't you think you could have used proportional reasoning to give a simple solution of the problem?"[4] Also keep in mind that, since most professional developers only know TSM, in-service mathematics PD cannot help but reinforce teachers' knowledge of TSM. Therefore, if teachers only know TSM, the upshot of any effort to foster lifelong learning and continuous education is likely to result in the creation of a robust environment for TSM to thrive and metastasize throughout the body of school mathematics education. This is not a consummation devoutly to be wished. For the good of school mathematics education, we cannot import (A) and (B) into the American system without first getting rid of TSM.

Before exploring the *content* issue any further, we must point out that it is impractical to make (B) a reality in the teaching culture without first introducing major changes in the present education system. If teachers are going to engage in mutual classroom visits, collaborative lesson planning, and attending professional development institutes as a way of life, they will need *time* in their job to engage in such time-consuming activities, valuable as these may be. It would appear that, as of 2017, the Chinese education system allows its teachers to do that but the American system does not, because:

> Chinese teachers have much larger classes: typically around twice the size of U.S. classes.... (National Research Council, 2010, p. 5) Chinese teachers have fewer classes than do U.S. teachers, typically just two or three per day, whereas U.S. teachers are in their classrooms for most, if not all, of their day. (*ibid.*, p. 6)

The long and short of the matter is that (B) will not materialize in the American system for free and, to make it happen, we will have to make some hard choices. First, we will have to choose between smaller classes and less opportunities for professional growth, or larger classes and more time for professional growth. The debate about the pros and cons of class size has been going on for a long time

295

(Mishel & Rothstein, 2002), and at some point, I hope teachers themselves will be able to make the right choice for themselves.

A second consideration is that if we opt for larger classes, the demand on each teacher's mathematical and pedagogical competence will likely increase. This will require an upgrade of the teaching corps. Such an upgrade will not happen unless we can make teaching a more attractive profession. Needless to say, that is a major social issue that is mostly beyond the concerns of this chapter. For example, even one of the easiest parts of this problem, raising teachers' salaries, already stirs up a political hornets' nest. However, there is at least one thing that is very much pertinent to the present discussion. One of the many reasons teachers are dissatisfied with their profession is that they are not given the respect due them as professionals. For example, teachers have little input into school decision-making and "have only limited authority over key workplace decisions" such as "which courses they are assigned—or misassigned—to teach" (Ingersoll, 1999, p. 34). Any changes along this line will require an overhaul of school management and the education administrative bureaucracy. Do we have the political will to face up to this challenge?

Let us return to the content issue and make a few comments on the feasibility of bringing about the necessary changes to make (A) and (B) a reality. To help pre-service teachers change their knowledge base from TSM to principle-based mathematics, institutions of higher learning across the land will have to be aware of the deleterious effects of TSM on school mathematics education and make the commitment to change. This will require cooperation between schools of education and departments of mathematics in these institutions as well as the willingness to make financial investments by the institutions' administrations. As of 2017, there seems to be no sign of such awareness and, in any case, the cooperation between the mathematics and education communities may not be easy to come by.

An additional difficulty in getting rid of TSM is the need for a default version of *principle-based mathematics* to demonstrate that there can be a systematic exposition of mathematics that toes the line of the school curriculum *and* is consonant with FPM.

This is because, although principle-based mathematics is a branch of mathematics, it is different from standard college mathematics, in the same way that although the basic theory of electrical engineering is part of physics, electrical engineering is *not* part of physics (see Wu, 2011b, section 3). The two demands on principle-based mathematics, (I) and (II) above, pull in opposite directions, so that something as simple as the concept of a fraction in *abstract algebra* easily satisfies (II) but is too sophisticated to be used in the elementary classroom (i.e., does not satisfy (I)). It is at times not entirely straightforward to find a way to develop mathematics that is consistent with FPM and also suitable for use in the school classroom. The very existence of principle-based mathematics therefore cannot be taken for granted.

At the moment, an overwhelming majority of teachers and university mathematics educators were themselves educated in TSM, and many have come to believe that TSM *is* mathematics. These teachers accordingly teach their own students TSM and these educators carry out their research in terms of TSM (e.g., a great deal of research effort was invested in the vain attempt to make fractions-as-pizzas teachable and learnable). Unfortunately, both actions end up imprinting TSM on the next generation and perpetuating a vicious cycle (see Wu, 2011c, p. 9 for a more detailed discussion of the vicious cycle). Having a default version of *principle-based mathematics* will make it possible to teach it to pre-service teachers and break this vicious cycle.

There is another benefit of having available a default version of principle-based mathematics that seems not to be fully appreciated, namely, it will facilitate the development of teachers' PCK as described, for example, in Chapters 4–7, 9 of this volume. This is because PCK is the *bridge* between teachers' *content knowledge* and their *pedagogical practices* in the classroom. At the moment, the education literature seems to be uncertain about exactly what this bridge is, or might be. The general idea (Chapter 6, p. 2) that PCK is needed to transform mathematics in "its academic form featuring rigorous deduction and logical reasoning"—which is "is precise but cold"—into something "interesting, beautiful, and easy to access" in the school classroom speaks to the misconception about what constitutes the "content" in *pedagogical content knowledge* (PCK). If we correctly understand this "content" to be *principle-based mathematics*, then the fact that this content is already conceived and developed in accordance with students' learning trajectories (i.e., condition (I) of principle-based mathematics) means that it will be, by design, already much easier to access than the supposedly "precise but cold" mathematics. Moreover, since the reasoning in principle-based mathematics is introduced at the level of the school curriculum for each grade, teachers and educators should have little doubt about whether it is appropriate for students to learn this "logical reasoning." Instead, they can concentrate their efforts on the pedagogical issues of how to facilitate their students' learning about this reasoning. Therefore, making principle-based mathematics the *content* of PCK adds clarity to what PCK is by closing the gap between content knowledge and pedagogical practices in the classroom.

Finally, we note that there has been an ongoing effort to write an exposition of principle-based mathematics for K-12; see Wu (2011b) for elementary school, (2016a, 2016b) for middle school, and (forthcoming) for high school.

SPECULATION ON CONTENT KNOWLEDGE PREPARATION

We will look into the content-knowledge preparation of mathematics teachers in China and propose a few related research directions.

In Chapter 9 by Pu et al. it is stated that school mathematics textbooks are very important to Chinese teachers:

> Mathematics textbooks are used in several important ways in the Chinese school education, including, as tools for teachers' professional development through studying textbooks; as self-paced learning materials for out-of-school children; and as the main resource for teaching and learning in the classroom.

> Both in urban and rural schools, mathematics textbooks and related teaching manuals are important resources for teachers' instructional and lesson planning.

Such an overt reliance on textbooks by teachers is natural, yet to someone from the U.S. who has experienced the devastation of TSM firsthand, this passage does raise some concerns: Could some incipient version of TSM be at work in Chinese school mathematics education? Such a question runs counter to the general tenor of the preceding chapters, and I raise it only because of my own sporadic and unorthodox encounters with Chinese school mathematics education.

I have visited China (essentially only Beijing) eight times and—except the one-week visit in 2010—each visit lasted between three and seven weeks. The first five visits, from 1976 to 1985, were entirely devoted to mathematics;[5] I went there in my capacity as a geometer. The last three, from 2006 to 2010, were wholly or partly related to school mathematics education. In seven of the eight visits, I got to meet with officials from the Ministry of Education. Now, during my visits up to 1985, I never once suggested that I knew anything about school education, yet each time, those officials would voluntarily bring up the question of what to do about what they called the "duck-stuffing" phenomenon (*tianya*, i.e., rote-learning) in school math classrooms. The fact that such an issue about *school education* would be discussed at all during my first visit in May of 1976, four months before the end of the Cultural Revolution, must be regarded, in retrospect, as nothing short of stunning. Given the wholesale disruption of all phases of education during the ten years of the Cultural Revolution, the raising of this question about *education* in that setting could not help but make an indelible impression on me. The officials' concern was obviously genuine, so the phenomenon itself had to have been deep-rooted. The fact that the same question—in the same terminology of *tianya*—would continue to be raised by different officials in each of my subsequent visits with them could only mean that the Ministry's search for a solution to the problem was still ongoing.

Before 1992, I was a professional mathematician full-time, and my knowledge of school mathematics education was nearly nonexistent. My answers to the Ministry officials' question about "duck-stuffing" in those days were at best *pro forma*, but the concept of "duck-stuffing" became permanently lodged in my mind. It is because of this *idée fixe* that I am now led to wonder about the possible presence of some form of TSM in Chinese school mathematics education. Obviously I do not have any evidence that Chinese school mathematics education is suffering from TSM, yet there are a few indicators—from my firsthand experiences and from what I read in the preceding chapters—that suggest that this idea may not be so outlandish after all but may actually be worthy of serious investigation.

Let me recount my limited firsthand experiences. In my 2011 visit to China, I once had a roundtable discussion with Ministry officials and local teachers (in Beijing). I brought up the subject of *reasoning* in teaching. Around that time, I happened to be wrestling with the teaching of the "laws of exponents" to American secondary teachers. In TSM, these laws are factoids to be memorized, and the exponential notation is just one more new notation like a host of other notations. In the TSM tradition, a main emphasis is to drill students on becoming fluent in re-writing radicals of numbers in exponential notation, and that is an end in itself. TSM treats these laws as *number facts* that have nothing to do with the real reason for learning about these laws—the amazing properties of the exponential function. I was faced with the problem of finding a way to make teachers aware of the real mathematical issues behind these laws, and to make my arguments sufficiently persuasive so that they would rethink their TSM-infused knowledge.[6] For this reason, I inquired specifically about how these laws were taught in Chinese classrooms. My memory of the teachers' responses is not too clear after all these years, but I think they said that their main emphasis in general was on teaching students how to use the laws to solve problems but not so much on why the laws are true or even what they mean. I was a bit surprised, so I asked whether, in their lessons, they would at least consider *defining* clearly the meaning of fractional exponents and *proving* certain special cases such as the ubiquitous identity: $x^{1/n} y^{1/n} = (xy)^{1/n}$ After a short silence, one teacher finally spoke up and explained that if she were to do this, students would ask whether this material would be on *gaokao*,[7] and if not, then they would simply tune her out. The other teachers nodded and murmured in agreement.

Some days after that meeting, I was given the opportunity to visit several classes in a high school and, afterwards, I had a short session with some teachers for an exchange of views. Very likely, something I heard in one of the classes made me bring up the need to clearly differentiate between a *definition* and a *theorem* in teaching. I said that, in America, our teachers were not always able to draw that distinction correctly and that, in Australia, many teachers also seemed to have the same problem. For example, I asked whether "$3^0 = 1$" was a definition or a theorem.[8] A teacher said confidently that it was clearly a theorem. This is an answer that the American TSM textbooks would emphatically agree with.

Because I had no firsthand knowledge of Chinese school mathematics textbooks, I reflected on these two anecdotes and wondered whether they were related, and also how students' preoccupation *gaokao* might impact textbooks. Would these books make any effort to discuss things not directly related to the procedures of solving problems on *gaokao*?

According to what I read in the preceding chapters, the answer to this question is more complex than I would have hoped. Unlike the American situation, textbooks *per se* seem not to be the main issue in Chinese mathematics education. Textbooks in China don't seem to include a lot of information about PCK and "a large amount of materials used in classroom instruction are literally added by teachers based on their experience" (Liang, Chapter 12, this volume). This is consistent with the information about the state of elementary mathematics education in Chapter 5, for example. In that chapter, one finds some reactions of pre-service elementary teachers to the rigorous *mathematics* course requirements for pre-service elementary teachers. One teacher said, "Some of the courses are so boring, also difficult." Another teacher said, "I don't know why we need to learn these. Useless" (Xie et al., Chapter 5, this volume). These comments point to an obvious gap between the mathematics taught in the required mathematics courses and the mathematical knowledge teachers actually *need for teaching in a school classroom*. As a matter of fact, one could have easily arrived at the same conclusion by inspecting the course requirements for elementary teachers in Chapter 5 and those for secondary teachers in Chapter 6 (see also Wu, 2011b, pp. 372–373, for an analogous discussion.). Some deans of normal universities acknowledged that this gap is substantial and real, and have tried to forge a closer connection between these mathematics courses and teachers' pedagogical needs (Xie et al., Chapter 5, this volume).

In this light, we now better understand the quote from Chapter 9 at the beginning of this section: the disconnect between the rigorous mathematics in university preservice courses and actual on-the-job pedagogical needs has driven teachers to rely on school textbooks and collaborative lesson planning (Chapter 10) to develop their own content knowledge and PCK for teaching. The burning question is whether teachers do so by isolating themselves and turning their collective backs on the mathematics community, and if such is the case, whether the mathematical knowledge so produced is consistent with FPM. In the case of the laws of exponents, for example, could experienced teachers' perception of the futility of teaching these laws *as principle-based mathematics* persuade new teachers to concentrate on teaching these laws as *procedures* because this is "what works" in a *gaokao*-fixated classroom? If so, might this tradition set up a vicious cycle so that such an approach to teaching the laws of exponents becomes a permanent fixture in Chinese mathematics education? Could similar classroom "realities" of this nature also come to inform the teaching of other mathematical topics in teachers' reservoir of PCK? Might not this kind of "reality-based" instruction be the source of the "duck-stuffing" phenomenon?

We need evidence-based answers to these questions. What we know from the American experience is that such isolation from the mathematics community was in fact a main cause that triggered the development of TSM—*Textbook* School Mathematics. In the U.S., the abandonment of FPM for pedagogical expediency was apparently what led teachers and educators to certain practices of TSM (e.g., fractions as pieces of pizza) which, when codified in textbooks, eventually became the orthodoxy in American school mathematics education. In this context, it may be useful to recall the two basic requirements of principle-based mathematics:

I. Principle-based mathematics closely parallels the school mathematics curriculum.
II. Principle-based mathematics is consistent with FPM.

Our present discussion of the isolation of teachers from the mathematics community puts in perspective why both requirements (I) and (II) are absolutely essential to a sound *content knowledge for teaching*, and why it is important for teachers to learn principle-based mathematics.

SOME REMARKS ON THE PREPARATION OF MATHEMATICS TEACHERS IN CHINA

Now, back to the main thread of our discussion: could the knowledge (content knowledge as well as PCK) created by mathematics teachers in such isolation have something to do with "duck-stuffing"?

If the concern of the Ministry of Education is to be taken seriously, real research effort should be devoted to settling this question one way or the other. The most direct approach would be to evaluate current textbooks to see whether they are consistent with FPM. This will require the cooperation of the mathematics and education communities. But, as we mentioned above, looking at textbooks is not enough, because we would need to collect a representative sample of actual classroom lessons to see the kind of mathematics instruction school students are actually getting. In year 2017, it should not be difficult to obtain video records of such lessons for a detailed study. Given the vast scope of such a study, perhaps one should break it down into three parts: elementary school, junior high school (middle school), and senior high school. Again, the video analysis will require a team of both mathematicians and educators.

Another study worth undertaking would be the effect of *gaokao* on mathematics education. Teaching-to-the-test is hardly a new phenomenon; it has galvanized the attention of American educators for a long time. Clearly, the *gaokao* is not going away, so there should be serious research on how to minimize its impact on general mathematics education. One school of thought believes that, if a standardized test is to serve a *useful* purpose for school mathematics education, it has to be low-stakes (see Wu, 2012, pp. 15–17). Since the *gaokao* is by definition a high-stakes test, it may be time to consider the possibility of having two tiers of mathematics education in the last two years of senior high school: students who intend to go to college take the *gaokao* but those who don't follow a different curriculum.

It is important to recognize that those who want to excel on *gaokao* will have the *obligation* to learn more than what the *gaokao* dictates. These advanced students must learn that excellence in the sciences and mathematics cannot be achieved solely by a superior ability to do *problems that are handed to them.* They will have to learn mathematics consistent with FPM, especially its reasoning and coherence, to nurture their creativity. We hope the (yet-to-be-written) curriculum for this group of students will leave no doubt about such an expectation. Classrooms in which principle-based mathematics is taught will certainly steer mathematics education away from "duck-stuffing." For the other group not going to college, being freed from the tyranny of the *gaokao* will allow them to settle down and learn about—not all the technical skills to negotiate the tricky problems in the *gaokao*—but the two components of mathematics education that validate the presence of thirteen years of mathematics in their school curriculum: the reasoning and critical thinking that are needed for decision making. These two qualities are indispensable to life in the high-tech age but they will not survive a duck-stuffing education. Because many of the "duck-stuffing" skills are increasingly being taken over by computers, students who do not go to institutions of higher learning have no choice but to learn to reason and think critically—things that computers have not yet mastered. This is not the place to enter into a detailed discussion of a curriculum that can promote reasoning and critical thinking, but Part II of Wu (2011d) can serve to give *some* idea of its potential.

Such a sea-change in the basic structure of school mathematics education cannot be carried out without detailed research into all the associated social questions it raises. For example, will such a two-tiered education be socially acceptable and, in fact, politically feasible? Fortunately, some nations (Japan, Germany, etc.) already practice variations of such a two-tiered system, so there is no need to reinvent the wheel. There should also be research into the creation of a complete series of school textbooks that are less *technically-oriented* but which nevertheless meet the high standards of FPM. Altogether, this would be a vast and Herculean undertaking, and until these preliminary hurdles have been overcome, it would be inadvisable to launch such a reform.

Finally, I will strongly suggest that all teachers be taught principle-based mathematics in their teacher preparation programs. Back in 1972, when Begle first looked into the disconnect between teachers' "mathematical" knowledge and the effectiveness of their teaching, he concluded:

> ... teachers should be provided with a solid understanding of the courses they are expected to teach. (Begle, 1972, p. 8)

Subsequent works on PD for teachers have basically borne out Begle's dictum (Wu, 2011b). Now if principle-based mathematics is to be made part of the mathematical preparation of teachers, then adjustments will have to be made to the rest of the required *mathematics* courses listed in section 1 (of this chapter). There is a need to lower the *technical* level and deepen the conceptual level of most of these courses by replacing them with *general surveys* for the benefit of teaching school mathematics. Recall one teacher's comment on these courses: "Some of courses are so boring, also difficult." For example, it is hard to imagine that many elementary teachers would be enthralled by the need to prove theorems in a course on Non-Euclidean Geometry or Projective Geometry when they will probably never come across the hyperbolic axiom or Desargues' Theorem in their professional life. Would not a single survey course on geometry that gives the history of the Parallel Postulate with some hands-on activities on hyperbolic geometry, spherical geometry, and projective geometry better serve elementary teachers? A similar comment can be made about the requirement for secondary teachers to take Complex Analysis, Abstract Algebra II, Differential Geometry, and Combinatorics and Graph Theory. However, to make a comprehensive reform possible, research will be needed to create a cohesive program that attends to the goal of broadening teachers' mathematical knowledge without sacrificing the relevance of this knowledge to their professional practice. This research will also have to include the creation of a series of textbooks for the new teacher preparation program. There is a tremendous amount of work ahead, but for the good of the next generation, it needs to be done.

NOTES

[1] Near the end of this chapter, there are a few comments on the appropriateness of some of these mathematical requirements.
[2] The same table gives 21.2 percent of the teachers reporting that they were worse, but the standard error for this figure is between 30 and 50 percent.
[3] And sometimes not even then.
[4] Proportional reasoning is a mainstay of TSM, but it is not correct mathematics. See section 7.2 in Wu (2016b, pp. 144–154) for an explanation of why it is not correct.
[5] My first visit was with a large mathematics delegation and one subgroup was entrusted with the task of reporting on Chinese mathematics education. If memory serves, I was not with that subgroup, but I am not absolutely certain.
[6] The eventual outcome of this effort is recorded in Wu (2016b, chapter 9).
[7] This is the examination that is a prerequisite for entering almost all higher education institutions in China at the undergraduate level. It is usually taken by students in their last year of high school.
[8] I raised this question because this is a definition that is usually presented as a theorem in TSM.

REFERENCES

Ball, D. L., Thames, M. H., & Phelps, G. (2008). Content knowledge for teaching: What makes it special? *Journal of Teacher Education, 59*, 389–407. Retrieved from http://jte.sagepub.com/cgi/content/abstract/59/5/389

Begle, E. G. (1972). *Teacher knowledge and student achievement in algebra* (SMSG Reports, No. 9). Pola Alto, CA: School Mathematics Study Group. Retrieved from https://eric.ed.gov/?id=ED064175

Bill & Melinda Gates Foundation. (2014). *Teachers know best: Teachers' views on professional development.* Retrieved from http://tinyurl.com/n22zhwh

Goldring, R., Taie, S., & Riddles, M. (2014). *Teacher attrition and mobility: Results from the 2012–2013 teacher follow-up survey* (NCES 2014–077). Washington, DC: National Center for Education Statistics, U.S. Department of Education. Retrieved from https://nces.ed.gov/pubs2014/2014077.pdf

Ingersoll, R. (1999). The problem of underqualified teachers in American secondary schools. *Education Researcher, 28*(2), 26–37. Retrieved from http://www.gse.upenn.edu/pdf/rmi/ER-RMI-1999.pdf

Ingersoll, R., Merrill, L., & May, H. (2014). *What are the effects of teacher education and preparation on beginning teacher attrition?* (Research Report #RR-82). Philadelphia, PA: Consortium for Policy Research in Education, University of Pennsylvania. Retrieved from http://www.cpre.org/prep-effects

Mishel, L., & Rothstein, R. (Eds.). (2002). *The class size debate.* Washington, DC: Economic Policy Institute. Retrieved from http://tinyurl.com/luj5f7e

National Research Council. (2010). The teacher development continuum in the United States and China. A. E. Sztein (Ed.), *U.S. national commission on mathematics instruction.* Washington, DC: National Academy Press. Retrieved from https://www.nap.edu/read/12874/chapter/1

NPR Ed. (2014). *The teacher dropout crisis.* Retrieved from http://tinyurl.com/kj2f896

Poon, R. C. (2014). *Principle-based mathematics: An exploratory study* (PhD. dissertation). University of California, Berkeley, CA: Retrieved from http://escholarship.org/uc/item/4vk017nt

Schoenfeld, A. H. (1988). When good teaching leads to bad results: The disasters of "well-taught" mathematics courses. *Educational Psychologist, 23*(2), 145–166.

Shulman, L. (1986). Those who understand: Knowledge growth in teaching, *Educational Researcher, 15*, 4–14. Retrieved from http://itp.wceruw.org/documents/Shulman_1986.pdf

Wu, H. (2011a). *Understanding numbers in elementary school mathematics.* Providence, RI: American Mathematical Society. (Chinese translation: Peking University Press, 2016.)

Wu, H. (2011b). The mis-education of mathematics teachers. *Notices American Mathematical Society, 58*, 372–384. Retrieved from https://math.berkeley.edu/~wu/NoticesAMS2011.pdf

Wu, H. (2011c). Bringing the common core state mathematics standards to life. *American Educator, 35*(3), 3–13. Retrieved from http://www.aft.org/pdfs/americaneducator/fall2011/Wu.pdf

Wu, H. (2011d). *Syllabi of high school courses according to the common core standards.* Retrieved from https://math.berkeley.edu/~wu/Syllabi_Grades9-10.pdf

Wu, H. (Spring-Summer 2012). Assessment for the common core mathematics standards. *Journal of Mathematics Education at Teachers College, 4,* 6–18. Retrieved from https://math.berkeley.edu/~wu/Assessment-JMETC.pdf

Wu, H. (2016a). *Teaching school mathematics: Pre-algebra.* Providence, RI: American Mathematical Society. Retrieved from http://tinyurl.com/zjugvl4

Wu, H. (2016b). *Teaching school mathematics: Algebra.* Providence, RI: American Mathematical Society. Retrieved from http://tinyurl.com/haho2v6

Wu, H. (2018). The content knowledge mathematics teachers need. In Y. Li, J. Lewis, & J. Madden (Eds.), *Mathematics matters in education: Essays in honor of Roger E. Howe.* Dordrecht: Springer.

Wu, H. (forthcoming). Volume I, *Rational numbers to linear equations.* Volume II, *Algebra and geometry.* Volume III, *Pre-calculus, calculus, and beyond.*

Hung-Hsi Wu
Department of Mathematics
University of California
Berkeley, California, USA

ABOUT THE CONTRIBUTORS

Sigrid BLÖMEKE, Professor of Educational Measurement at University of Oslo, Norway, and Director of the Centre for Educational Measurement (CEMO). Blömeke holds a PhD from University of Paderborn, Germany, and was Associate Professor at University of Hamburg, Professor at Humboldt University of Berlin (both Germany) and Visiting Professor at Michigan State University, USA. Her areas of expertise include research on the development of teacher competence and educational effectiveness research. Blömeke was the National Research Coordinator of TEDS-M in Germany.

Wei CHEN is professor and dean of School of Education, Harbin University. Dr. Chen is the member of the National Elementary Teacher Education Committee and National Elementary Education Committee. Her research interests include children development and education, teachers' professional development, elementary teacher education. In recent years, she has published nearly 20 academic articles and 3 books. She has been involved in 7 projects as the director. She has served in elementary teacher education research and practice for 12 years.

Martina DÖHRMANN, full professor for mathematics education at the University of Vechta since 2010. She studied mathematics and physics at the University of Bremen and received the second state examination for teaching in 2005. She completed her PhD with a study on the improving of data skills. Her areas of research are teachers' professionalism and the design of learning environments.

Chia-Jui HSIEH, assistant professor at National Taipei University of Education, and the secretary-general of Taiwan Association for Mathematics Education. He received his Ph.D. in mathematics education from National Taiwan Normal University. Dr. Hsieh had been a middle school mathematics teacher for three years before pursuing higher degrees. After receiving Ph.D. degree, he taught mathematics pedagogy courses for pre-service mathematics teachers at National Taiwan Normal University. At National Taipei University of Education, he focuses on primary teacher preparation. Dr. Hsieh's research interests include mathematics teacher education, remedial teaching, and concept image for mathematics teaching.

Feng-Jui HSIEH, professor of the Mathematics Department at the National Taiwan Normal University, and the president of Taiwan Association for Mathematics Education. She received her M. Sc. in Mathematics and Ph. D. in Instruction and Curriculum at Purdue University. Her research focuses on mathematics learning and teaching, and pre- and in-service teachers' professional development. She served as Taiwan's NRC of two international studies about mathematics teacher education,

ABOUT THE CONTRIBUTORS

TEDS-M 2008 granted by IEA and MT21 granted by NSF in USA, and the Chair of ICMI-EARCOME8 conference.

Rongjin HUANG is a professor at Middle Tennessee State University, USA. His research interests include mathematic classroom instruction, mathematics teacher education, lesson study, and comparative mathematics education. He has completed several research projects and published scholarly work extensively. His recently published books include, *Prospective mathematics teachers' knowledge of algebra: A comparative study in China and the United States of America* (Springer, 2014) and *Learning and teaching mathematics through variations: Confucian heritage meets western theories* (Sense, 2017). Dr. Huang has served as a guest editor for *ZDM Mathematics Education* and *International Journal for Lesson and Learning Studies*. He has taken leadership in various professional organizations by chairing conference secessions and serving on professional committees such as AERA and ICME.

Xianhan HUANG, assistant professor at the University of Hong Kong. Her research interests include teacher professional development, curriculum design, as well as washback effects of high-stake testing. yxhhuang@hku.hk

Xingfeng HUANG is an Associate Professor at Shanghai Normal University. He obtained his Ph.D. in mathematics education from East China Normal University. Dr. Huang taught middle and high school mathematics for ten years. Then, he taught mathematics pedagogy courses for pre-service teachers, and also taught advanced mathematics for engineering at Changshu Institute of Technology. At Shanghai Normal University, Dr. Huang has focused on primary teacher preparation. Dr. Huang research interests include mathematics teacher education.

Gabriele KAISER holds a master's degree as a teacher for mathematics and humanities for lower and upper secondary level. She completed her PhD in mathematics education in 1986 with a study on applications and modelling and her post-doctoral study in pedagogy on international comparative studies at the University of Kassel. Since 1998, she is full professor for mathematics education at the Faculty of Education of the University of Hamburg. Her areas of research are empirical studies on teacher education and teachers' professionalism, modelling and applications in school, international comparative studies, gender and cultural aspects in mathematics education.

Since 2005 she serves as Editor-in-chief of ZDM Mathematics Education (formerly Zentralblatt fuer Didaktik der Mathematik), published by Springer. She is Convenor of the 13th International Congress on Mathematics Education (ICME-13), which took place in July 2016 at the University of Hamburg with 3,500 participants from all over the world.

ABOUT THE CONTRIBUTORS

Huk-Yuen LAW is an adjunct assistant professor at The Chinese University of Hong Kong (CUHK). He obtained his PhD in mathematics education from University of East Anglia. Dr. Law taught secondary school mathematics for 23 years. Then, he taught mathematics pedagogy courses for both pre-service and in-service teachers, and also taught action research for post-graduate as well as undergraduate education students at CUHK. His research interests include mathematics teacher education, action research in education, communication in the teaching and learning of mathematics, and values in mathematics education. hylaw@cuhk.edu.hk

Yeping LI is a Professor of Mathematics Education at the Department of Teaching, Learning, and Culture at Texas A&M University, USA. He has also been named by Shanghai Municipal Education Commission as an "Eastern Scholar" Chair Professor at Shanghai Normal University, China, since 2016. His research interests focus on issues related to mathematics curriculum and teacher education in various education systems and understanding how factors related to mathematics curriculum and teachers may shape effective classroom instruction. He is the editor-in-chief of the *International Journal of STEM Education* and *Journal for STEM Education Research* published by Springer, and is also the editor of a monograph series *Mathematics Teaching and Learning* published by Brill | Sense. In addition to co-editing over 10 books and special journal issues, he has published more than 100 articles that focus on mathematics curriculum and textbook studies, teachers and teacher education, and classroom instruction. He has also organized and chaired many group sessions at various national and international professional conferences, such as ICME-10 (2004), ICME-11 (2008), and ICME-12 (2012). He received his Ph.D. in Cognitive Studies in Education from the University of Pittsburgh, USA.

Su LIANG is an Associate Professor in Department of Mathematics at California State University, San Bernardino. She received her Ph.D. in mathematics from University of Connecticut in 2010. Dr. Liang has been teaching mathematics content courses for pre- service teachers for seven years. Her research interest includes mathematics teacher preparation, teacher professional development, scholarly teaching, and applying technology in mathematics teaching.

Sin-Sheng LU, associate professor at Shanghai Normal University. He obtained his Ph.D in mathematics education from Tokyo Gakugei University. Dr. Lu taught foundations of mathematics and mathematics pedagogy courses for pre-service teachers at Nantong Normal College. At Shanghai Normal University, Dr. Lu has focused on middle school teacher preparation. His research interests include mathematics teacher education, Instructional Materials Development, educational technology in math teaching, and origamis.

ABOUT THE CONTRIBUTORS

Yu-Jen LU is a curriculum and instruction supervisor in Yilan County in Taiwan. He had served for many years as an elementary school mathematics lead teacher for the County Teacher Training Center and a member of curriculum and instruction consulting team for mathematics teachers. His major interests are in the areas of classroom practices and in-service teacher professional development in mathematics. hitachi6@gmail.com

Yun-peng MA is professor and former dean of School of Education, Northeast Normal University. Dr. Ma also serves as the co-director of the Chinese Curriculum Studies Committee, the co-director of the Chinese Mathematics Education Committee, the member of the National Basic Education Curriculum Materials Committee, and vise director of Instruction Advisory Board of Primary Teachers Training in Higher Education. His research interests include curriculum implementation and assessment, curriculum reform in K-12 education, and K-12 mathematics education. Dr. Ma has published more than 20 books and multiple research articles. He is a recipient of various awards at the national, provincial, regional and university levels.

Despina POTARI is a professor of mathematics education at the Mathematics Department of the National and Kapodostrian University of Athens. She has been visiting professor in different universities in Europe and US and currently at Linnaeus University in Sweden. Her research interest is mainly on the development of mathematics teaching and learning and teacher development and in particular on the role of different contexts and tools in the classroom setting as well on teacher knowledge and teaching practices. She has publications in international research journals, conference proceedings and book chapters. She is an editor in chief of the Journal of Mathematics Teacher Education, member of editorial boards and teams and reviewer for international journals and conferences.

Shuping PU is a professor at Chongqing Normal University. She obtained her Ph.D. in mathematics education from East China Normal University, and now she is conducting her postdoctoral research at Southwestern University. Dr. Pu taught mathematics pedagogy courses for pre-service teachers. At Chongqing Normal University, Dr. Pu has been focusing on primary teacher preparation. Dr. Pu research interests include mathematics teacher education & HPM.

Tim ROWLAND has spent most of his professional life at the University of Cambridge, UK. His research in the last 20 years has focused on the ways that teachers' mathematics-related knowledge is made evident in their classroom performance. In August 2012, he was appointed Professor in Mathematics Education at the Norwegian University of Science and Technology (NTNU), in Trondheim, Norway, and Honorary Professor of Mathematics Education at the University of East Anglia, Norwich, UK. He has held Visiting Professor positions in several countries, and was Vice President of the International Group for the Psychology of Mathematics Education (PME) until 2013.

Gloria Ann STILLMAN is an Associate Professor at Australian Catholic University, Ballarat, and a researcher in the Learning Sciences Institute Australia. She obtained her PhD in mathematics education from The University of Queensland. Dr Stillman taught secondary school mathematics before entering academia. She has taught mathematics, statistics, instructional design, education studies and computer studies to pre-service and in-service teachers in mathematics education and computing courses at several Australian universities. At Australian Catholic University, Dr Stillman has focused her teaching on primary and early childhood teacher preparation. Her research interests include the teaching and learning of mathematical modelling at primary and secondary school levels, metacognition, and teacher professional knowledge.

Xuhua SUN is an assistant professor, faculty of education, university of Macau. Her research focus on variation problems, Chinese lesson study, teacher professional knowledge. She publishes in both Chinese and English. She is the author of spiral variation: the Chinese logic to develop curriculum and instruction (in Chinese). She is co-chair of International Commission for Mathematical Instruction (ICMI) Study 23: primary mathematics study on whole number (2012–2018).

Hak Ping TAM, associate professor at the National Taiwan Normal University. His research interests include mathematics curriculum, assessment and applied statistics. He is currently developing several learning progressions in mathematics and a knot theory curriculum for secondary school students. Meanwhile, he is designing a curriculum to learn plane geometry concepts through paper folding. t45003@ntnu.edu.tw

Shu-Jyh TANG, teacher at Taipei Municipal Bailing High School. He obtained his Ph.D in mathematics education from National Taiwan Normal University. Dr. Tang has taught junior and senior high school mathematics for twenty years. At the same time, he has focused his research interests on secondary mathematics teacher preparation, life-long learning, and performance evaluation.

Ting-Ying WANG, project assistant professor at National Taiwan Normal University. She teaches mathematics pedagogy courses and develops programs for fostering IB teachers for the Mathematics Department at the National Taiwan Normal University. Her research has focused on mathematics teacher education and secondary school mathematics teachers' teaching competences. She is currently cooperating with researchers from Mainland China to investigate preservice and in-service teachers' perspectives of ideal mathematics teaching behaviors, and also with researchers from Germany to investigate teachers' noticing and pedagogical reasoning.

ABOUT THE CONTRIBUTORS

Hung-Hsi WU is Professor Emeritus of Mathematics at the University of California at Berkeley where he taught from 1965 to 2009. He got into mathematics education in 1992 and worked with the State of California in 1997–2005 in all aspects of mathematics education. From 2000 to 2013, he gave three-week, content-based summer professional development institutes every year for teachers in K-8. His long-term project is to reform the mathematical education of teachers and educators so that all school mathematics teachers and mathematics educators are taught the correct content knowledge they need for their work. To this end, he is finishing a series of six volumes that cover the mathematics of K-12.

Yingkang WU is an Associate Professor at East China Normal University. She obtained her Ph.D. in mathematics education from National Institute of Education, Nanyang Technological University of Singapore. Dr. Wu has taught mathematics pedagogy courses for pre-service secondary teachers, and also taught calculus for college students. Her research interests include mathematics teacher education, mathematics classroom teaching, mathematics assessment, as well as teaching and learning of school statistics.

Shu XIE is lecturer of School of Education, Northeast Normal University. Dr. Xie's research areas focus on elementary teacher education, pedagogical content knowledge, and curriculum and pedagogy. She has involved in 17 projects. While 3 of them serve as director are national level, which respectively are, study on evaluation of elementary mathematics teachers' pedagogical content knowledge, study on elementary mathematics teachers' curriculum design, and reciprocal study on elementary mathematics classroom teaching between China and Canada. She has published 33 articles.

Yudong YANG is a Professor at Shanghai Academy of Educational Sciences, Deputy Secretariat of Shanghai Committee of School Mathematics Teaching. He obtained his Ph.D. in mathematics education from East China Normal University. His major research interests are in in-service mathematics teacher education.

Zhiqiang YUAN is Associate Professor of Mathematics Education at the College of Mathematics and Computer Science, Hunan Normal University, Changsha, Hunan. He received his PhD in mathematics education from East China Normal University in 2012, and MSc and BSc degrees in mathematics education from South China Normal University in 2003 and 2000, respectively. He is a member of the Executive Committee of the Mathematics Education Research Association of China. His current research interests are in mathematics teacher education, technology in mathematics education, and teaching and learning of probability and statistics.

Bo ZHANG is an Associate Professor at Yangzhou University. She obtained her Ph.D. in mathematics education from East China Normal University. Dr. Zhang taught mathematics pedagogy courses for pre-service teachers, and also taught advanced mathematics for engineering at Yangzhou University. Dr. Zhang has focused on secondary pre-service teacher preparation. In recent years, Dr. Zhang's research interests include mathematics teacher education and mathematics teaching design.

INDEX

A

Algebra, 6, 11, 12, 20, 21, 48, 67–70, 77, 78, 88, 93, 97, 99, 110, 112, 116, 119, 123–131, 156–158, 182n3, 215, 249, 290, 296, 302

Ancient China, 38–41, 44, 45, 47, 49–52, 53n1

Apprenticeship, 6, 57, 138, 241–260, 264, 265, 270, 271

Assessment, 15, 17, 23, 26, 39, 46, 65, 78, 94, 117, 120, 139, 157, 180, 181, 188, 279, 281

Attitude, 41, 43, 45, 95, 96, 105, 120, 157, 173, 174, 176, 179, 187, 256

B

Ball, D., 60, 61

Bass, H., 132, 263

Baumert, J., 63, 64, 111, 284

Beliefs, 15, 46, 53n5, 59–62, 64, 66, 67, 85, 86, 99, 100, 113, 138–141, 147, 148, 157, 158, 166, 187–190, 192, 205, 219, 221, 268

Blömeke, S., 3, 5, 57–79

Bransford, J., 138

Burkhardt, H., 166

C

Case study, 6, 7, 94, 100, 148, 159, 167, 168, 181, 185–206, 209–222, 241–260, 264–271

Chen, W., 85–105

Chinese societies, 37, 39–46, 48, 51–53

Classroom discussion, 120

Classroom observation, 100, 211, 230, 232, 247

Classroom teaching, 6, 12, 94, 95, 102, 103, 111, 114, 116, 117, 120, 139–141, 143, 145–149, 157, 179, 185–187, 192, 201, 203, 225, 226, 232, 233, 237, 267

COACTIV, 63, 64, 66, 284

Coaching practice, 244, 248–258, 265

Cobb, P., 270

Common content knowledge, 51, 60, 87, 263, 270, 279, 280, 283

Competence, 37, 39, 41, 44, 46, 52, 53, 63, 64, 66, 138, 160, 185, 186, 205, 206, 249, 259, 296

Confucian tradition, 37, 42–46, 50

Critical point, 52, 210, 220, 221, 268, 269, 283

Cultural context, 3, 7, 64

Curriculum knowledge, 57, 86, 87, 89, 90, 94–98, 100, 111, 119, 120

Curriculum reform, 10, 11, 18, 23–26, 85, 109, 185, 243

Curriculum standard for teacher education, 11, 109, 110, 113–118, 131

Curriculum structure, 5, 18, 21, 91–95, 113, 114

D

Data analysis, 90, 124, 189, 204, 250, 251

Decision-making, 203, 219, 296, 301

Deductive, 281

Definition, 65, 67, 68, 78, 127, 128, 169–172, 174, 175, 180, 267, 285, 289, 293, 294, 299, 301, 303n8

Diagrams, 67, 169, 170, 174

Difficult point, 120, 174, 210, 214, 215, 217, 220, 221, 245, 247, 258, 268, 269, 283, 284

Dimensions, 37, 58, 59, 64, 65, 71, 72, 87, 88, 113, 121, 153, 173, 175, 226, 284–287
Discourse, 201
Diversity, 12, 31, 121, 153
Döhrmann, M., 3, 5, 57–78, 289
Domains, 11, 15, 21, 64–66, 68, 70, 77, 78, 172, 187, 210, 263, 266, 269, 271, 279, 280

E

Educational practice, 31, 92, 100, 105, 113, 114, 137, 146
Education in Chinese regions, 4, 18, 26, 37, 39, 46, 53, 85, 86, 101, 113, 140, 158, 160, 161, 166, 225, 297, 303n7
Elementary mathematics, 85–105, 131, 154–156, 181, 290, 299
Elementary mathematics teacher, 85–105, 131, 154–156, 290
Elementary school, 60, 90, 97, 99, 100, 102–104, 155, 251, 252, 263, 266, 267, 283, 297, 301
Equality, 73, 293
Equity, 13
Evaluation, 15, 23, 26, 39, 67, 78, 94, 98, 117, 120, 121, 168, 171, 173, 176–181, 219, 227, 231, 257, 264, 278
Exemplary lesson, 133, 138, 225–228, 233–236, 245, 259, 268
Expectation, 14, 18, 110, 156, 157, 230, 231, 257–259, 271, 301
Experienced teachers, 6, 114, 138, 177, 185–206, 210, 211, 225, 227, 229–232, 234, 245–247, 258, 267, 268, 282, 290, 291, 300

F

Feedback, 45, 103, 160, 206, 211, 229, 233, 234, 250, 278

Field observation, 6, 137–143, 146–149, 158, 159, 171

G

Geometry, 11, 20, 21, 48, 61, 67, 68, 71–73, 77, 78, 93, 116, 119, 124, 125, 127, 131, 155, 158, 169, 176, 190, 195, 199, 215, 230, 249, 282, 290, 294, 302
Goals, 19, 26, 39, 78, 99, 104, 109, 120, 154, 155, 160, 185, 186, 190, 191, 200, 202–204, 214, 219, 221, 236, 257, 270, 277, 302
Group lesson preparation, 167
Group processes, 90, 228

H

Hiebert, J., 137, 153, 209
Hill, H., 3, 279, 280
Historical perspective, 37–53
Hong Kong, 4, 5, 9–32, 53n2, 277, 282
Hsieh, C.-J., 28, 29, 37–53
Hsieh, F.-J., 3, 5, 28–30, 37–53
Huang, R., 3–7, 105, 109–133, 137–149, 153, 159, 185, 191, 205, 206, 209, 228, 236, 241–260, 264, 265, 267, 268, 270, 290, 291
Huang, X., 9–32, 185–206

I

ICMI, 153
Impact, 23, 37–39, 85, 86, 96, 98, 101, 104, 105, 126, 127, 132, 140, 148, 149, 153, 154, 158–160, 201, 225, 229–231, 255, 256, 263, 271, 299, 301
Implications, 26
In-service mathematics teachers, 9, 14, 24–27, 29–31, 167, 209–222, 282, 283, 295

Instruction, 4, 11, 14, 15, 17, 18, 21–27, 30–32, 41, 48, 50, 51, 60, 62, 64, 118, 119–121, 138–140, 145, 165–169, 171–181, 182n1, 185, 187, 188, 192, 193, 209, 210, 219–222, 229, 241–244, 247, 250–253, 255–260, 267, 268, 270, 271, 280, 283–285, 289, 291, 298–301
Instructional planning, 6, 64, 166–168, 173–181
International perspective, 4, 57–79, 277

K
Kaiser, G., 3, 5, 57–79
KCS, 51, 60, 62, 87, 111, 263, 265, 267–271, 279
KCT, 51, 61, 62, 87, 111, 263, 265, 267, 269–271
Key point, 120, 174, 210, 214, 215, 218–221, 254, 268, 269, 283
Knowledge acquisition, 3–7, 57, 234
Knowledge development, 5, 6, 78, 96, 131, 180, 190, 255
Knowledge Quartet (KQ), 62, 63, 284, 286
Knowledge requirement, 38, 41, 46–49
Krainer, K., 28, 139

L
Lampert, M., 209
Law, H., 5, 9–32
Learning community, 138, 139, 186, 187, 232, 237, 247, 248, 258, 260, 270, 282, 292
Learning mathematics, 3, 139, 287, 293
Leinhardt, G., 209
Lesson design, 139, 159, 176, 217, 236, 237, 246, 248–252, 254, 255, 271
Lesson explanation, 171, 172, 176
Lesson study, 27, 28, 32, 133, 139, 159, 185, 205, 206, 209, 220, 243, 259, 268, 283

Li, Y., 3–7, 109, 137, 138, 148, 153, 165–182, 185, 187, 205, 206, 209, 225, 228, 231, 236, 241–260, 263, 265–267, 270, 271
Liang, S., 153, 225–237, 264, 265, 269, 290, 291, 299
Lifelong learning, 7, 104, 105, 156, 291, 295
Literacy, 5, 15, 19, 37–53, 85, 115, 117, 118, 266
Lu, S.-S., 37–53

M
Ma, L., 4, 64, 165–167, 181, 205, 209, 241, 258
Ma, Y.-P., 85–105
Macau, vii
Mainland China, 4, 5, 7, 9–32, 39, 85, 86, 137, 167, 182, 211, 241–260, 264, 270, 271
Massive Open Online Courses (MOOC), 103, 104, 156
Master teacher workstation, 6, 185–206, 243, 264, 265, 267, 268, 282, 291
Master teachers, 6, 117, 138, 148, 168, 173, 176, 180, 182n1, 185–206, 211, 218, 225, 227, 234, 235, 243, 244, 248, 259, 260n1, 264, 265, 267, 270, 282, 285, 287, 291
Mathematical activities, 158, 159, 175, 181, 209, 278
Mathematical beliefs, 67, 99, 139, 148, 188, 190, 192, 219
Mathematical discussion, 60, 62, 120, 147–149, 153, 160, 161, 181, 182, 265, 292, 300, 301
Mathematical knowledge, 3–7, 51, 60, 61, 65, 67, 87, 90, 95–98, 100–104, 110, 153, 154, 156, 171, 205, 237, 263–271, 279, 280, 289

315

INDEX

Mathematical Knowledge for Teaching (MKT), 3–7, 51, 60, 61, 65, 67, 87, 90, 95–98, 100–104, 110, 153, 154, 156, 171, 205, 237, 263–271, 279, 280, 289

Mathematical learning, 3, 4, 6, 7, 12, 13, 30, 49–51, 74–77, 85, 88, 96, 104, 120, 132, 137, 139, 156, 160, 166, 167, 173, 175, 176, 186, 192, 210, 245, 282–287, 292–294

Mathematical perspective, 3, 4, 37–53, 57–79, 99, 110–112, 137, 237, 266, 277, 300

Mathematical practices, 9–11, 14, 18, 20, 24, 27, 28, 30, 37, 38, 52, 57, 65, 85, 100, 102, 103, 132, 137, 155, 159, 182, 188, 209

Mathematical proficiency, 14, 67

Mathematical thinking, 93, 155, 176, 178, 236

Mathematician, 30, 31, 47, 48, 52, 96, 155, 298, 301

Mathematics Content Knowledge (MCK), 28, 30, 49, 62, 65–68, 71, 73–78, 110, 115, 119, 131, 137, 139, 154, 155, 157, 158, 187, 236, 237, 280

Mathematics educators, 6, 46, 51, 53, 85, 97, 99, 140, 155, 158, 160, 296

Mathematics Pedagogical Content Knowledge (MPCK), 28–30, 49, 65–68, 73, 75–78, 87, 98, 102, 122, 137, 139, 172

Mathematics teacher literacy, 38, 47–52

Mathematics teachers, 3–7, 9–32, 37–53, 57, 64, 65, 67, 70, 73, 77, 78, 85–105, 109–133, 137–149, 153–161, 167–169, 171, 173, 176, 177, 180–182, 185, 209–222, 226, 229, 231, 232, 236, 241, 242, 245, 263, 264, 267, 277–283, 285–287, 289–302

Mathematics teachers' professional development, 5–7, 23–27, 31, 85, 113, 140, 147, 153, 166, 167, 263–271, 277, 285, 287, 289, 297

Mathematics teaching, 23, 25, 27, 37, 49, 52, 60, 73, 74, 89, 90, 92–97, 99, 103, 104, 117, 119, 120, 139, 145, 148, 153, 155, 157, 160, 172, 173, 177, 179, 188–190, 192, 219, 236, 237, 243, 245, 258, 270, 277, 282, 283, 285

Mathematics textbooks, 26, 48, 49, 71, 99, 165–169, 278, 297–299

Mentors, 5, 15, 19, 41, 90, 98, 100, 102–104, 105n1, 138, 145, 146, 149, 159, 185, 186, 190, 225, 227, 229, 231, 232, 242, 244–248, 258, 259, 265, 268, 270, 271, 291

Methods, 15, 17, 20–22, 26, 41, 44–46, 49–52, 53n9, 74, 87–90, 92, 93, 99, 100, 102, 117, 120, 124, 125, 127, 129, 139–141, 144, 146, 147, 155, 168–180, 200, 203, 210, 211, 221, 226, 227, 230, 233, 234, 244, 250–252, 254–258, 266, 280, 283, 285

Methodological issues, 86–90, 188, 189

Microteaching, 6, 20, 89, 94–96, 116, 138–140, 143–149, 158, 159

Misconceptions, 51, 64, 72, 75, 101, 127, 145, 231, 236, 237, 297

Models, 18, 21, 24, 37, 38, 42, 43, 47, 52, 57, 59, 60, 62–66, 68, 78, 86, 90, 92–96, 99, 104, 105n1, 109, 111, 112, 116, 138, 142, 145, 147, 154, 155, 158, 172, 185, 187, 190, 191, 193, 195, 196, 199, 202, 211, 219, 225, 227, 234, 243, 245, 266, 284, 286

Motivation, 43, 50, 64, 66, 101, 247

Multiplication, 101, 280

N

Nature of mathematics, 99, 139, 155, 180, 191, 219
Noticing, 141, 159, 231
Novice teacher, 19, 62, 68, 132, 185, 186, 189, 206, 210, 211, 228, 229, 231, 232, 244, 245, 247, 259, 268–270, 285
Number, 11, 18, 21, 26, 40, 47, 49, 51, 67–71, 73, 77, 78, 86–88, 93, 97, 99, 101, 102, 105n1, 116, 124, 125, 127, 128, 139, 140, 153, 159, 167, 182n3, 185, 212–218, 249, 251, 254, 259, 277, 280, 284, 290, 299

O

Orientations, 11, 66, 78, 92, 97, 209, 210, 222, 241

P

Pedagogical Content Knowledge (PCK), 6, 24–26, 28, 29, 49, 51, 58–60, 62–64, 85–87, 89, 90, 95–98, 100–103, 105, 111, 112, 114, 117–119, 122, 131–133, 138, 139, 141–144, 147, 148, 155–158, 166, 167, 171, 172, 180, 181, 209, 210, 219–222, 236, 263, 267, 268, 279, 280, 283, 290–292, 297, 299–301
Pedagogical issues, 52, 117, 297
Pedagogical knowledge, 22, 25, 30, 52, 57, 58, 64, 66, 86, 94, 110, 131–133, 139, 141–143, 148, 172, 237, 256, 279, 290
Pedagogical training, 5, 6, 137–149
PISA, 26, 277, 278
Potari, D., 6, 153–161
Practice, 5–7, 9–11, 13, 14, 18, 20, 22, 24, 26–32, 37–40, 43, 44, 46, 52, 53, 53n5, 57, 60, 61, 65, 85, 92, 94–97, 100–105, 111–114, 116–118, 121, 131–133, 137–143, 145, 146, 148, 149, 153–161, 165, 166, 168, 175, 177, 178, 181, 182, 185–189, 192, 200, 202, 204–206, 209, 210, 222, 225, 227, 228, 231, 233–235, 241–260, 263–268, 270, 271, 277, 282–286, 290, 291, 297, 300, 302
Practicing teachers, 9, 133, 137–139, 141–143, 147, 148, 186–188, 248, 249
Preschool, 78
Pre-service mathematics teachers, 7, 21, 30, 85–87, 110, 117, 167, 279–283
Probability, 11, 20, 68, 73, 77, 88, 93, 97, 116, 119, 138, 155, 182n3, 249, 290
Problem solving, 30, 59, 78, 110, 117, 119, 124, 154, 174, 178, 191, 199, 201, 247, 252, 254, 266
Professional competence, 63, 138, 186, 205
Professional development, 5–7, 9, 11, 23–27, 31, 62, 64, 85, 99–102, 104, 105, 113, 114, 132, 138–140, 147, 148, 153, 158–160, 165–167, 171, 182, 185–188, 205, 225, 226, 228, 229, 233, 235, 237, 241–243, 250, 258–260, 263–271, 277, 282, 285, 287, 289–292, 295, 297, 302
Professionalism, 63
Professional standards for teachers, 109
Proficiency, 67
Proficient performance, 177
Proof, 69–71, 77, 78, 190, 197, 281, 294
Prospective mathematics teacher, 9, 66, 110, 123, 131, 132, 137–149, 154–160, 166, 167, 280, 285
Pu, S., 6, 165–182, 265, 266, 297

INDEX

Public lesson, 6, 139, 147, 148, 167, 185, 186, 188–190, 200, 201, 203, 225–237, 245, 255–258, 264, 265, 268–270, 282, 285, 287, 290, 291

R

Reasoning, 15, 30, 44, 50, 68, 70, 72, 112, 118, 127, 154, 178, 191, 257, 293–295, 297, 298, 301, 302, 303n4

Reflections, 5, 7, 27, 44, 63, 65, 78, 100, 104, 105, 156, 159, 160, 166, 176, 187, 191, 193, 195, 199–201, 205, 206, 210, 218–222, 234, 235–237, 244, 246, 255, 256, 268–271, 283, 290, 291

Representations, 59, 62, 64, 74, 89, 100, 102, 103, 119, 124, 125, 127, 129–131, 145, 158, 172, 180, 209, 213, 269, 285

Research lesson, 160, 226, 227, 232–234, 245

Resources, 23, 24, 27, 61, 100, 102, 103, 138, 145, 155, 156, 165, 166, 177, 193, 202, 220, 226, 231, 232, 237, 250, 266, 269, 278, 297, 298

Role of beliefs, 61, 62, 64, 66, 67

Rowland, T., 7, 62, 63, 263, 277–287

S

Same Content Different Designs (SCDD) activity, 6, 139, 143, 144, 158

Schoenfeld, A., 62

Secondary mathematics teacher, 6, 109–133, 156–161, 279

Sense making, 192, 193

Sherin, M. G., 133

Shulman, L., 57–60, 62–66, 86, 87, 102, 105, 111, 172, 173, 209, 210, 219, 229, 263, 264, 279, 280, 284, 289, 290

Social-political development, 5

Standards, 11, 31, 40, 42, 43, 48, 52, 73, 78, 88, 92, 101, 113–118, 130, 132, 139, 156–158, 160, 165, 168, 173, 177, 182n2, 192, 225, 234, 242, 271, 290–292, 296, 301, 302, 303n2

Statistics, 11, 73, 77, 88, 93, 97, 116, 119, 249, 257, 278, 292

Stein, M. K., 165, 283

Stillman, G. A., 7, 263–271

Student teaching, 6, 95, 96, 110, 114, 116, 117, 131, 137–140, 143–149, 158, 159, 290

Student thinking, 166, 187

Subject matter knowledge, 51, 58, 60, 62, 63, 85–90, 93–104, 111, 156, 172, 209, 263, 267, 279, 280–282, 291

Sun, X., 165–182

T

Taiwan, 4, 5, 9–32, 40, 42, 43, 46, 48, 49, 53n2, 69, 71, 73, 77, 132, 277

Tam, H. P., 5, 9–32

Tang, S.-Z., 5, 28, 29, 37–53

Teacher education, 4, 5, 9–32, 38–40, 43, 46, 53, 57, 62, 65, 67, 68, 77, 85, 86, 92, 101–104, 109, 111, 113–118, 140, 153–161, 241, 258, 259, 269, 283, 287n1, 290

Teacher education programs, 9, 10, 16, 17, 22, 24, 27, 31, 32, 32n3, 48, 49, 90, 102, 103, 110, 111, 113–118, 154, 156–158, 160, 280

Teacher preparation, 3–6, 9, 13–17, 40, 43, 46, 49–53, 85–105, 109–133, 137, 140, 148, 153–161, 167, 236, 277, 279, 285, 290, 302

318

Teacher professional development, 9, 24, 113, 114, 148, 165, 167, 171, 185, 186, 241–243
Teacher professional development standards, 185, 242
Teacher qualification, 14–16, 31, 38–41, 43, 46, 52, 53, 118–123
Teacher qualification examinations, 14, 15, 17, 46, 53, 110, 118–123, 157, 160
Teachers' expertise, 6, 138, 241–260
Teachers' knowledge, 3–7, 18, 30, 57, 59, 60, 85, 87, 97, 110–112, 123–131, 137, 141, 145, 154, 156–159, 172, 173, 176–178, 180–182, 187, 219, 220, 229, 236, 242, 243, 260, 263, 295
Teachers' learning, 6, 7, 94, 133, 139, 147–149, 161, 166, 187, 188, 206, 226, 229–237, 258, 260, 264, 282–286
Teaching competition, 185, 189, 202, 205, 227, 228, 234–236
Teaching mathematics, 23, 25, 27, 28, 37, 49, 50, 52, 60, 73, 74, 89, 90, 92, 93, 95–97, 99, 103, 104, 111, 117, 119, 120, 124, 126, 127, 139, 145, 148, 153, 155, 157, 160, 172, 173, 177, 179, 188–193, 219, 226, 228, 229, 231, 233, 235–237, 243, 245, 258, 263, 266, 270, 277, 282, 283, 285
Teaching process, 24, 49, 112, 120, 192, 230
Teaching research activities, 177, 178, 181, 182, 185, 186, 189, 209, 243–245, 249, 259

Teaching Research Groups (TRG), 6, 138, 141, 146–148, 158, 159, 167, 168, 173–182, 209–222, 225, 243, 264, 265, 267–269, 282, 283, 285, 287, 291
Teaching Research Office (TRO), 167, 210
TEDS-M, 5, 28–31, 49, 57–79, 132, 137, 138, 153
Textbook studies, 6, 165–182, 265, 266
Theory, 9, 10, 17, 20, 21, 24, 26–32, 46, 52, 53, 57–67, 70, 71, 73, 77, 78, 86, 89, 92–94, 97–101, 103–105, 110–113, 115, 116, 118, 120, 138–140, 155, 156, 159, 186–188, 211, 222, 232–234, 263, 277, 278, 284, 290, 296, 302
TIMSS, 67, 68, 109, 153

V

Vision, 138, 139, 148, 149, 155, 254, 285

W

Wu, H.-H., 289–302
Wu, Y., 109–133

X

Xie, S., 5, 85–105, 142, 290, 300

Y

Yang, Y., 110, 138, 167, 181, 186, 209–222, 259, 268, 270, 283
Yuan, Z., 137–149

Z

Zhang, B., 186, 190, 209–222